THE OBSERVER'S
ARMY VEHICLES DIRECTORY TO 1940

RESEARCH AND EDITING BY
BART H. VANDERVEEN
OLYSLAGER ORGANISATION BV

FREDERICK WARNE & CO LTD   London

FREDERICK WARNE & CO INC.   New York

CONTENTS

FOREWORD	3	HUNGARY	235
ACKNOWLEDGEMENTS	3	ITALY	238
KEY TO TECHNICAL DATA	4	JAPAN	262
ABBREVIATIONS	5	NETHERLANDS	268
		POLAND	273
AUSTRALIA	6	RUSSIA	277
AUSTRIA	8	SWEDEN	284
BELGIUM	27	SWITZERLAND	290
CANADA	32	USA	294
CZECHOSLOVAKIA	34	OTHER COUNTRIES	364
DENMARK	62		
FINLAND	63	INDEX	373
FRANCE	64		
GERMANY	115		
GREAT BRITAIN	174		

Published by FREDERICK WARNE & CO. LTD LONDON ENGLAND 1974
© OLYSLAGER ORGANISATION BV and B. H. VANDERVEEN 1974
Printed in Great Britain by BAS PRINTERS LTD. WALLOP, HAMPSHIRE
LIBRARY OF CONGRESS CATALOG CARD
No. 74-80615
ISBN 0 7232 1540 5
814,474

FOREWORD

When in the late 1800s the self-propelled vehicle began to develop into a machine of increased reliability and it became obvious that it had certain distinct advantages over the then customary prime movers, horse and mule, there were several advocates for the application of motor vehicles for military purposes. These people, however, were few and far between and met considerable resistance from the old school, especially in cavalry circles. Nevertheless, they persevered and managed at least to get motor vehicles entered in military manoeuvres. By about 1910 it had become clear that there were definite roles for motorcycles, cars, trucks, tractors and the like in military organizations. Various armies began to purchase them in limited quantities and to institute subsidy schemes so that in the event of hostilities they had a fleet of commercial transport vehicles readily available. The wisdom of this move was demonstrated in World War I (then called the Great War) when considerable numbers of standardized or at least unified vehicles were at those armies' immediate disposal. This war also became known as the mechanized war, mainly because it saw the birth of the tank. The internal combustion engine had proved its worth and had come to stay.

Of course, after this 'war to end all wars' most of the military motor vehicles were disposed of and the horse regained much of its lost ground, but many individuals, military as well as civilian, kept both the interest and further development going. On the whole, however, the 1920s was a decade of many new designs but little procurement, partly for monetary reasons. It was not until the 1930s and particularly the second half of this period when war clouds were looming again, that military motorization got in full swing. This book describes the majority of transport vehicles built and used during the first four decades of our century and includes many which saw active service in the Second World War. Also included are tractors and other special purpose vehicles, as well as typical examples of armoured cars. Those countries which produced many if not most of their military vehicles themselves, are dealt with in separate sections, in alphabetical order. Other nations, which relied largely or totally on imports are listed at the end of the book.

Together with *The Observer's Fighting Vehicles Directory—World War II* and *The Observer's Military Vehicles Directory—from 1945* this volume covers military motor vehicles, quantity-produced models as well as many experimental prototypes, which appeared during a time span of 75 years.

Piet Olyslager, MSIA, MSAE, KIVI

ACKNOWLEDGEMENTS

This book is largely based on the private collection of the editor, who would like to thank the many individuals, manufacturers, government agencies and other organizations who have assisted him in assembling his military transport reference library over the last 30 years. Particular mention should be made of the considerable help provided by the following fellow collectors/historians:
Brian S. Baxter, Sven Bengtson, Günter Buchwald, Peter Chamberlain, Colonel Robert J. Icks, Aimé van Ingelgom, Jean-Gabriel Jeudy, William F. Murray, Yasuo Ohtsuka, Walter J. Spielberger, B. T. White, and Laurie A. Wright.
The editor would also like to stress that without the encouragement and co-operation of his employer, Mr. Piet Olyslager, it would not have been possible to compile either this Directory or the similar volumes covering World War II and after.

Note: Inasmuch as the value of a work of this kind is determined by the degree of accuracy of its contents, the editor would like to ask readers and users to call to his attention any errors and omissions they may detect. Constructive criticism and supplementary material, factual and/or pictorial, will always be welcomed. (Editorial Productions Division, Olyslager Organisation BV, Book House, Vincent Lane, Dorking, Surrey, RH4 3HW, England).

KEY TO TECHNICAL DATA

Vehicle nomenclature: A uniform system has been used, wherever practicable, throughout the book, regardless of vehicle's country of origin. Example: 'Truck, 3-ton, 6 × 6, Cargo' means six-wheel, six-wheel drive cargo truck with rated payload capacity of three tons. '4 × 2' indicates a four-wheeled vehicle with two-wheel drive, etc. For this purpose dual tyres count as single wheels. A cross-country truck's rated payload capacity is always lower than its actual carrying capacity when used on roads. The vehicle type/nomenclature is followed by the make and usually the model designation in parentheses.

Year of manufacture: The year given indicates the year when the vehicle was first produced or, when used in others than the country of origin, the year when it entered military service there. Therefore indication of a single year does not necessarily mean that the vehicle was in production during that year only. In some cases the dates are estimated.

Engine: Engine make is usually the same as the vehicle make, unless stated otherwise. Fuel is petrol (gasoline) unless stated otherwise. Power output is given in bhp where known; HP is used where only the taxable horsepower (treasury rating) was available. Cubic capacity is given in cubic inches or centimetres, depending on usage in the country of origin. Engine type in 'Technical Data' is indicated by a letter code, e.g. I-L-W-F = cylinders in-line, side valves (L-head), water-cooled, front mounted, V-I-A-R = cylinders in-Vee, valves in head, air-cooled, rear-mounted, etc.

Transmission: Number of forward speeds and reverse and, if applicable, number of ratios in auxiliary gearbox or transfer case. Example: 5F1R × 2 indicates five-speed main gearbox with two-speed transfer, providing total of ten forward and two reverse ratios.

Wheelbase: Distance between centres of front and rear wheels, or, in the case of tandem axles, the centre of the bogie. '(BC)' indicates the wheelbase of the tandem bogie. Thus, a wheelbase of 3.40 (BC 1.10)m on a six-wheeler indicates a distance of 3.40 m between centre of front wheel and centre of rear bogie and 1.10 m between centres of rear wheels. This could alternatively be indicated as 2.85 + 1.10 m.

Dimensions: Overall length, width and height are given, where available, in the units generally used in the vehicle's country of manufacture. Where only two dimensions are given they refer to length and width. A figure in parentheses following the height figure indicates the minimum (reducible) height.

Weights: Vehicle weights, where known, are approximate net kerb weights (unladen) unless stated otherwise.

Conversions:

Length
1 inch (1″ or 1 in) = 25.40 mm
1 foot (1′ or 1 ft) = 12 in = 304.8 mm = 0.30 m
1 mile = 1760 yards = 1609 m
1 millimetre (mm) = 0.039 in
1 centimetre (cm) = 10 mm = 0.39 in
1 metre (m) = 1000 mm = 39.37 in = 3.28 ft
1 kilometre (km) = 1000 m = 0.62 mile

Weights
1 pound (1 lb) = 0.45 kg
1 quarter = 28 lb = 12.70 kg
1 cwt = 4 qtr = 112 lb = 50.80 kg
1 long ton = 20 cwt = 1.12 short tons = 1016 kg
1 short ton = 2000 lb = 0.89 long ton = 907.2 kg
1 kilogramme (kg) = 2.2 lb
1 metric ton = 1000 kg = 0.98 long ton = 1.1 short tons

Capacities
1 cubic inch (1 cu. in) = 16.39 cc
1 Imperial gallon = 1.2 US gallons = 4.55 litres
1 US gallon = 0.833 Imp. gal = 3.79 litres
1 cubic centimetre (1 cc or cm³) = 0.061 cu. in
1 litre (1 l) = 1000 cc = 61.025 cu. in

ABBREVIATIONS

A	Austria	IC	internal combustion (engine)	qv	quod vide (which see)
AA	Anti-Aircraft	IFS	independent front suspension	RAF	Royal Air Force (GB)
A/C	Armoured Car	IRS	independent rear suspension	RASC	Royal Army Service Corps (GB)
AFV	Armoured Fighting Vehicle	IWM	Imperial War Museum	RFC	Royal Flying Corps (GB)
AOP	Armoured Observation Post	LHD	Left-hand drive	RHD	right-hand drive
APC	Armoured Personnel Carrier	L-head	side valves (engine)	RN	Royal Navy (GB)
APG	Aberdeen Proving Ground (US)	*Lkw*	*Lastkraftwagen* (D; truck)	rpm	revolutions per minute
ASC	Army Service Corps (GB)	LWB	long wheelbase	SAE	Society of Automotive Engineers
AT	Anti-Tank	mech.	mechanical	SC	Sidecar
aux.	auxiliary	mfr(s)	manufacturer(s)	SP	Self-propelled
BC	bogie centres	MG	machine gun	S-T	semi-trailer
bhp	brake horsepower	MWEE	Mechanical Warfare Experimental Establishment (GB)	std	standard
BxS	bore and stroke			SV	side valves (L-head)
c	circa (approximate date)	NA	not available	SWB	short wheelbase
carb(s)	carburettor(s)	NC	normal control	trans.	transmission
cc	cubic centimetres	NL	Netherlands	US(A)	United States (of America)
CID	cubic inches displacement	OD	overdrive	USSR	Union of Soviet Socialist Republics
CKD	completely knocked down	OHC	overhead camshaft (engine)	wb, WB	wheelbase
COE	cab over engine (FC)	OHV	overhead valves (engine)	WD	War Department (GB)
CR	compression ratio	*Pkw*	*Personenkraftwagen* (D; Passenger car)	wh.	wheel(ed)
CS	Czechoslovakia			WO	War Office (GB)
cu. in	cubic inch(es)	PTO	power take-off	Wt	weight
D	Germany				
diff(s)	differential(s)				
Dim.	dimensions				
D/S	dropside (body)				
exp.	experimental				
F	France				
FAT	Field Artillery Tractor				
FC	forward control				
F/R	front/rear				
GB	Great Britain				
GCW	gross combination weight				
Govt	government				
GS	General Service				
GVW	gross vehicle weight				
HO	horizontally-opposed				
How.	howitzer				
HP	horsepower (treasury rating)				
hyd.	hydraulic				
I	Italy				

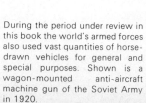

During the period under review in this book the world's armed forces also used vast quantities of horse-drawn vehicles for general and special purposes. Shown is a wagon-mounted anti-aircraft machine gun of the Soviet Army in 1920.

AUSTRALIA

The first recorded use of a motor vehicle in Australian military exercises was, reputedly, in 1908 when a Tarrant car (made in Melbourne; 1901–07) performed so well that an Australian Volunteer Automobile Corps was formed. Its staff consisted of officers who used their own cars. About 1910 the Caldwell-Vale Motor and Tractor Co. Ltd of Auburn, NSW, constructed some four-wheel drive vehicles but these were not adopted by the military. In World War I mainly British and American vehicles were used. During the 1930s American (and some British) vehicles were assembled and part-manufactured, the latter being mainly of Canadian origin. Bodywork was usually of Australian design and manufacture. Ford and General Motors (which in 1931 had purchased Holden's Motor Body Works) were particularly active. When World War II broke out large numbers of civilian-type vehicles were taken into service, soon supplemented by Canadian military types (see *The Observer's Fighting Vehicles Directory—World War II*).

Truck, 3-ton, 4×2, Cargo (Lacre 38 HP) Supplied in 1914. During World War I various types of British vehicles were used (in Europe), e.g. Douglas motorcycles, Dennis and Halford trucks, Rolls-Royce armoured cars, etc. US makes included FWD and Peerless trucks, Holt Caterpillar tractors, etc.

Ambulance, Medium, 4×2 (Morris 25 HP) At the outbreak of World War II many civilian-type vehicles were put into service. This RAN ambulance was based on a 1938/39 British Morris Series III chassis. Army had 1938/39 Chevrolet, Dodge, Fargo, Ford, Hudson, Studebaker, etc.

Truck, 12-cwt, 4×2, Cargo (Ford V8 68) Typical Australian Coupé Utility ('Ute'), known officially as GS Van. Similar pickup trucks were based on Chevrolet chassis. Front end was North American pattern. Specimen shown was experimentally fitted with a producer gas unit.

Truck, 15-cwt, 4×2, Wireless (Chevrolet) Light truck chassis with Australian body and roadster type cab. Known officially as 'Van, 15-cwt, Battery Staff, Wireless (Aust.), Chevrolet'. Basically a 1937 utility truck, fitted with side extensions to carry aerial poles, etc.

Truck, 2-ton, 4×2, Cargo (Fordson BBE) This truck was of British origin (1934–38) and had the 30 HP Ford V8 engine and 6.00-20 tyres. Vehicle shown was fitted with Charcoal Gas Plant, supplied by High Speed Gas (Australia) Pty Ltd of Melbourne for experimental use. c. 1939.

Truck, 30-cwt, 4×2, Gun Portee (Ford 917T) Early type of 2-pounder Gun Portee, improvised on 1939 Ford V8 truck. The 1938/39 style cab, with triple ribs on the sides, was continued on Australian-made Ford NC trucks for several years (N. American 1940–47 models had two, wider, ribs).

Tractor, Artillery, 4×4, Hathi Mk II (Thornycroft) Produced in Great Britain (qv) in the mid-1920s, several of these tractors found their way to Australia where they were utilized for towing field and AA guns until well into World War II, albeit not on active service.

AUSTRIA

Motorization in the Austro-Hungarian armed forces commenced as early as 1898 when, following tests with German Daimler (Marienfelde) trucks in 1897/98, a licence-produced Daimler 5-ton truck was taken into service. It was soon followed by other vehicles and it was not long before some very advanced four-wheel drive types, including an armoured car with revolving turret, made their appearance. Later came the legendary Austro-Daimler 4 × 4 heavy artillery tractors, the Daimler-Landwehr petrol-electric road/rail trains and others. Responsible for these early developments was a relatively small group of people, including some enthusiastic army officers and automotive designers such as Paul Daimler and Ferdinand Porsche. Examples of their efforts can be seen in the following pages.

At the end of World War I the Austro-Hungarian Empire was divided into what are now the countries of Austria, Czechoslovakia and Hungary and those vehicles which were made in the latter two parts of the old empire are dealt with in the sections 'Czechoslovakia' and 'Hungary'.

The Austrian Army has always worked very closely with the motor industry, particularly with the Steyr-Daimler-Puch concern which was formed in 1934/35 when Austro-Daimler and Puch (who merged in 1928) joined forces with the old armaments firm of Steyr. When in 1938 Austria was annexed by Germany, many vehicles were taken over by the *Wehrmacht* and other Austrian military vehicles were used by Balkan and other countries. From 1938 until the end of World War II the Austrian motor industry produced large quantities of cars, trucks, half-tracks, etc. for the *Wehrmacht* (see *The Observer's Fighting Vehicles Directory—World War II*).

Standard postal truck chassis (PABZ), fitted with military bodywork for use in World War I.

Who's Who in the Austrian Automotive Industry

Note: pre-1940; for pre-1918 see also Czechoslovakia and Hungary.

Austro-Daimler	Österreichische Daimler Motoren AG,* Wiener-Neustadt (originally Bierenz, Fischer & Co.).
Austro-Fiat	Austro-Fiat AG, Wien XXI (originally a branch of FIAT of Turin, Italy; later became Österreichische Automobil-Fabriks-AG).
Austro-Tatra	Austro-Tatra Werke AG, Wien.
Berna-Perl	Berna-Automobilfabrik Ing. Perl AG, Liesing, Wien (licence-production of Swiss Berna trucks).
Bock & Holländer	Bock & Holländer, Wien III (taken over by WAF, *qv*).
Fross-Büssing	A. Fross-Büssing, Fabrik für Nutzkraftwagen KG, Wien XX (formerly Ferdinand Götz und Söhne; licence-production of German Büssing trucks).
Gräf & Stift	Wiener Automobilfabrik AG vorm. Gräf & Stift, Wien XIX (later: Gräf & Stift Automobilfabrik AG, Döbling und Liesing, Wien).
PABZ	Postauto-Betriebszentrale, Wien XXI, Stadlau.
Puch	Puchwerke AG,* Graz
Saurer	Österreichische Saurer Werke AG, Wien X (originally licence-production of Swiss Saurer trucks).
Steyr	Steyr-Werke AG* (formerly Österreichische Waffenfabriks-Gesellschaft), Steyr und Wien.
WAF	Wiener Automobil-Fabrik GmbH, Wien XVI.

*these firms later combined to form the Steyr-Daimler-Puch AG.

AUSTRIA

CARS, 4×2 and MOTORCYCLES

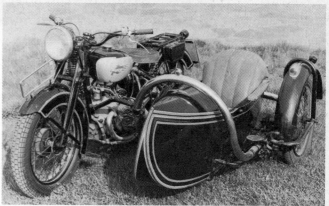

Makes and Models: *Cars:* Austro-Daimler 16, 27, 28, 35, 45 and 80 PS (between 1904 and 1920), ADV, ADM and ADR (between 1920 and 1930), etc. Austro-Fiat 24, 27, 35 and 40 PS (between 1909 and 1920), 1001 34 PS (1932), etc. Austro-Tatra 57A (1934–38). Bock & Holländer 8/24 (1902). Gräf & Stift 32, 45 and 50 PS (between 1912 and 1918), SP5 (1930), etc. Puch VIII (1915). Steyr VII, XII, XVI and XX (between 1926 and 1930), 30, 430, 530, 50/55, 120(S), 220 and 250 (between 1930 and 1940), etc. WAF (1916), etc.

Motorcycles: various commercial types, notably Puch.

General Data: In addition to the above makes there were products of Laurin-Klement, Nesselsdorfer, Praga and RAF. These were made in the parts of the Austro-Hungarian Empire which after 1918 became Czechoslovakia (*qv*).
The same applies to the MAG and the Rába, for which see Hungary. The first military car in Austria was a 1902 Bock & Holländer, used mainly for driving instruction. The cars listed above were in the main civilian types with four-door open tourer type bodywork and were used by field and convoy commanders, high-ranking officers, etc. Some were equipped as mobile repair units, others had special bodywork, e.g. ambulance, balloon servicing, etc.
During World War I practically all motor vehicle manufacturers in Austro-Hungary were engaged in vehicle production and most of them produced cars as well as trucks. Total output of all types of vehicles, including trailers, in 1916, was about 830 units per month. At the beginning of the war the Army had 54 passenger cars and 9 artillery-battery-commander cars. The latter, known as *Batterie-Kommandantenwagen der schw. Art./mot.* had special bodywork with rear entrance to the rear body, on a specially strengthened car chassis and were supplied mainly by Gräf & Stift, Praga, Puch and Rába. During the 1930s several special command cars were produced on light 6×4 truck chassis (*qv*).

Vehicle shown (typical): Motorcycle, with sidecar (Puch 800).

Technical Data:
Engine: Puch 4-cylinder, V-L-A-C, 792 cc (60×70 mm), 20 bhp at 4000 rpm, CR 5.0:1. (Cylinders set under slight 'V').
Transmission: 4F. Clutch in rear hub.
Brakes: mechanical.
Tyres: 4.00-19 (27×4).
Wheelbase: 1.43 m.
Overall l×w×h: 2.18×0.83×1.05 m (solo machine).
Weight: 290 kg. GVW: 540 kg.
Note: 550 produced, 1936–38. In military service mostly with sidecar. Max. speed 125 km/h (w/SC 95 km/h) approx. Weight and GVW for solo machine 195 and 365 kg respectively.

Car, Heavy, 4×2, Postal Service (Austro-Daimler 18 PS)
Until 1908/09 postal vehicles were operated by the military authorities. This 4-cyl. model entered service in Dec. 1905 and was the second supplied by Austro-Daimler (the first was a 16 PS model in 1904). Other *Militär-Postwagen* were supplied by Gräf & Stift and Saurer.

Car, 4-Seater, 4×2 (Austro-Daimler) 4-cyl., 30/35 bhp, 4F1R, wb 3.00 m. *Stabswagen* (staff car) with 'tulip-type' bodywork. Basically civilian *Alpenwagen*, produced 1911–14. Rudge wire-spoke wheels with 880×120 tyres. 2212-cc (80×110 mm) L-head engine. Similar cars were supplied by several other manufacturers.

Car, 4-Seater, 4×2 (Steyr XII) 6-cyl., 30 bhp, 4F1R, wb 3.00 m, 1085 kg. 1560-cc (61.5×88 mm) OHC engine. First mass-produced Steyr car with swing-axle independent rear suspension. Standard civilian type (11,124 produced during 1926–29). Maximum speed 85 km/h. Tyres 775 or 745×145.

Car, 5-Seater, 4×2 (Steyr 250) HO-4-cyl., 25.5 bhp, 4F1R×2, wb 2.60 m, 4.10×1.68×1.87 m, 1050 kg. Tyres 7.00-18. *Mannschafts-Geländewagen* produced for Austrian and Axis forces (77 in 1938, 975 in 1939, 148 in 1940). Lockable rear differential. Derived from model 150 light truck.

AUSTRIA

TRUCKS, 4×2

Makes and Models: Austro-Daimler 12 PS (1901), 12/14 PS (w/Winch, 1902), 12/14 PS (3-ton, 1903), ADN (3–3½-ton, 1936), ADTK (1-ton, 1936–37), etc. Austro-Fiat Wz-06 (2-ton, 1906), TV2 (2-ton, 1910–18), FT1 (1-ton, 1914–18), TS3* (3-ton, 1913–18), V2 (2-ton, 1917–22), AFN (2-ton, 1930–33), AFL (1¼-ton, 1935–38), AF25 (3-ton, 1930–32), AFH (3-ton, 1935–37), HRM5 (3-ton, 1936–38), etc. Berna-Perl BP3* (3-ton, 1913–20). Fross-Büssing FN2 (2-ton, 1910–18), FN3* (3-ton, 1911–18). Gräf & Swift 45Sp (2½-ton, 1909–12), Wz09, 12 and 14 (2-ton, 1909–15), 3 Sub.* (3-ton, 1912–26), 5-Sub. (5-ton, 1914–26), WzW16 (5-ton, 1916–18), V2 (ambulance, 1920–28), V6 (3½-ton, 1928–32), 335D (3-ton, 1932–38), etc. PABZ ET11 and 13 (2-ton, 1911–14). Puch VIII (ambulance c. 1915), X (F,D,VIII) (2½-ton, 1913), XB and XC (VIII) (2½-ton, 1914–17). Saurer 2TC (2-ton, 1912–20), 3TK* (3-ton, 1914–18), 5TK (5-ton, 1914–18), 3BH (3-ton, 1928–33), 3BOD (3½-ton, 1933), 5BHw (5-ton, 1935–36), 4BTDw (5½-ton, 1938–40), etc. Steyr III (2½-ton, c. 1925), XIIN (ambulance, 1927–29), 150 (¾-ton, 1938–39), etc. WAF 35 PS (2-ton, 1914–18), 45 PS* (3-ton, 1914–18), etc.

* Subsidy-types.

Vehicle shown (typical): Truck, 3-ton, 4×2, Cargo (Austro-Fiat TS3)

Technical Data:
Engine: Austro-Fiat 4-cylinder, I-L-W-F, 45 bhp.
Transmission: 4F1R. Chain final drive. Differential lock.
Brakes: Mechanical (on rear wheels only).
Tyres: Solid rubber; front: 820 or 920 × 120, rear: 1040 × 120 (dual).
Wheelbase: 4.20 m.
Overall l × w × h: 5.30 × 1.80 × 2.50 m approx.
Weight: 3620 kg.
Note: subsidy type, employed for heavy transport columns, towing four-wheeled 2-ton trailer. Winching drums on rear wheels. Carried three pairs of chain drive sprockets (11, 16 and 21 teeth) for different operating conditions.

General Data: The very first vehicle supplied for Austrian military service was a 5-tonner produced under German Daimler licence by Maschinenfabrik Bierenz, Fischer & Co. in Wiener-Neustadt. It was delivered in March 1898 and named *Dromedar*. The second was a 4-ton German Marienfelder-Daimler, in 1900. It was named *Hyäne* and was the prototype for two more, the 2-ton *Iltis* (Daimler) and 3-ton *Känguruh* (Reinickendorfer). Before the splitting-up of the Austro-Hungarian Empire in 1918 there were several other makes in service; in this book these are dealt with under Czechoslovakia (Laurin-Klement, Nesselsdorfer, Praga) and Hungary (Danubius, MAG, Marta, Rába). There were also Adlers and Dürkopps from Germany and Fiats from Italy. Most of these manufacturers supplied medium trucks during the 1914–18 period. As in several other countries before WWI, Austria had a subsidy scheme for commercial users of certain vehicles which were suitable for military use and which had to be available for impressment if required. This happened on a large scale in 1914. During WWI many captured Italian vehicles were also used. In 1914 a number of Studebaker car chassis were modified for ambulance use.

Truck, 3-ton, 4×2, Cargo (Austro-Daimler 12/14 PS) 4-cyl., 12–14 bhp, 4F1R. Shaft drive to rear axle. Used by the Court of Archduke Friedrich. (Text on bodysides: *Obersthofmeisteramt Seiner k.u.k. Hoheit des durchlauchtigsten Herrn Erzherzogs Friedrich*). Third Austro-Daimler *Militärlastwagen*, 1903.

Truck, 3-ton, 4×2, Cargo (Fross-Büssing FN3) 4-cyl., 45 bhp, 4F1R, wb 4.20 m. Chain final drive. Subsidy type. Normally on solid rubber tyres (front: 920×120, rear: 1040×120) but shown converted for use on Carpathian railroads in WWI. As such these 'trains' had a capacity of 40–45 tons.

Truck, 3-ton, 4×2, Cargo, with Trailer, 2-ton 4-wh., Cargo Typical example of standardized 3+2-ton *Subventionslastzug* of 1914. They were made by Austro-Fiat (shown). Berna-Perl, Fross-Büssing, Gräf & Stift, Saurer, WAF, etc. Later production had soft-top 'half cab'. A 10-ton combination was also utilized.

Truck, 2½-ton, 4×2, Cargo (Steyr 12/34 PS Typ III) 6-cyl., 40 bhp, 4F1R, wb 3.70 m. Tyres 955×155 (optional: 920×120 solid tyres on steel spoke wheels). Commercial truck; small number used by Army. 3325-cc (80×110 mm) OHC engine. Worm-drive rear axle. Mechanical brakes, on rear wheels only. 1922–28.

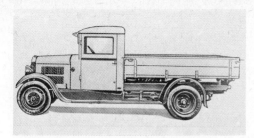

Truck, 1¼-ton, 4×2, Cargo (Austro-Fiat AFL) 4-cyl., 35 bhp, 4F1R, wb 3.25 m, 5.00×1.80×2.10 (w/tilt) m, 1590 kg approx. 2224-cc (85×98 mm) SV engine. Tyres 5.50-20. Also available with 3.50 m wb. Used as *Kolonnen-Rüstwagen* (weapons/equipment carrier). Produced in 1935.

Truck, 1-ton, 4×2, Cargo and Personnel (Austro-Daimler ADTK) 4-cyl., 18 bhp, 4F1R, wb 1.75 m, 3.81×1.51×1.46 m, 1045 kg. Multi-purpose *Tross-Karren*, used for carrying troops, field kitchen, etc. and as light tractor. FB 12/20 air-cooled 2312.2-cc engine. Tyres 5.25-18. Hyd. brakes. IFS. 159 made, 1936–37.

Truck, 3-ton, 4×2, Cargo (Saurer 3BH) 4-cyl., 65 bhp, 4F1R, wb 4.20 m. Tyres 32×6. Produced 1928–33, then superseded by diesel-engined model 3BOD 3½-tonner. Similar trucks were produced by Austro-Fiat (AF25, AFH). Military Saurer 3BOD and 5BHw (5-tonner) had hard-top cab with 'half doors'.

Truck, 3-ton, 4×2, Cargo (Austro-Fiat AFH) 4-cyl., 60 bhp, 4F1R, wb 4.00 m. Tyres 34×7. OHV engine. Soft-top cab with 'half doors'. Low chassis frame height (frame upswept over rear axle, underslung semi-elliptic rear springs). Fitted with radiator guard. Dropside body. Produced during 1935–37.

AUSTRIA
TRUCKS, 6×4

Makes and Models: Austro-Daimler ADG (2-ton, 1935–36), ADGR (2–3-ton, 1936–40). Steyr 140/40D (2-ton, 1932–33), 340/*BH Geländewagen* (2-ton, 1933), 440 (1½-ton, 1935–37), 640 (1½-ton, 1937–41), 640/643 (ambulance, c. 1940).

General Data: The Austro-Daimler *Geländewagen* (ADG) chassis made its debut in 1931 and one was exhibited at the Olympia Show in London in November of that year. It was beautifully designed and like several contemporary Austrian cars had a beam type front axle and swinging type rear axles. The chassis frame featured longitudinal frame members at front, converging behind the gearbox to form a central box section girder extending to the rear and forming the support for the rear wheel bogies. The 7F3R gearbox had single-lever control and was separately mounted behind the overhead-camshaft 3915.4-cc (85 × 115 mm) engine and single dry-plate clutch. Behind the gearbox there was a transfer box with differential unit from which two shafts, one on each side of the central frame members, drove the rear wheels. The wheels were mounted on swinging half axles with forked inner ends journalled on housings containing bevel gearing by which the axle shafts were driven. Intermediate propeller shafts conveyed the drive to the rearmost axles. Inverted semi-elliptic leaf springs were mounted lengthwise on trunnions and their ends were spherically jointed to the axles. The spare wheels were mounted in line with the dashboard and were free to revolve on stub axles, thus assisting the vehicle in riding over high obstacles. The ADG went into quantity production in 1935 when 66 were made, followed by a further two in 1936. In 1936 the ADGR version was introduced. This was a slightly modified model, mainly for export to Rumania. 361 of these were made until 1940. The French firm of Laffly (*qv*) produced a range of cross-country vehicles based on the ADG design, but mostly with additional drive to the front wheels. The Steyr 6 × 4 was not unlike the ADG(R) but had a normal ladder-type chassis frame and a single propeller shaft to the first rear axle differential/transfer box. It was also smaller and lighter. Steyr production figures were as follows: Model 140: 15 (1932–33), Model 340: 5 (1933), Model 440: 713 (1935–37), Model 640: 3780 (1937–41).

Vehicle shown (typical): Truck, 1½-ton, 6 × 4, Command (Steyr 440)

Technical Data:
Engine: Steyr 430 6-cylinder, I-I-W-F, 2078 cc (70 × 90 mm), 45 bhp at 3800 rpm (40 at 3000). CR 5.75 :1.
Transmission: 4F1R × 2.
Brakes: hydraulic.
Tyres: 7.00-18.
Wheelbase: 3.03 m, BC 1.06 m.
Overall l × w × h: 5.24 × 1.65 × 2.28 m.
Weight: 2450 kg. GVW: 3690 kg.
Note: IRS with swing axles. Spare wheels on idler hubs. 11-seater *Kommandeur-* or *Kommandowagen* body with folding top. Also with cargo/personnel and signals van bodywork. Produced 1935–37.

Truck, 2-ton, 6×4, Chassis (Steyr 140/40D) 6-cyl., 40 bhp, 4F1R×2, wb 3.03 (BC 1.06) m. Tyres 7.00-18. Prototype chassis with test body. Engine of Model 30 car. Commercial design, intended for military use also. First of over 4500 Steyr *Geländewagen* produced during 1932–41 (see General Data).

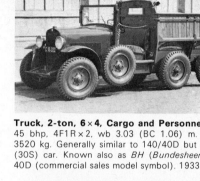

Truck, 2-ton, 6×4, Cargo and Personnel (Steyr 340) 6-cyl., 45 bhp, 4F1R×2, wb 3.03 (BC 1.06) m. Tyres 7.00-18. GVW 3520 kg. Generally similar to 140/40D but engine of Model 230 (30S) car. Known also as *BH (Bundesheer) Geländewagen* and 40D (commercial sales model symbol). 1933.

Truck, 1½-ton, 6×4, Cargo and Personnel (Steyr 440) 6-cyl., 45 bhp, 4F1R×2, wb 3.03 (BC 1.06) m, 5.33×1.73×2.33 m, 2360 kg. Shown is the *Mannschaftswagen* (personnel carrier) version which had seating for 14 troops, incl. driver. Engine as contemporary model 430 Steyr cars. Note 'belly support wheels'.

Truck, 1½-ton, 6×4, Signals (Steyr 440) 6-cyl., 45 bhp, 4F1R× 2, wb 3.03 (BC 1.06)m. 2820 kg. *Funkwagen* version with coach-built house-type van body for radio communications equipment. Crew 6, including driver. Of the Model 440 chassis a total of 713 was made during 1935–37.

Truck, 1½-ton, 6×4, Cargo and Personnel (Steyr 640) 6-cyl., 55 bhp, 4F1R×2, wb 3.03 (BC 1.06) m, 5.33×1.73×2.33 m, 2400 kg. GVW 4000 kg. Used by Austrian and Axis forces. Last of Steyr 6×4 *Geländewagen*, produced 1937–41. Note bracket for Notek blackout driving light on left-hand front wing.

Truck, 1½-ton, 6×4, Ambulance (Steyr 640/643) 6-cyl., 55 bhp, 4F1R×2, wb 3.03 (BC 1.06) m, 4.88×1.80×2.56 m, 2885 kg. *Krankenkraftwagen* bodywork with double rear doors. Used by German *Wehrmacht* with military designation *Kfz.* 31. Accommodation for 4 stretcher cases or 8 sitting patients.

Truck, 2-ton, 6×4, Cargo (Austro-Daimler ADG) 6-cyl., 65 bhp, 7F3R, wb 3.75 (BC 1.10) m, 5.96×1.74×1.64 m, GVW 5700 kg. Tyres 7.50-20. Payload on roads 3 tons. Commercially available. IRS with swing axles and semi-elliptic leaf springs. Spare wheels on idler hubs. 3915-cc (85×115 mm) OHV engine.

Truck, 2-ton, 6×4, Command (Austro-Daimler ADG) *Kommandowagen* on same chassis as shown on left. 429 chassis produced, 1935–40 (incl. ADGR version for Rumania). Some had dual rear tyres. Other body variants included cargo/personnel, tanker, ambulance, searchlight. First announced in 1931. Also with LWB for bus bodies.

AUSTRIA
TRACTORS

Makes and Models: Austro-Daimler 20 PS (4×2, 1905), 50 PS (4×4, 1905), 90 PS M09 (4×4, 1909), 100 PS M12 (4×4, 1912–15), 100 PS M12/16 (4×4, 1916–17), 100 PS M12 *Seilwindenauto* (4×4, 1912–15), 100 PS M12/16 *Seilwindenauto* (4×4, 1916–17), 80 PS M17 (4×4, 1916–18), KP II (2×2, 1917), ADZK (4×4, 1937), ADAZ (6×6, 1935). Austro-Daimler/Landwehr *Elektro-Train* (road and road/rail trains, 1912–15). Austro-Daimler/Skoda *C-Zug* ('Gigant', 1915–18). Gräf & Stift 70 *Spezial* (4×2, 1912–18), Wz-09/12/14 and WzW-16 (4×2, 1909–18), etc. Holt Caterpillar 75 (60 PS) and 120 (115 PS) (Half-track, 1914, US).

General Data: Development of special artillery (howitzer and mortar) tractors in Austria commenced as early as 1904/05 when the Austrian branch of the German Daimler concern, under the direction of Gottlieb Daimler's son Paul, produced a 20 and a 50 HP *Militärzugwagen*. At about the same time the Austro-Hungarian Army ordered similar machines from Reinickendorf in Berlin and Daimler in Marienfelde, both in Germany, with engines of 60 and 80 HP respectively. Meanwhile, Archduke Leopold Salvator Habsburg-Lothringen, Inspector of the Artillery, designed (or initiated) a 4×4 tractor with separate drive shafts from a central differential to each wheel. This tractor was completed in 1905, tested and subsequently modified to have solid rubber tyres instead of iron wheels and other improvements. Another officer, the later General von Tlaskal-Hochwall, had designed a 60 HP tractor with four-wheel drive and steering, towing five self-tracking four-wheel steer trailers. This 'train' was produced by Rába in Gyor in 1904. Several more experimental tractors followed and in 1909 a series of 36, produced by Austro-Daimler (M09), entered service. They all bore names of beasts of prey. Until 1918 a variety of impressive Porsche-designed 4×4 tractors, notably the M12 and M16, were produced in quantity. The latter, with 30.5 Skoda mortars, were borrowed by the Germans in World War I in the siege of some Belgian forts. In fact, these vehicles were built by Skoda, who during World War I were linked with Austro-Daimler. Also of interest was a series of petrol-electric tractors with trailers, initiated by General Landwehr von Pragenau and executed by Ferdinand Porsche and his design team at Austro-Daimler. These machines culminated in the famous *C-Zug*. The American Holt Caterpillar tractor had been widely used for agricultural purposes since 1912. Its track bogies were copied by the Germans for the A7V tank, etc.

Vehicle shown (typical): Tractor, 4×4, Heavy Artillery, MZW M12 (Austro-Daimler *Zugwagen M12*)

Technical Data:
Engine: Austro-Daimler 6-cylinder, I-I-W-F, 20,180 cc (140×220 mm), 100 bhp at 800 rpm. CR 4.5:1.
Transmission: 4F1R.
Brakes: mech. ext. contracting (handwheel-operated on rear wheels; foot-operated on trans.).
Tyres: solid rubber, front: 1060×160 mm, rear 1500×160 mm (dual).
Wheelbase: 3.50 m.
Overall l×w×h: 6.80×2.12×3.20 m.
Weight: 10,140 kg. Payload 5 tons.
Note: central diff, twin propeller shafts front and rear. Twin carbs. Magneto ignition. Acetylene lighting. Power winch at front. Some had iron-shod (1500×500 mm) wooden rear wheels.

B

Tractor, 4×4, Heavy Artillery, MZW MO9 (Austro-Daimler MO9) 6-cyl., 90 bhp, 4F1R, wb 3.80 m, 4500 kg. Power winch at front. Used for towing 30.5-cm mortar (separate 4-wheel carriages for mortar barrel, cradle and base; overall length 25 metres), 36 produced in 1909. Carried 3 tons of ballast.

Tractor, 4×4, Winch, MO9 (Austro-Daimler 'Goleath') 6-cyl., 90 bhp, 4F1R, wb 3.80 m. *Seilwindewagen* (winch vehicle) version of MZW MO9. Powerful engine-driven Wolf'scher vertical drum winch at rear. Small self-recovery winch at front. Max. speed 26 km/h. Note sprag (hill holder) just in front of rear wheels. 1909.

Tractor, 4×4, Heavy Artillery, MZW M12/16 (Austro-Daimler M12/16) 6-cyl., 100 bhp, 4F1R, wb 3.50 m, 6.80 × 2.30 × 3.20 m, 11,240 kg. Differed from M12 (see previous page) mainly in having iron-shod rear wheels, rectangular bodysides, single carburettor, electric lighting and starter (Bosch). 3-ton winch at rear. 127 produced (200 M12s).

Tractor, 4×4, Winch, M12 (Austro-Daimler) 6-cyl., 100 bhp, 4F1R, wb 3.50 m, 6.80 × 2.40 × 3.20 m, 12,000 kg. (10,760 kg if equipped with solid rubber rear tyres). 1-ton winch at front, large 5-ton Skoda winch at rear, both engine-driven. Early models had Petravic main winch. 32 produced (plus 30 on M12/16 chassis). MZW M12 also appeared as wheel-cum-track tractor (exp.).

Tractor, 4×4, Heavy Artillery, MZW M17 (Austro-Daimler M17) 4-cyl., 80 bhp, 4F1R, wb 3.00 m, 6.35 × 2.22 × 2.45 (min.) m, 9700 kg. 1460-mm diam. metal wheels. Some were later fitted with solid rubber tyres. 13.5-litre (140 × 220 mm) OHV engine. Used for 30.5-cm mortar and trailers. 3-ton rear winch. Max. speed 14.5 km/h.

Tractor, 4×4, Heavy Artillery, MZW M17 (Austro-Daimler M17) Similar in many respects to M12/16 which it superseded, but 4-cyl. engine and same wheel size front and rear. Model shown had riveted wheels. By July 1918, 138 had been made. Some were still in use in Czechoslovakia and Italy in 1938.

Tractor, 4×4, Artillery, MZW M05 'Löwe' chassis with four-wheel drive, designed by Archduke Leopold Salvator, produced by Skoda in 1905. Engine (4-cyl. 45-bhp) and clutch were supplied by Gräf & Stift, transmission and 4-wh. steering by Austro-Daimler. Later modified (fitted with solid rubber tyres, etc.).

Tractor, 4×2, Searchlight, M12 (Gräf & Stift 70 Spezial) 4-cyl., 70 bhp, 4F1R, wb 4.80 m. Solid rubber tyres, front 820 × 120, rear 950 × 120 (dual). Tractor-cum-generator unit, known as *120-cm Auto-Scheinwerferzug M12*. Produced 1912–18. OHV engine. Similar type made by Nesselsdorfer/Tatra (see 'Czechoslovakia').

Road Train, Petrol-Electric (Austro-Daimler/Landwehr) First of several road trains produced by Austro-Daimler from an idea of General Landwehr von Pragenau (*Elektrotrain System Daimler-Landwehr*). Ferdinand Porsche was involved in its design Two of this 15-ton 10-trailer type were ordered in 1909/10 and delivered in 1912. The tractor had a 6-cyl. 100-bhp engine, driving a 250-volt 63-(or 70-)kW generator which powered electric motors in the wheels of every other trailer. The tractor rear wheels were driven mechanically. Knorr air brakes throughout.

Road/Rail Train, Petrol-Electric (Austro-Daimler/BE-Zug) The third type of the petrol-electric *Landwehr-Trains* could, like the second type (*B-Zug*), be used on roads and on railway tracks after changing the wheels. The *B-Zug* was rather similar in appearance to the type shown but the front axle was not mounted so far forward. On roads they towed up to five 4/5-ton trailers, on rails 12 to 15. Both had 100-bhp engine (like MZW M12) and 93-kW generator (engine output also quoted as 150 bhp).

Tractor-Trailer, Petrol-Electric (Austro-Daimler C-Zug) The *C-Zug*, also known as 'Gigant', was in production during 1915–18. It was designed by Ferdinand Porsche for transporting heavy Skoda howitzers and mortars on roads and rails (normal and Russian gauge). Road transport of one 38-cm M16 Howitzer required five *C-Zuge* (i.e. tractors plus four-axle 30-ton trailers) viz. one each for the left and right base, one for the mount, one for the barrel and one for ammunition (shown). It was also used for haulage of other heavy loads.

Tractor, Petrol-Electric (Austro-Daimler C-Zug) 6-cyl. 120 (or 150) bhp with 70 (or 93) kW generator, powering 135-hp electric motors in tractor rear wheels (and all 8 trailer wheels). Wb 3.25 m, length 5.00 m, wt 8900 kg approx. Shown being converted from rail to road operation. Vacuum brakes. Rear-mounted winch.

Tractor, 4×2, Mobile Workshop, MZW M16 (Gräf & Stift WzW-16) 4-cyl., 55/70 bhp, 12F3R, wb 4.20 m. Solid rubber front tyres. 1.8-m diameter iron rear wheels. Winch at front. Introduced 1916. Towed fully-equipped 4-wh. workshop trailer. One of various models produced by Gräf & Stift from 1909. Incorporated electric generator.

Tractor Unit, 4×2 (Austro-Daimler 'Motorprotz') This 'mechanical horse' (*Motorpferd*) was designed by Ferdinand Porsche for towing field guns, etc. It was also used for towing agricultural ploughs. At the rear was a tailpiece with two small metal wheels, one of which is visible. Note seats at front.

Tractor Unit, 2×2, M17 (Austro-Daimler KP II) 4-cyl., 15 bhp, 4F1R. Used in conjunction with 8-cm field gun (shown) or field-kitchen unit. Air-cooled 1500-cc engine. Steering by articulation. Used from March 1917. Produced until 1918. Also known as '*Daimler-Pferd*' (Daimler-horse).

Tractor, Light, 4×4 (Austro-Daimler ADZK) 4-cyl., 60 bhp, 4F1R×2. *Zugwagen* (tractor) with air-cooled engine. Three produced, 1937/38. Crew 1+7 IFS/IRS. Similar chassis used for ADSK rear-engined armoured car. A few were licence-produced in Britain by Morris (models Q armoured car, QW tractor).

Tractor, Heavy, 6×6 (Austro-Daimler ADAZ) 6-cyl., 150 bhp, Voith hyd. transmission with 3-speed auxiliary gearbox, wb 2.80 (BC 1.05) m, 5.73×2.43×2.25 m, 7840 kg. Tyres 8.25-20. IRS. 10-ton winch. 30 produced, 1935–37. Chassis also as AA/AT gun carriage. 8×8 edition used for armoured cars (*qv*).

AUSTRIA

HALF-TRACK and WHEEL-CUM-TRACK VEHICLES

During the 1930s several Austrian firms designed wheel-cum-track vehicles, i.e. vehicles which could run on tracks for cross-country use but quickly converted to run on wheels for operation on roads. Of these, the Austro-Daimler ADMK *Maschinengewehr-Traktor* or *Motor Karette* was the most numerous. It was popularly known as 'Adamek' and proved itself to be a versatile go-anywhere vehicle on the rough terrain of the Austrian Alps. It could transport three men, a heavy machine gun with 5000 rounds and equipment plus a tow-load of 1000 kg. The ADMK, as well as the Saurer RK7 (also known as RR7) were later used by the German *Wehrmacht*, the latter with various body types, including armoured (see following section). In 1914 Archduke Leopold Salvator had filed a patent application for a wheel-cum-track vehicle (tracks located between front and rear axles; US Patent No. 1,355,853, 19 Oct. 1920).

Tractor, Half-Track, Artillery (Austro-Fiat AFRS) 6-cyl., 80 bhp, 4F1R × 2. *Raupenwagen* or *Zwitterfahrzeug*, produced in 1935/6. Rubber tracks similar to those of the French Citroën-Kégresse P19 (which was also tested). Skis could be fitted to the front wheel stub axles. Intended for field artillery.

Carrier, Wheel-cum-Track, Machine Gun (Austro-Daimler ADMK) 4-cyl., 18–20 bhp, 4F1R, wb (on wheels) 1.77 m, 2.77 × 1.06 m (on wheels: 3.57 × 1.50 m), 1450 kg. Tyres 5.25-18. Speed 15 km/h (on wheels 45 km/h). Production: 12 in 1935, 302 in 1936, 4 in 1937, 16 in 1938. Note front axle assembly in swung-up position.

Tractor, Wheel-cum-Track, Artillery (Austro-Daimler ADAT) 6-cyl., 80 bhp, 4F1R, wb 3.20 m, 5.28 (on tracks 4.70) × 2.03 m. Tyres 230-18. Payload 1600 kg. *Raupentraktor* variant of ADMK carrier with 6- vs. 4-cyl. engine. Seating for crew of 9. Manufactured in 1937 but no quantity production.

Carrier, Wheel-cum-Track, Machine Gun (Austro-Daimler ADMK) Variant of the standard ADMK *Maschinengewehr-Traktor* with large dual wheels replacing the smaller two-wheeled rocking bogies. Like the standard model it had the AD FB12/20 4-cyl. 2312.2-cc (80 × 115 mm) engine with forced air cooling.

Carrier, Wheel-cum-Track (Austro-Daimler ADMK) 4-cyl., 20 bhp, 4F1R, wb 1.77 m, 3.68 × 1.54 × 1.46, 1730 kg. Combat weight 2300 kg. This is how the 'Adamek' appeared in the German *Wehrmacht* vehicle catalogue (D600), with open-top bodywork for crew of 6. The vehicle was also employed as a light gun tractor.

Tractor, Wheel-cum-Track (Saurer RK7) Prototype for *Räder-Kettenwagen*, developed by Saurer. It had a 75/90-hp 6-cylinder diesel engine and changing from wheel to track operation was possible with the vehicle moving at low speed. Small-scale production during 1937–40, mainly for the *Deutsche Wehrmacht*.

Tractor, Wheel-cum-Track (Saurer RK7) 4-cyl. diesel, 70 bhp, 5F1R × 2, wb 2.40 m, 4.50 × 1.96 × 2.20 m, 5300 kg. Tyres 8.25-20. Payload 1500 kg. This 1937–40 edition appeared with various body types, incl. AOP and with V-blade snow plough. Shown: *Instandsetzungskraftwagen* (maintenance/repair vehicle).

AUSTRIA
ARMOURED VEHICLES

In 1903, Ing. Paul Daimler of the Austrian Daimler firm, following an idea of the then Captain Ludwig von Tlaskal-Hochwall, designed a four-wheel drive chassis with armoured bodywork. It became the world's first purpose-built armoured car and was demonstrated to Emperor Franz Joseph during Army manoeuvres. Reportedly, the first visible effect was that all the nearby horses bolted and a celebrated general was thrown from his mount. Upon this the Emperor made a gesture of discontent and although some more demonstrations followed this was the premature end of a machine which was much ahead of its time. The first Austrian armoured vehicles to see active service were mainly rather crude truck-based models and it was not until the 1930s that some new purpose-built designs made their appearance.

Armoured Car, 4×4 (Austro-Daimler 40/45 PS) 4-cyl., 40 bhp, 4F1R, wb 3.50 m, 4.60×1.76 m, 3000 kg approx. Revolving turret. Crew 4–5. Driver occupied conventional position but could lower his seat to disappear from view. Produced in 1903/04. Permission to sell the vehicle to France was granted in late 1906.

Armoured Car, 4×2 (Fiat-Ansaldo M1930) 4-cyl., 40 bhp, 4F1R, 5.40×1.80×2.40 m, 3800 kg approx. Crew 6. Armament 2 heavy machine guns. 1930. Other 4×2 A/Cs (*Strassenpanzer*) included truck-mounted Heigl, Junovic and Romfell designs and some captured from the Italians and Russians in WWI (Austin, Rolls-Royce).

Armoured Car, 4×4 (Austro-Daimler ADSK) 4-cyl., 60 bhp, 4F1R×2, wb 2.25 m, 3.72×1.95×1.70 m (1937 prototype: wb 2.00 m, 3.70×1.65×1.57 m), 3200 kg (gross). Tyres 10.50-16. Air-cooled 3620.5-cc (98×120 mm) OHV engine at rear. Crew 1+4. IFS/IRS with transversal leaf springs. Five produced, 1938.

Armoured Car, 6×6 (Austro-Daimler ADKZ) 6-cyl., 85 bhp, Voith hyd. trans. w/3-speed auxiliary gearbox, wb 2.67 (BC 1.20) m, 4.76×2.40×2.42 m, 8000 kg (gross). Tyres 210-20. AD M650 4994-cc (96×115 mm) OHV engine, developing 100 bhp max. Exp. model, with certain components in common with ADAZ and ADGZ. 1937.

Armoured Car, 8×8 (Austro-Daimler ADGZ/P) 6-cyl., 150 bhp, Voith hyd. trans. w/3-speed auxiliary gearbox, wb 1.85+ 1.05+1.85 m, 6.26×2.16×2.56 m, 12,000 kg (gross). Front and rear wheel steering. Crew 8. 8.25-20 bullet-proof tyres. Armament: 3.5-cm and 2-cm guns and 4 heavy MGs. Prototype, 1931.

Armoured Car, 8×8 (Austro-Daimler ADGZ) Of the ADGZ production model, which had dual tyres on the centre axles and other improvements, 27 were built during 1935–37 for the Army and *Polizei*. Some were used by the German military police in Danzig in 1939 (shown). A further 25 were made in 1942.

Armoured Car, Wheel-cum-Track (Saurer RR7/2) 4-cyl., 70 bhp, 5F1R×2, wb 2.40 m, 4.50×2.47×2.02 m, 6400 kg (gross). Used by German *Wehrmacht* as AOP *Sd.Kfz.254* (captured vehicle shown). Gräf & Stift produced a Maybach V12-engined tank with similar wheel-cum-track system in 1938. (IWM photo E12273).

BELGIUM

In World War I the Belgian forces used many imported trucks including American Overland (300 were acquired from Willys-Overland in London in 1914) and Kelly-Springfield, British Commer, French Panhard, Italian Diatto, etc., in addition to vehicles (incl. armoured cars) of their own manufacture, e.g. Bovy, Metallurgique, Minerva, etc. In the late 1920s the Verviers firm of Dasse supplied a number of trucks with American 8-cyl. Lycoming engines.

During the inter-war period most Belgian motor vehicle manufacturers disappeared. In 1936/37 large-scale motorization of the armed forces began and the surviving firms of Brossel, FN, Miesse and Minerva supplied various types, supplemented by imported and locally-assembled vehicles, the latter including Chevrolet, Dodge, Ford, GMC, Latil, etc. Military type motorcycle combinations, with sidecar wheel drive if required, were produced by FN, Gillet and Sarolea. Ambulances were supplied on Dodge, Ford and Minerva chassis. Many of these vehicles were captured by the German *Wehrmacht* in 1940.

Motorcycle, Solo (FN) The Fabrique Nationale des Armes de Guerre SA (now Fabrique Nationale Herstal SA) of Herstal, Liège, produced motorcycles from 1902. Many saw service in the First World War, incl. these 285-cc models (introduced in 1912) used by Australian dispatch riders.

Motorcycle, with Sidecar, 3×1 (FN M86 mil.) Basically a civilian model (Estafette 600 cc, OHV) this machine was used by the Belgian Army with and without sidecar. Sidecar shown was standard type with FN/Browning MG. There were also ammunition carrier, wireless and other variants. 1936.

Motorcycle, with Sidecar, 3×2 (FN M12) Special military model with 1000-cc (80×78 mm) 22-bhp flat-twin SV engine, driving rear and sidecar wheel through a gearbox with four forward and reverse speeds (with normal and low ratio). Weight 240 kg. Max speed 100 km/h. Produced during 1937–39.

Motorcycle, with Sidecar, 3×2 (Gillet 750) The Herstal firm of Gillet produced this 2-cyl. 2-stroke 750-cc combination in 1938 for the *Gendarmerie*. The Army version (1938–40) had cross-country tyres and a military pattern sidecar. Both had 4F1R transmission, driving rear and sidecar wheel.

Motorcycle, with Sidecar, 3×2 (Sarolea 1000) Like FN and Gillet, Sarolea during the late 1930s built heavy combinations with sidecar wheel drive. 978-cc (88×80 mm) HO-2-cyl. SV engine developing 20 bhp at 3000 rpm. 3F×2 and reverse transmission (with planetary reduction). Weight 525 kg approx. Speed 80–85 km/h.

Motorcycle, with Sidecar, 3×1, Armoured (FN M86 mil.) One of the variants of the militarized FN M86 combination was this *Groupe blindé* version of *c.* 1937. Top, rear and left-hand side of sidecar were open. As on the non-armoured sidecars there were MG mountings at front and rear. Also supplied to Argentina.

Motortricycle, 3×2 (FN Tricar) This multi-purpose three-wheeler was derived from the FN M12 sidecar combination and could carry 600 kg or a crew of five, including the driver. Ground clearance was 225 mm. Transmission 4F1R×2. Disc wheels at rear. It was introduced in 1939.

Car, Heavy, 4×2, Command (FN 8 Coloniale) Developed in the mid-1930s this 8-cylinder car was intended mainly for colonial use. It had IFS, dual servo brakes (Dewandre) and good ground clearance. Weight 2600 kg. Wheelbase 3.35 m. Track 1.55 m. Max. road speed 90 km/h. Fording depth 0.60 m.

Truck, 3-ton, 4×2, Cargo (FN 8 Cyl. 3T) Typical Belgian Army truck (*Camion type Armée Belge*) on 8-cyl. FN chassis. Similar military cab and GS bodywork was mounted on Brossel, Miesse and other 4×2 chassis, including American Chevrolet and Ford (which were assembled in Antwerp). *c.* 1936.

Truck, 3-ton, 4×2, Searchlight (Miesse) Mobile searchlight unit for air section of Belgian Army engineers' division. Bodywork by Carrosserie Jonckheere. 20 supplied in 1932. Miesse trucks, buses and tractors were made at Buysinghen, Brussels. From 1932 many had licence-produced Gardner diesel engines.

Truck, 2½-ton, 4×4, Cargo (GMC) American four-wheel drive 1938 GMC chassis with Belgian cab and bodywork. Belgium also bought light and medium 4×4 Ford/Marmon-Herrington chassis and were the latter company's first customer for the all-wheel drive converted Ford ½-ton truck in 1936.

Tractor, 4 × 4, Light Artillery (FN 47) Twin-engined prototype for low-silhouette cross-country artillery tractor, designed early in 1937. Independent front and rear suspension with exceptional wheel travel. Shared certain components with FN 63C truck. Vehicle was not put into production.

Tractor, 4 × 4, Medium Artillery (Latil TL6) Several types of 4 × 4 and 6 × 6 Latils were in service. The chassis were assembled in Belgium and had Belgian bodywork. Shown is all-wheel steer tractor with 75-mm gun. Earlier wheeled artillery tractors included Chenard-Walcker (F) and Pavesi (I).

Tractor, 4 × 4, Heavy Artillery (Brossel TAL) One of a series of *Tracteurs Artillerie Lourde* built in 1938 by Brossel Fréres of Brussels. 6-cyl. petrol engine. Centrally mounted power winch. All-wheel steering. Brossel also supplied diesel-engined military cargo trucks and truck-mounted revolving cranes.

Tractor, 4 × 4, Heavy Artillery (FN 63C/4RM) Produced during 1938–40 this tractor was developed from the conventional FN 63C truck. It had a 6-cyl. 65-bhp 3.94-litre SV engine, 4F1R × 2 transmission, lockable differentials in transfer box and both axles, centrally-mounted winch. 5.68 × 2.30 m, wb 2.70 m.

Tractor, Half-Track, Heavy Artillery (FN/Kégresse) 6-cyl. 60 bhp, 4F1R × 2, wb 2.71 m, 5.13 × 1.90 × 2.12 m, 4050 kg. Kégresse-Hinstin bogies, produced under Citroën licence. FN 63T 4-litre engine. 1935. Preceding model (1934 Minerva-FN-Kégresse) had Minerva 36 3-litre 55-bhp sleeve-valve engine.

Tractor, Full-Track, Light Artillery Ford 4-cyl., 52 bhp, 4F1R, 2.41 × 1.63 × 1.44 m, 1960 kg. Crew 3 (Model 1) or 1 (Model 2). *Tracteur 'Utility B'*, used for towing 47-mm AT gun. One of several vehicles built by SA des Ateliers de Construction de Familleureux under licence from Vickers-Armstrongs Ltd (GB).

Armoured Car, 4 × 2 (Minerva) During World War I the Belgian forces used various types of improvised armoured cars on Minerva (1914 model shown), SAVA and other car chassis. Improved developments of these were still in use in the 1930s, together with more modern all-wheel drive types by Berliet (F).

Tractor, Armoured, 4 × 4 (Ford/Marmon-Herrington) One of several body types on Antwerp-assembled V8-engined 1-ton truck chassis. The armoured hulls were built in Belgium and the vehicles were destined for towing the 47-mm Anti-Tank gun. Other body styles (unarmoured): command car, radio and cargo truck. 1939/40.

CANADA

The Canadian automotive industry, situated mainly in Ontario, has always been closely linked with that of neighbouring Michigan, USA. In 1914 the six major producers/assemblers were Ford (by far the biggest), Studebaker, Reo, McLaughlin-Buick, Hupp and Gramm. Military vehicles used in World War I and during the inter-war period were mainly US types, supplemented by certain British vehicles. In 1915 the Drednot Motor Truck Co. of Montreal exhibited a 46 HP armoured car built for the Russian Government. About the same time a fleet of Autocar trucks, made in the US, was acquired. These included eight open-topped armoured vehicles which saw service in France with the 1st Canadian Motor Machine Gun Brigade; they had been designed by M. Raymond Brutinel, a French engineer who resided in Canada. In 1939/40 a family of standardized trucks was developed in accordance with British WD specifications. Made by Ford and General Motors these were subsequently supplied in large quantities to the armed forces of the British Commonwealth (see *The Observer's Fighting Vehicles Directory—World War II*).

Car, 4-seater, 4×2, Command (Autocar) Battery Commander bodywork on 1914 US-built 2-cyl. 2-ton truck chassis. Other body styles: Cargo, Ambulance, Ammunition and Fuel Carriers, Armoured Car (see USA section). Touring cars included Briscoe, Cadillac (standardized), Chalmers, Ford, Russell.

Truck, 3-ton, 4×2, Cargo (Kelly-Springfield) In 1914 the Canadian Government ordered some 150 trucks from United States factories, incl. Autocar, Jeffery Quad, Kelly-Springfield, Packard, Peerless and White. More followed and most were shipped to Europe for active service with the Canadian Contingent.

Trucks, 1½-ton, 4×2, Cargo During the 1930s most Army transport vehicles were supplied by the Canadian subsidiaries of Ford, General Motors and Chrysler. Illustrated are some typical examples: 1936 Ford and 1937 Chevrolet and GMC. They were commercial chassis with general service bodies, operated by the RCASC.

Truck, 1½-ton, 6×4, Wireless (Crossley 20/60 HP) In the 1930s the Dominion of Canada acquired a limited number of military trucks and tracked MG carriers from Great Britain. Vehicle shown was based on WD-type Crossley 30-cwt 6×4 chassis. Normal-control Leyland six-wheelers were also used.

Truck, 2½-ton, 6×4, Cargo (Gotfredson) 4-cyl., 28.9 HP (RAC), 4F1R×2. Weight 9184 lb. Payload on roads 5 tons. Commercial truck, fitted with auxiliary gearbox and Hendrickson rear bogie. Made by Gotfredson Motor Co. of Walkerville, Ontario, in 1927. Note skis at front, chains at rear.

Armoured Car, 4×2 (Case) Built on a Case passenger car chassis and presented to the Overseas Battalion of Canadian Volunteers of Victoria by a citizen of British Columbia. It reputedly saw active service in France in 1915. Case cars were made in Racine, Wisconsin, USA, during 1910–27.

Armoured Truck, 4×4 (Russell/Jeffery) 40 built in 1915 on US Jeffery Quad chassis. Featured integrated fighting compartment, turret-mounted MG, dual controls, 4-wheel steering. Dim. 17′10″ × 5′2″ × 8′9″. Shipped to Britain in 1916 but not used operationally. In 1917 they were passed on to India.

C

CZECHOSLOVAKIA

The republic of Czechoslovakia was formed on 28 October 1918, mainly from Bohemia, Moravia and Slovakia. Before 1918 these territories had been part of the Austro-Hungarian Empire (see also Austria and Hungary). The country's motor industry has a long tradition and in particular Tatra, at one time Nesselsdorfer, ranks with the oldest motor vehicle manufacturers in Central Europe. Most Czech motor manufacturers have at one time or other supplied military vehicles but the majority of special vehicles were products of Praga, Skoda and Tatra. The following table shows the growth of the number of motor vehicles in service with the Czech armed forces from 1918 until World War II.

Type	Quantity and date				
	Nov. 1918	Dec. 1922	Dec. 1934	March 1938	March 1939
Motorcycles	—	83	357	937	1238
Passenger Cars	107	186	260	1557	2110
Trucks	360	1196	825	4560	7176
Tractors	1	180	274	349	403
Special Vehicles	—	114	410	889	829
AFVs	—	17	196	491	490
Total	468	1776	2322	7783	12246

In 1938 Hitler-Germany annexed the Sudetenland (border regions in the North) and on 15 March 1939, the Nazis occupied Bohemia and Moravia. Czechoslovakia had been well prepared from the military point of view but was let down by her allies. As a result the *Wehrmacht* went in virtually without a fight. They confiscated all military equipment and much of it was given to other Axis powers. Others, particularly AFVs, were used by the *Wehrmacht* itself, mainly in Poland and France (a Praga RN 6×4 truck was encountered by the editor in France as late as 1971). Production was not discontinued in 1939, however.

Under Nazi rule the following production programme was set up:

Cars: Aero 1—1.5-litre model, Praga 1.5—2-litre model, Skoda 2—3-litre model; Trucks: Praga 3- and 4½-ton models, Skoda 6½-ton model; Motorcycles: Autfit 250-cc, CZ 125- and 350-cc, Jawa 125-, 250- and 350-cc models. Tatra was considered to be German and came under the *Reichsdeutsche Bauprogramm*. Tracked AFVs continued in production by CKD/Praga. Later, of course, several German designs were co-produced by various Czech firms (see also *The Observer's Fighting Vehicles Directory—World War II*).

Who's Who in the Czechoslovak Automotive Industry

Aero	Aero, Továrna Letadel, Praha-Vysočany.
CKD	(see Praga).
CZ	Česka Zbrojovka, Strakonice.
Jawa	Zbrojovka Ing. F. Janeček, Praha.
L & K, Laurin & Klement	Laurin & Klement AS, Mladá Boleslav (also in Praha and Plzen) (taken over in 1925 by Skoda of Pilsen, *qv*).
Nesselsdorfer	Nesselsdorfer Wagenbau-Fabriks-Gesellschaft (Vorm. k.k. priv. Wagenfabrik Ignaz Schustala & Co.), Nesselsdorf (later Kopřivnice); in 1923 merged with Ringhoffer-Waggonfabriken-Konzern of Prague to become Tatra (*qv*).
Praga	Ceskomoravské Továrny na Stroje, Praha; later Ceskomoravska-Kolben-Danek AS, Praha-Karlin (also Böhmisch-Mährische Maschinenfabriken AG, Praha-Lieben).
RAF	Reichenberger Automobilfabrik, Reichenberg (later Liberec) (merged with Laurin & Klement in 1913).
Skoda	Akc. Spol. drive Škodovy Závody, Plzen and Skoda-ASAP, Mladá Boleslav.
Tatra	Závody Tatra AS (Tatra-Werke AG), Kopřivnice (from 1935: Ringhoffer-Tatra; originally Nesselsdorfer, *qv*).
Z	Ceskoslovenská Zbrojovka AS (also Zbrojovka Závody Jana Švermy), Brno.

CS

CARS, 4×2
and MOTORCYCLES

Makes and Models: *Cars:* Aero 30 (1934–39), 50 (1936–40). Jawa Minor I (1936–39). Laurin & Klement S 24PS (1913–18), etc. Nesselsdorfer S4 30 PS (1906–11), S4T 45 PS (*c.* 1911), S6 50 PS and S6U 65 PS (1913–18), etc. Praga Alpha 12 PS (1914–18), Mignon 25 PS (1913–18), Grand (1913–18), Piccolo (1932), Alfa (1932–38), etc. RAF SV70 PS (1915–18). Skoda Popular 1000 (1935–38), Popular 1100 (1939–41), etc. Tatra T (1924), 30 (1930), 57 and 57a (1932–41), etc. Z Z18 (1927–35), Z4 (1936), etc.

Motorcycles: BD 500 cc (1926), CZ 175 (1935–37), etc. Itar (Walter) 744 cc (1921). Jawa 175 (1937), 350 SV (1934), etc. Ogar 4 (1937), etc. Praga (1928), etc.

General Data: The pre-1919 cars listed were used in the Austro-Hungarian Army. Of these, the Praga Grand was also made by Rába (see Hungary). The Aero 30 was a civilian type sports car. The military versions, two- and four-seaters, were used as command cars and differed from the civilian models mainly in having a water pump in the cooling system (rather than thermosyphon). The Aero 50 command/staff car was a civilian type also. It had a four-in-line water-cooled two-stroke engine with two carburettors. In 1940 a final series of 40 was supplied to the German *Wehrmacht*. The Jawa Minor was basically a civilian car but had special military bodywork. Like the Aeros it had a water-cooled two-stroke engine, front wheel drive and independent suspension with transversal leaf springs. The Skoda Popular 1000 was based on the civilian 1935–38 chassis and like the Jawa had military bodywork. Its successor, the 1939 Popular 1100 (with 1098-cc OHV engine) was produced with soft-top four-door bodywork for the German Army until 1941. The Tatra (T)57 was a light but rugged car designed by the late Dr. Techn. h.c. Hans Ledwinka. It was known popularly as the 'Hadimirschka' and like the original twin-cylinder 11 and 12 models featured a backbone chassis with the propeller shaft enclosed in the central tube. The flat-four engine unit, attached to the forward end of the tube, carried the transversal leaf independent front suspension; the swing axles at the rear were also sprung by a transversal leaf spring, anchored on the top of the differential housing which was at the other end of the chassis tube. This chassis was in production from 1932 until 1948 and was also made in Austria (Austro-Tatra) and Germany (Röhr Junior, Stoewer Greif). Motorcycles were basically commercial machines and some of the heavier types were used with sidecar.

Vehicle shown (typical): Car, 4-seater, 4 × 2 (Aero 30)

Technical Data:
Engine: Aero 2-cylinder, I-T-W-F, 998.6 cc (85 × 88 mm) 30 bhp at 4000 rpm. CR 5.2 :1.
Transmission: 3F1R. Front-wheel drive.
Brakes: mechanical.
Tyres: 5.25-16 or 130 × 40.
Wheelbase: 2.515 m. Track, front and rear 1.20 m.
Overall l × w × h: 3.78 × 1.43 × 1.45 m.
Weight: 850 kg.
Note: independent front and rear suspension with transversal leaf springs. Max. speed 100 km/h. Also with two-seater bodywork. 1934–39. Trucks in background are Praga model RV 6 × 4 2-tonners (see page 51).

Motorcycle, Solo (CZ 175) 1-cyl., 5.5 bhp, 3F, wb 1.30 m, 2.00 × 0.77 m, 97 kg. 172-cc (60 × 61 mm) two-stroke engine. Tyres 3.00-19. Maximum speed 80 km/h. Amal carburettor. Bosch ignition and electrical equipment. 1935–37. Other military motorcycles used at this time included Jawas and Ogars.

Car, 4-seater, 4 × 2 (Aero 50) 4-cyl., 50 bhp, 3F1R, wb 2.94 m, 4.40 × 1.60 × 1.51 m, 1235 kg. Tyres 5.75-16. 1997-cc (85 × 88 mm) two-stroke twin-carburettor engine, driving the front wheels. Independent front and rear suspension. Track, front 1.18 m, rear 1.24 m. Speed 125 km/h. 1936–40.

Car, 4-seater, 4 × 2 (Jawa Minor I) 2-cyl., 19 bhp, 3F1R, wb 2.41 m, 3.60 × 1.35 m, 700 kg. Tyres 4.75-16. 616-cc (70 × 80 mm) two-stroke engine, driving front wheels. IFS, IRS. Mechanical brakes. Track, front 1.05 m, rear 1.07 m. Speed 95 km/h. Basically a civilian model, used as command car, 1935–38.

Car, 4-seater, 4 × 2 (Skoda Popular 1000) 4-cyl., 22 bhp, 3F1R, wb 2.43 m, 3.77 × 1.36 × 1.52 m, 750 kg. Tyres 5.50-16. 995-cc (65 × 75 mm) side-valve engine, driving rear wheels. IFS, IRS (swing axles at rear). Track, front 1.05 m, rear 1.10 m. 350 two- and four-seaters were supplied as command cars, 1935–38.

Car, 6/10-seater, 4×2 (Praga/Rába Grand) 4-cyl., 40–45 bhp, 4F1R, wb 3.20 m. Tyres 880×120 or 895×135. Shaft drive. Reinforced chassis. Winching drums on rear wheels. Used as battery commander's car in heavy motorized artillery units of Austro-Hungarian Army, 1914–18. Also 6-seater staff car with 3.30-m wb.

Car, 4-seater, 4×2 (Praga Piccolo Normand) 4-cyl., 22 bhp, 3F1R, 3.87×1.45 m, 895 kg. Tyres 720×120. 995-cc (60×88 mm) SV engine. Max. speed 70 km/h. Two-door convertible bodywork with wooden rear section, based on car-derived civilian pickup truck, c. 1932. Used by *Gendarmerie* (police forces).

Car, 5/6-seater, 4×2 (Praga Alfa) 6-cyl., 34 bhp, 4F1R, wb 3.00 m, 4.35×1.53×1.72 m, 1320 kg. 1795-cc (65×90 mm) side-valve engine. Track, front 1.25 m, rear 1.27 m. Max. speed 90 km/h. Tyre size 5.00-19. Wooden rear body. 1932. Note the revolving arrow type direction indicator.

Car, 6/8-seater, 4×2 (Praga Alfa) 6-cyl., 55 bhp, 4F1R, 5.30× 1.62 m approx. Police car, based on civilian Praga Alfa car chassis. One of a batch produced in 1938. 2490-cc (75×94 mm) side-valve engine. Four-door soft-top bodywork with removable side screens.

Car, 6-seater, 4×2 (Tatra T30) 4-cyl., 24 bhp, 4F1R, wb 3.17 m, 4.10×1.65×1.65 m, 1090 kg. 1680-cc (75×95 mm) air-cooled flat-four ('boxer') engine. CR 4.9:1. Track, front and rear 1.30 m. IFS, IRS. Mechanical brakes. Maximum speed 90 km/h. 1930.

Car, 2-seater, 4×2 (Tatra T57a) 4-cyl., 20 bhp, 4F1R, wb 2.55 m, 3.72×1.56 m, 750 kg approx. Tyres 5.25-16. 1160-cc (70×75 mm) engine with horizontally-opposed air-cooled cylinders and 5.0:1 CR. Basically civilian two-seater, used as command car. 1935.

Car, 4-seater, 4×2 (Tatra T57a) 4-cyl., 20-bhp, 4F1R, wb 2.55 m, 3.72×1.56 m. Tyres 5.25-16. 1160-cc OHV air-cooled flat-four engine. Soft-top command car bodywork with removable side screens. 1935. One of various body types on this civilian T57 chassis, designed by Hans Ledwinka.

Car, 2/4-seater, 4×2 (Tatra T57a) Open version of the T57a *vojensky* (military) car, produced during 1936–41. Hoops, tilt and side curtains were provided. Some had rear seats. Subsequent production (for German *Wehrmacht*) had four-door *Kübelwagen* bodywork (T57K). Bonnet-cum-wings assembly was hinged at scuttle.

CARS, 4×4, 6×4 and 6×6

Makes and Models: Praga AV (6×4, 1936–39). Skoda 6LT6 (6×4, c. 1935), 903 (6×4, 1935–37), etc. Tatra V750 (4×4, 1937), T79/V799 (4×4, 1938), V809 (4×4, 1940), T26 (6×4, 1925–27), T26/30 (6×4, 1927–33), T71 (6×4, 1932), T72 (6×4, 1933–35), T82 (6×4, 1935–38), T93 (6×6, 1939–40), etc. Z and Z-T2 (4×4, prototypes, 1936 and 1938).

General Data: Four- and six-wheel drive cars were not as numerous as 6×4 types. The latter were simpler and less costly, being based on existing light 6×4 truck chassis. Most were of Tatra manufacture and designed by the firm's chief designer Dr Hans Ledwinka. They featured a tubular backbone chassis, like the contemporary Tatra civilian cars. The first were built on the T26 chassis, a four-cylinder introduced in 1925 and developed from the 4×2 T13, which, in turn, was a 1-ton truck version of the T12 car chassis. The T26 was further developed and engine power was increased several times during its production span throughout the early 1930s. Instead of the T13's single differential unit with the transversal leaf spring above it, the tandem-drive T26 had a combined housing for the two final drive units and longitudinal inverted semi-elliptical leaf springs suspending the four swinging type half-axles. Front wheel suspension was not independent. The final models in this range were the T72 of 1933–35 and the T82 of 1935–38. The latter had a restyled rounded front end (still designed as one assembly and hinged at the scuttle for maximum engine accessibility) and some of them had 'belly support wheels' at the centre of the running boards. The Tatra T93 was based on the T93 6×6 truck chassis.

The Tatra V750 was a 4×4 chassis and appeared with different bodies. One took part in British Army vehicle trials in North Wales in May, 1937. It developed into the V799 of 1938 and eventually the V809 of 1940 which the German *Afrika Korps* used. The 'V' in these designations stands for *Vojensky* (military).

Vehicle shown (typical): Command Car, 4-seater, 4×4 (Tatra T79/V799).

Technical Data:
Engine: Tatra V799 4-cylinder, I-I-A-F, 2191 cc (80×109 mm), 50 bhp at 3000 rpm.
Transmission: 4F1R×2.
Brakes: hydraulic.
Tyres: 5.50-18.
Wheelbase: 2.60 m. Track 1.30 m.
Overall l×w×h: 3.70×1.60×1.90 m approx.
Weight: N.A.
Note: supplied in 1938 as command car for artillery units. Independent front and rear suspension with swing axles. Overhead camshaft engine with Scintilla magneto ignition.

Car, 4-seater, 4×4 (Tatra V750) 4-cyl., 30 bhp, 4F1R×2, wb 2.43 m, length 3.31 m, track front/rear 1.30/1.25 m. Air-cooled 1688-cc (80×84 mm) OHV flat-four engine. There was a variant with 2.60-m wheelbase, four doors and different front end styling. Prototypes, 1937.

Car, 4-seater, 4×4, Radio (Z-T2) 4-cyl., 42 bhp, 4F1R×2, wb 2.53 m, 3.90×1.59×1.75 m, 1423 kg. Air-cooled 2-litre (81.5× 94 mm) two-stroke twin-carb. in-line engine with magneto ignition. Diff. locks front and rear. IFS/IRS with swing axles and transversal leaf springs. Track 1.24m. 1938.

Car, 6-seater, 6×4, Command (Praga AV) 6-cyl., 70 bhp, 3F1R×2, wb 2.96 (BC 0.92) m, 5.10×1.75×1.85 m, 2200 kg. Water-cooled 3462-cc (80×115 mm) side-valve engine. Tyres 5.50-18. Track, front 1.39 m, rear 1.41 m. IFS, IRS. Max. speed 92 km/h. Just under 400 made, 1936–39.

Car, 6-seater, 6×4, Command (Skoda 903) 6-cyl., 66 bhp, 4F1R×2, wb 2.93 (BC 0.92) m, 5.15×1.80×1.90 m, 2000 kg approx. Payload 700 kg. 3140 cc (80×104 mm) engine. Tyres 5.50-18. Track 1.40 m. Speed 90 km/h. Also soft-top variants (*Faeton, Break*; spare wheels in front wing wells). 1935.

Car, 6-seater, 6×4, Command (Tatra T26/30) 4-cyl., 24 bhp, 4F1R×2, wb 2.66 (BC 0.85) m. Succeeded T26 of 1925/26. T71 of 1932 was similar but had 30-bhp engine. All had air-cooled flat-four power units and tubular backbone chassis with IRS. They were basically light truck chassis. Max. speed 70 km/h. 1931.

Car, 6-seater, 6×4, Command (Tatra T72) 4-cyl., 32 bhp, 4F1R×2, wb 2.66 (BC 0.92) m, 2200 kg approx. Also with 3.36 (BC 0.92) m wheelbase. Various body types, incl. bus. 1910-cc (80×95 mm) flat-four air-cooled engine. LWB model had full-length soft top and spare wheels between doors and rear mudguards. 1933.

Car, 6-seater, 6×4, Command (Tatra T82) 4-cyl., 55 bhp, 4F1R×2, wb 3.36 (BC 0.92) m. Last in the line of Tatra 6×4 cars, produced from 1935 until World War II. Engine was enlarged to 2490 cc (90×98 mm, air-cooled flat-four). Hydraulic brakes. Four-door bodywork. Twin spare wheels at rear. Also with full-length body.

Car, 6-seater, 6×6, Command (Tatra T93) V-8-cyl., 70 bhp, 4F1R×2, wb 3.27 (BC 0.94) m. 3980-cc (80×99 mm) OHV engine with forced air-cooling. Backbone chassis with IFS/IRS (similar to that of T93 *Rumänian* truck) and 6.00-20 tyres, single rear. Also supplied with four-door bodywork as shown on T82 (*qv*). 1939/40.

CS

BUSES and PERSONNEL CARRIERS
4×2

Over the years various types of basically civilian buses and coaches were used for troop transport. During World War I these were mostly impressed types, including postal service vehicles. The Laurin & Klement shown on this page was used in Montenegro (now part of Yugoslavia) and was designed about 1910 specifically for use on mountain roads. In order to provide a wheelbase short enough to enable the vehicle to negotiate hairpin bends the jackshaft and sprockets of the chain final drive were located behind the rear axle. There was provision for towing a four-wheeled cargo trailer. The Praga personnel carriers shown were commercial chassis with special bodywork. The windscreen was fixed, rather than of the folding type. Note the difference in the body panelling; the early type's rear body was of composite wood and sheet metal construction.

Truck, 1½-ton, 4×2, Postal Coach (Laurin & Klement) Cab-over-engine chassis used mainly for postal service in Montenegro with interchangeable coach (shown) and box van bodies. Picture shows one of these vehicles impressed into military service in 1916. Four-cylinder 24-bhp engine (105×130 mm).

Truck, 1½-ton, 4×2, Personnel (Praga PN) 4-cyl., 25 bhp, 3F1R, wb 3.00 m. Tyres 6.50-18. 1447-cc (70×94 mm) engine. Track, front 1.30 m, rear 1.32 m. Max. speed 60 km/h. Commercial truck chassis of c. 1930. Special bodywork with full-length folding top. Note long overhang at rear.

Truck, 2-ton, 4×2, Personnel (Praga AN-6) 6-cyl., 40 bhp, 4F1R, wb 3.50 m. Tyres 32×6.50. 2493-cc (75×94 mm) engine. Track, front 1.40 m, rear 1.38 m. Max. speed 60–65 km/h. Commercial truck chassis of mid-1930s. Special composite bodywork with full-length folding top.

CS

AMBULANCES 4×2

Illustrated here are some typical examples of ambulances based on civilian chassis. The Praga Mignon was a *Gebirgs-Sanitäts-Kraftwagen* of the *Deutschen-Ritter-Ordens* used in WWI.

The Tatra was typical of the late 1920s and 1930s and was used by military and other government authorities. There were several other Tatra ambulances, including a six-wheeler (T92 6×4). The T13 shown was based on a one-ton truck chassis which was a direct development from the Ledwinka-designed T11 and T12 cars. It differed from the cars mainly in having reinforced suspension. These twin-cylinder vehicles were extremely robust and long-lasting; some reached over 1,000,000 km. They were popularly known as 'Tatritschek'.

During 1936–39 Skoda supplied 62 Model 206 ambulances with 2.7-litre engine.

Truck, Light, 4×2, Ambulance (Praga Mignon) 4-cyl., 21 bhp, 4F1R, wb 3.10 m. 1850-cc (70×120 mm) side-valve engine. Pneumatic tyres, size 875×105. Track 1.35 m. Max. speed 70 km/h. 1911–14. Later models (4th–6th Series) had 30-bhp 2296-cc (75×130 mm) engine; 10 of these were supplied in 1920.

Truck, Light, 4×2, Ambulance (Tatra T20/T20 Sp) 4-cyl., 45 bhp, 4F1R, wb 3.37 m. Track 1.28 m. Tyres 820×120. 3560-cc (90×140 mm) water-cooled OHC engine. Max. speed 90 km/h. Mechanical brakes on rear wheels only. Four stretchers. Also with flat radiator and dual rear tyres. 1925.

Truck, Light, 4×2, Ambulance (Tatra T13) 2-cyl., 12–14 bhp, 4F1R, wb 3.06 m. 1056-cc (82×100 mm) air-cooled horizontally-opposed OHV engine with 4.8:1 CR. Speed 45–50 km/h. 1-ton truck chassis with backbone chassis and independent rear suspension, 1925–1933.

CS

TRUCKS, 4 × 2

Makes and Models: Laurin & Klement DL (2-ton, 1909–15), LK2, MS (2-ton, 1914–17), etc. Nesselsdorfer/Tatra NL2, TL2 (2-ton, 1915–23), TL4 (4-ton, 1916–24), etc. Praga Charon (0.6-ton, 1909–11), V Series (3-4-ton, 1911–22), L Series (2–3-ton, 1912–31), R Series (2-ton, 1913–26), N Series (5-ton, 1915–31), RN and RND (2½-ton, from 1936), etc. Skoda 500, 506, 540, 550 (3-ton and Skoda-Martin tractors for S-T, 1924–35), 104 (1¼-ton, 1938), 254 (3-ton, 1937), 256B (2½–3-ton, 1940), 706 (7-ton, 1940), etc. Tatra T20/20 Sp (radio van, 1925), T23 (4-ton, 1927–31), T27, T27a, T27b (3-ton, from 1930), T43 (1½-ton, from 1928), etc.

General Data: Most of the earliest trucks listed above were used by the Austro-Hungarian Army during World War I. Pragas, produced in the Bohemian capital of Prague, were particularly popular in Army supply columns. There were several types and many were licence-produced by Rába (pronounced Rava) in what in 1918 became Hungary (*qv*). The Praga V or PS3 truck was designed for military use in 1911 and before the First World War was sold to civilian customers under a special subsidy scheme. Like several similar pre-1914 Austro-Hungarian trucks, e.g. Austro-Fiat and Gräf & Stift, they had an open cab with folding top. From 1914 a semi-closed cab, i.e. one with half-doors, was fitted and from about 1920 those which were still in production had a hard-top cab with fixed windscreen.

Some of the Praga trucks had a very long production span. The Praga L, for example, was first introduced in 1912 as a 2–2½-tonner; three series of these were produced during 1912–13. The fourth to eighth series were made during 1914–23. The ninth to 13th series (3-ton) were in production during 1925–26, followed by the 14th to 16th during 1927–28. The latter had pneumatic tyres as standard equipment, as did the modernized 17th to 24th (and last) series which ran from 1929 until 1931.

Praga trucks were also exported, for example to Yugoslavia, Rumania and Turkey. During the 1930s the Tatra T27 was widely used as a military cargo truck. After 1938 these and many Praga and Skoda 4 × 2 trucks went into service with the German and Axis forces. Among imported 4 × 2 trucks was a number of British Tilling-Stevens petrol-electric searchlight vehicles, delivered in 1925.

Vehicle shown (typical): Truck, 2-ton, 4 × 2, Cargo (Nesselsdorfer TL2).

Technical Data:
Engine: Nesselsdorfer 4-cylinder, I-I-W-F, 3552 cc (90 × 140 mm), 35 bhp at 1600 rpm.
Transmission: 4F1R. Shaft drive.
Brakes: mechanical.
Tyres: 920 × 140, solid; dual rear.
Wheelbase: 3.70 m. Track 1.45 m.
Overall l × w × h: 5.38 × 1.80 × 2.45 m.
Weight: (chassis) 2100 kg.
Note: overhead-camshaft engine. Speed 35 km/h. Fuel tank capacity 140 + 20 litres. Used during WWI by Austro-Hungarian Army. Nesselsdorfer was later renamed Tatra.

Truck, 2-ton, 4×2, Cargo (Laurin & Klement MS, LK2) 4-cyl., 40–45 bhp, 4F1R, wb 4.50 m. 5911-cc (112 × 150 mm) engine. Tyres, solid 920 × 120, later *'Sembusto-Bereifung'* (resilient metal and rubber type, as illustrated). L & K trucks were also bought by the Japanese Army (Model DL 2-ton, 4508-cc, 1909–15).

Truck, 4-ton, 4×2, Cargo (Nesselsdorfer TL4) 4-cyl., 35 bhp, 4F1R, wb 4.00 m, 6.06 × 1.80 × 2.50 m, 3000 kg (chassis) approx. Engine as 2-ton model TL2 (*qv*) but stronger wheels, chassis, etc. Originally used by Austro-Hungarian Army, continued from 1919 until 1924 as commercial truck (shown).

Truck, 3–4-ton, 4×2, Cargo (Praga V) 4-cyl., 37 bhp, 4F1R, wb 4.00 m, 5.90 × 1.80 × 2.85 m, 3850 kg. Chain final drive. 6840-cc (110 × 180 mm) engine. Three produced in 1911, 10 in 1912, 42 in 1913, 75 in 1914, 90 in 1915, 180 in 1916, 602 in 1917 and finally 390 in 1920–22.

Truck, 5-ton, 4×2, Cargo (Praga N) 4-cyl. 50 bhp, 4F1R × 2, wb 4.20 m, 6.80 × 2.00 × 2.75 m, 4800 kg. Shaft drive. 7478-cc (115 × 180 mm) engine. Solid tyres, 140 × 880 front, 160 × 1050 dual rear. 124 produced during 1915–18. Known also as NV5. Used with Praga Model Q 5-ton trailer.

Truck, 2-ton, 4×2, Cargo (Praga R) 4-cyl., 35–45-bhp, 4F1R, wb 3.30 m, 2000 kg. Solid tyres, 910×100. 3824-cc (90×150 mm) engine. Late type shown, with fixed cab roof and windscreen. Used by Austro-Hungarian and Turkish, later also by Czech Army. 1913–26.

Truck, 3-ton, 4×2, Workshop (Praga L) 4-cyl., 45 bhp, 4F1R, wb 3.60 m, 5.90×1.90 m approx. 3824-cc (90×150 mm) engine. Shaft drive. Used in Czech Army. One of Praga L 9–13th Series, which was optionally available with pneumatic tyres (max. speed 35 vs 25 km/h). *c.* 1925.

Truck, 3-ton, 4×2, Cargo (Praga L) 4-cyl., 40 bhp, 4F1R, wb 3.60 m, 5.91×1.89×2.30 m, 2750 kg approx. Pneumatic tyres, size 36×6, dual rear. Side-valve engine. Max. speed 38 km/h. Late production had 12-volt Bosch electrical lighting. Note direction indicator on windscreen.

Truck, 3-ton, 4×2, Cargo (Praga L) 4-cyl., 40 bhp, 4F1R, wb 3.60 m, 5.56×1.95×2.29 m, 2750 kg approx. Payload 3000–3500 kg. Military version of one of the last production batches of the Praga L (1930/31). 1929 model had square cab. All had Bosch magneto ignition.

Truck, 3-ton, 4×2, Tractor (Skoda Martin 550) 4-cyl., 50 bhp, 4F1R, wb 4.00 m. Used with ballast body for towing full-trailers and with fifth wheel coupling for semi-trailers. Trailer shown is semi-trailer with dolly. 6786-cc (120×150 mm) engine. Maximum speed 30 km/h. Produced 1926–28.

Truck, 3-ton, 4×2, Tractor (Skoda Martin 540) 4-cyl., 40 bhp, 4F1R, wb 3.20 m. Shown with 5½-ton artillery semi-trailer (wb, from king pin, 3.00 m), carrying field gun, full-track tractor and crew. 5911-cc (112×150 mm) engine. Solid tyres, 920×120 front, 920×140 dual rear. 1924–29.

Truck, 3-ton, 4×2, Tractor (Skoda Martin 540) 4-cyl., 40 bhp, 4F1R, wb 3.20 m. Shown with fuel tanker semi-trailer (wb, from king pin, 3.00 m). Wheel track 1.40, 1.43 and 1.53 m. Solid tyres, 920×120 front, 920×140 dual rear (tractor and semi-trailer). Max. speed 20 km/h. 1924–25.

Truck, 3-ton, 4×2, Tractor (Skoda Martin 500) 4-cyl., 35 bhp, 4F1R, wb 2.90 m. Shown with signals semi-trailer, *Vzor* 24. Pneumatic tyres. Semi-trailer wheelbase (from king pin) 3.05 m. 4700-cc (100×150 mm) engine. Max. speed 40 km/h. Track, tractor, 1.60 m, semi-trailer 1.92 m. 1924–26.

Truck, Light, 4×2, Radio Van (Tatra T20/20Sp) 4-cyl., 45 bhp, 4F1R, wb 3.37 m. 3560-cc (90×140 mm) engine with overhead camshaft. Dual rear tyres. House-type body. Rigid axles with semi-elliptic leaf springs. Similar vehicle used as ambulance (*qv*). c. 1925.

Truck, 4-ton, 4×2, Cargo (Tatra T23) 4-cyl., 65 bhp, 4F1R, wb 4.34 m. Tyres 40×10.5. 7480-cc (115×180 mm) engine. Speed 55 km/h. Tubular backbone chassis. IFS with coil springs. IRS with oblique-mounted semi-elliptic cantilever leaf springs. Civilian bodywork shown. 1927–31.

Truck, 3-ton, 4×2, Cargo (Tatra T27) 4-cyl., 52 bhp, 4F1R, wb 3.80 m, 6.30×2.10×2.40 m, 3280 kg. Tyres 36×8. 4260-cc (95×150 mm) OHV engine. Tubular backbone chassis. IFS with transversal leaf spring, IRS with oblique-mounted quarter-elliptic leaf springs. Civilian bodywork shown. 1930.

Truck, 3-ton, 4×2, Cargo (Tatra T27) 4-cyl., 60 bhp, 4F1R, wb 3.80 m, 6.50×2.14×2.39 m, 3550 kg. approx. Tyres 9.75-20. 4712-cc (100×130 mm) water-cooled OHV engine. Tubular backbone chassis with independent suspension front and rear. Commercially available. Military bodywork shown. 1938.

CS

TRUCKS
6 × 4 and 6 × 6

Makes and Models: Praga RV, RVR (2-ton, 6 × 4, 1935–39), SV (4-ton, 6 × 4, 1935), etc. Skoda L (2–2½-ton, 6 × 4, 1932–35), 6LT6-L, 6L (2-ton, 6 × 6, 1935–37), 6ST6-T, H, HD (4–5-ton, 6 × 4, 1936–39), 6ST6-L, S, SD (4-ton, 6 × 4, 1935–39), 6STP6, 6S, 6STP6-LD, 6SD (4-ton, 6 × 6, 1936–39), 6V, 6VD (5-ton, 6 × 6, 1935–39), 6VTP6-T, 6K (11-ton, 6 × 6, 1935–39), etc. Tatra T22 (3-ton, 6 × 4, 1936–37), T24 Series (6–10-ton, 6 × 4, 1925–38), T26 (1½-ton, 6 × 4, 1925–27), T26/30 (1½-ton, 6 × 4, 1927–33), T28 (3-ton, 6 × 4, 1932–35), T29 (8-ton, 6 × 4, 1930–35), T72 (1½-ton, 6 × 4, 1933–35), T81 and T81H (6½-ton, 6 × 4, 1939–42), T82 (2-ton, 6 × 4, 1935–40), T85, T85A (5-ton, 6 × 4, 1936–41), T85/91 and T98 (4–5-ton, 6 × 4, 1936–38), T92 (2-ton, 6 × 4, 1937–40), T93 (2-ton, 6 × 6, 1937–40), etc.

General Data: Best known of the above six-wheeled trucks were the Ledwinka-designed light and heavy Tatras. The first of these were the 1½-ton 6 × 4 T26 and the 6–8-ton 6 × 4 T24. They were developed in 1925 and derived from the contemporary T13 and T23 4 × 2 trucks. The conversion was relatively simple, the vehicles retaining their backbone chassis but with a further set of swinging-type half-axles added at the rear. In the case of the T26 a common final drive unit was used for the rear axles (at the same time forming a rear extension of the chassis tube). The T24 had two final drive/differential units, connected by a further tubular section. Both types were further developed during the 1930s, culminating in the T92/T93 and T81. Developed from the latter were the wartime T111 and the post-war T138 and T148 ranges.

The Praga RV was used in relatively large numbers and like the Tatra T93 was also exported, notably to Rumania. Some of the Skoda types were exported to Sweden and many trucks of all three makes were used by the German and other Axis forces during the Second World War.

Vehicle shown (typical): Truck, 2-ton, 6 × 4, Cargo (Tatra T82).

Technical Data:
Engine: Tatra 4-cylinder, H-I-A-F, 2490 cc (90 × 98 mm), 55 bhp at 3200 rpm.
Transmission: 4F1R × 2.
Brakes: hydraulic.
Tyres: 6.00-20.
Wheelbase: 3.36 m, BC 0.92 m.
Overall l × w × h: 5.60 × 2.00 × 2.45 (1.95) m.
Weight: 3150 kg (w/winch 3220 kg).
Note: lockable differential. Semi-elliptic leaf springs, transversal front, longitudinal rear. 2½-ton winch optional. Also with soft-top command car bodywork (single rear tyres). Some had higher cab and bonnet.

D

Truck, 1½-ton, 6×4, Cargo (Tatra T72) 4-cyl., 32 bhp, 4F1R×2, wb 3.36 (BC 0.92) m. Track 1.30 m. Tyres 5.50-18. Speed 60 km/h. Gradability 34%. Horizontally-opposed 1910-cc (80×95 mm) air-cooled engine. 1933. Later appeared with single rear tyres and more modern cab.

Truck, 2-ton, 6×4, Photographic (Tatra T82) 4-cyl., 55 bhp, 4F1R×2, wb 3.36 (BC 0.92) m. This house-type van was used in the Czech Army as a photographic laboratory in 1938. Altogether 325 Tatra T82s were built, with various body types.

Truck, 2-ton, 6×4, Cargo (Tatra T92) V-8-cyl., 70 bhp, 4F1R×2, wb 3.27 (BC 0.94) m. 5.49×2.00×2.61 (2.13) m, 3580 kg. Tyres 6.00-20. 3-ton winch. 529 produced during 1937–38. A 6×6 version, model T93, was produced chiefly for Rumania (*qv*). All had Lockheed hydraulic brakes.

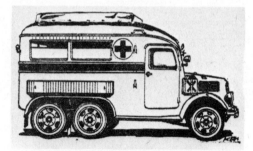

Truck, 2-ton, 6×4, Ambulance (Tatra T92) V-8-cyl., 70 bhp, 4F1R×2, wb 3.27 (BC 0.94) m. GVW 5580 kg. Field ambulance bodywork on T92 truck chassis with single rear tyres. Max. speed 70 km/h. Radius of action 550 km. 3981-cc (80×99 mm) OHC air-cooled engine with 5.5:1 CR.

Truck, 2-ton, 6×4, Cargo (Praga RV) 6-cyl., 68 bhp, 4F1R×2, wb 3.56 (BC 0.92) m, 5.69×2.00×2.50 (2.09) m, 3810 kg. 3-ton winch at rear. Max. trailer wt 2-ton. 3468-cc (80×115 mm) engine. Tyres 6.00-20. Torsion bar IFS. 1936. Also signals van (Model RVR; 57 units, 1936–39).

Truck, 4-ton, 6×4, Cargo (Praga SV) 6-cyl., 93 bhp, 4F1R×2, wb 4.12 (BC 1.15) m, 6.82×2.10×2.48 m. 7800-cc (105×150 mm) OHV engine with 5.0:1 CR. Tyres 9.00-20, single. IFS with coil springs. IRS with swing axles and leaf springs. Low chassis frame. 1935.

Truck, 2–2½-ton, 6×4, Cargo (Skoda L) 6-cyl., 66 bhp, 4F1R×2, wb 3.36 (BC 0.92) m, 5.78×2.00×2.38 m, 3700 kg. Max. trailer weight 3½-tons. 2½-ton winch. 3140-cc (80×104 mm) engine. Tyres 6.50-20. Stub-axle-mounted spare wheels. Produced during 1932–35.

Truck, 2-ton, 6×6, Cargo (Skoda 6L) 6-cyl., 66 bhp, 4F1R×2, wb 3.36 (BC 0.92) m, 5.78×2.00×2.38 m, 4000 kg. 2½-ton winch. Basically similar to Skoda Model L but with all-wheel drive. Tyres 6.50-20. Max. speed 63 km/h. Early models (6LT6-L) had 3-speed gearbox. 1935–37.

Truck, 4-ton, 6×4, Cargo (Skoda 6ST6-L) 6-cyl., 100 bhp, 4F1R×2, wb 4.07 (BC 1.15) m, 7.20×2.11×2.86 m, 7510 kg. Tyres 9.75-20. 5½-ton winch. Air brakes. Front suspension with transversal leaf spring. 8271-cc (112×140 mm) engine. 1937. Modernized and redesignated Skoda S (and SD diesel) in 1938.

Truck, 4–5-ton, 6×4, Cargo (Skoda 6ST6-T) 6-cyl. diesel, 100 bhp, 4F1R×2, wb 4.07 (BC 1.15) m, 7300 kg. Maximum trailer weight 5-tons. Speed 60 km/h. Tyres 10.50-20. 5-ton winch. 8550-cc (110×150 mm) diesel engine. Modernized and redesignated Skoda HD (and H petrol) in 1938.

Truck, 4-ton, 6×4, Cargo (Skoda H) 6-cyl., 100 bhp, 4F1R×2, wb 4.12 (BC 1.25) m, 7.10×2.16×2.65 m, 7800 kg. Tyres 10.50-20. 7-ton winch. Similar chassis were used for specialist bodies, incl. mobile workshop and crane. 1938–39. Used until 1950s. Model HD had diesel engine.

Truck, 4-ton, 6×4, Workshop (Skoda H or HD/APD) 6-cyl., 100 bhp, 4F1R×2, wb 4.12 (BC 1.25) m, 7.44×2.25×3.10 m, 7000 kg. Payload 3800 kg. Max. trailer weight 5-tons. 4-ton winch. Speed 60 km/h. Also supplied with diesel engine. Towed 4-wheel workshop trailer (various types). Produced in 1938.

Truck, 4-ton, 6×6, Cargo (Skoda 6SD) 6-cyl. diesel, 100 bhp, 4F1R×2, wb 3.73 (BC 1.25) m, 6.45×2.15×2.65 m, 7700 kg. Tyres 9.75-20. 5-ton winch. Air brakes. Tilt cab. 8550-cc (110× 150 mm) diesel engine. 75 produced, 1936–39. Early production designated 6STP6-LD.

Truck, 11-ton, 6×6, Tank Transporter (Skoda 6VTP6-T) 6-cyl., 175 bhp, 3F1R×2, wb 4.37 (BC 1.25) m, 7.50×2.50× 2.85 m, 11,000 kg. Tyres 10.50-20, dual all round. 10-ton winch. Max. trailer weight 17-tons. 12,920-cc (140×140 mm) engine. 1935–39. Late models designated Skoda 6K.

Truck, 3-ton, 6×4, Cargo (Tatra T22) 4-cyl., 60 bhp, 4F1R×2, wb 3.78 (BC 1.15) m. Tyres 36×8 (9.75-20). Auxiliary wheels with pneumatic tyres at front. Payload on roads 5-tons. Max. trailer weight 3-tons. 4712-cc (100×150 mm) OHV engine. Gradability, with trailer, 30%. 18 produced, 1934–35.

Truck, 3-ton, 6×4, Cargo (Tatra T28) 4-cyl., 63 bhp, 4F1R×2, wb 3.75 (BC 1.10) m. Tyres 8.25-20. 4712-cc (100×150 mm) water-cooled OHV petrol engine. Max. speed 48 km/h. Typical Tatra backbone chassis with independent suspension with leaf springs front and rear, 1931–32.

Truck, 5-ton, 6×4, Cargo (Tatra T85) 4-cyl., 80 bhp, 4F1R×2, wb 4.09 (BC 1.18) m, 7.30×2.35×2.60 m, 6000 kg. IFS, IRS. Track, front 1.70 m, rear 1.90 m. Max speed 60 km/h. Knorr air brakes. Also used as prime mover. 90 produced, 1936–38.

Truck, 5-ton, 6×4, Fuel Tanker (Tatra T85) 4-cyl., 80 bhp, 4F1R×2, wb 4.09 (BC 1.18) m, 7.30×2.35 m approximately. Tyre size (fuel tanker and cargo truck) 9.75-20 (8×20), single rear. 8180-cc (120×180 mm) water-cooled OHV petrol engine. 224 produced, 1936–38.

Truck, 5-ton, 6×4, Cargo (Tatra T85A) 4-cyl., 80 bhp, 4F1R×2, wb 4.09 (BC 1.18) m, 7.30×2.35×2.60 m, 6000 kg. Track, front 1.70 m, rear 1.90 m. Tyres 9.75-20. Max. speed 60 km/h. Generally similar to T85 which it superseded. Produced until 1941, also with bus bodywork.

Truck, 4–5-ton, 6×4, Cargo (Tatra T85/91) 6-cyl. diesel, 85-bhp, 4F1R×2, wb 4.07 (BC 1.15) m. Track front/rear 1.70/1.90 m. 7300-cc (105×140 mm) water-cooled diesel engine, developing 85 bhp at 1900 rpm. Air brakes. Tyres 9.75-20. Known also as T98. 1936–38.

Truck, 6–8-ton, 6×4, Cargo (Tatra T24) The T24 was developed in 1925 from the 4×2 T23. Early models had a four-cylinder petrol engine, later a six-cylinder petrol or diesel (MAN). Developments (civ.) included T24/58, T24/59, T24/63 and T24/67. Shown is a 1929 model undergoing cross-country tests.

Truck, 8-ton, 6×4, Cargo/Artillery (Tatra T29) 4-cyl., 65 bhp, 4F1R×2, wb 4.67 (BC 1.30) m. 7478-cc (115×180 mm) OHV petrol engine. Fitted with winch and special body for carrying gun, full-track tractor and crew. 112 produced, 1934–35, also with other body types.

Truck, 8-ton, 6×4, Workshop (Tatra T29) 4-cyl., 65 bhp, 4F1R×2, wb 4.53 (BC 1.30) m. Mobile workshop van with dropsides on T29 chassis, 1934. Chassis was similar to late-production T24 but equipped with 4-cyl. engine. Tyres 40×10.5. A balloon-winch was also based on this truck.

Truck, 6½-ton, 6×4, Cargo (T81H) V-8-cyl., 150 bhp, 4F1R×2, wb 4.56 (BC 1.22) m, 9.03×2.50×2.75 m, 8000 kg. 14,726-cc (125×150 mm) OHV engine. Tyres 9.75-20. Speed 65 km/h. 220 T81 and T81H trucks were made during 1939–42. Early production (T81) had 12,464-cc (115×150 mm) V-8 diesel engine.

CS

TRACTORS, WHEELED

After the splitting up of the Austro-Hungarian Empire at the end of World War I, Czechoslovakia acquired the successful Austro-Daimler M17 4×4 heavy artillery tractors. Some of these were still in use in the Czech Army as late as 1938. Praga and Skoda produced similar tractors during the early 1920s, viz. the U and Z models shown in this section. Skoda also produced some types of steam-propelled trucks and tractors under British Sentinel licence. During World War II Skoda produced the large Porsche-designed *Ostradschlepper* which, for that time, was unusual in having large all-metal wheels like the World War I designs. During the 1930s the Tatra T25 6×6, a development from the T24 6×4, was also used. The original version (1927) had solid rubber tyres. In addition many Praga full-track types (see following section) were in service.

Tractor, 4×2, Searchlight (Nesselsdorfer M) 4-cyl., 25 bhp, 4F1R, wb 3.35 m, length 5.08 m. Track, front/rear 1.25/1.37 m. Chain final drive. 4.9-litre (110×130 mm) SV engine. 900-mm diameter wheels with solid tyres. Towed searchlight trailer, carried crew and generating set. 5 built, 1909.

Tractor, 4×2, Artillery (Skoda/Sentinel) Produced under licence-agreement with Sentinel of Great Britain. Two-cylinder steam engine, developing 70 bhp at 250 rpm. Cylinder bore and stroke 170×230 mm, piston displacement 10,440 cc. Winch at rear. 1925.

Tractor, 4×2, Artillery (Skoda U) 4-cyl., 50 bhp, 4F1R, wb 3.80 m. Track, front/rear 1.80/1.70 m. Payload 5000 kg. Max. trailer weight 6000 kg. 7480-cc (115×180 mm) petrol engine. Max. speed 13 km/h. Prototype manufactured by Praga, series production by Skoda.

Tractor, 4×2, Artillery (Skoda Z) Generally similar to Skoda Model U but powered by a 6-cylinder petrol engine of 11,200-cc cubic capacity (115×180 mm), developing 80 bhp at 1000 rpm. Weight 11,900 kg. Max. drawbar pull 10,000 kg. 1921–23. Designed by Praga.

Tractor, 4×2, Artillery (Skoda Z) Generally similar to tractor shown on left but with solid rubber tyres instead of all-metal wheels. All models were fitted with a powerful engine-driven winch. Fuel consumption was considerable, namely 350–410 litres per 100 km.

Tractor, 6×6, Artillery (Tatra T25) 6-cyl., 120 bhp, 4F1R×2, wb 3.63 (BC 1.25) m, 10,000 kg. 5-ton winch. 12,210-cc (120× 180 mm) petrol engine. Developed from T24 in 1926. 28 produced during 1933–34. 1926 prototype ('Bulldog') had 115-mm bore (11,200-cc) 110-bhp engine.

Tractor, 6×6, Artillery (Tatra T84) One of three prototypes, 1935. Production licence sold to Lorraine in France. 4-cyl. 65-bhp 7478-cc (115×180 mm) OHV petrol engine behind cab. 4F1R×2 transmission. Wheelbase 3.78 (BC 1.15) m. Track 1.70 m. Tyres 9.75-20. Max. speed 55 km/h. 2½-ton winch.

TRACTORS, FULL-TRACK and WHEEL-CUM-TRACK

The Czech motor industry, mainly Praga and Skoda, produced a large variety of tracked tractors and many were exported. Praga customers included the Netherlands East Indies, Peru, Portugal, Rumania, Sweden and Turkey (chiefly variants of the T6 model). Principal models were the Praga BD25 (1925, 1927), T3, T4, T5, T6, T7, T8 and T9 (from 1937; also designated T III, T IV, etc) and the Skoda STH (from 1934), SH and SHD (from 1935), MH and MTH (from 1935). Tatra produced only a few types. Their V740 was rather similar in appearance to the Swedish Landsverk 131 and 132. The Skoda KH60 was basically the chassis of the KH60 wheel-cum-track tank, which was designed by Vollmer and used by the Czech, Italian, Yugoslavian and Russian Armies. 'KH' stands for *Kolo-Housenka* (wheel-tracks).

Tractor, Wheel-cum-Track, Artillery (Skoda KH60) 4-cyl., 60 bhp, 4F1R. Track, wheels 1.80 m, tracks 1.15 m. Max. speed on tracks 18 km/h, on wheels 45 (cross-country 30) km/h. Track width 300 mm. Gradability 45°, ground clearance 300 mm. Engine at rear. Prototype, 1924.

Tractor, Wheel-cum-Track, Artillery (Tatra KTT) Experimental vehicle of 1930. Overall length and width 6.10 × 2.10 m. 6-cylinder 12,210-cc (120 × 180 mm) water-cooled OHV petrol engine, developing 120 bhp at 1500 rpm. Truck type bodywork. Known as *Kolopásový Traktor*.

Tractor, Full-Track, Artillery (Skoda STH) Produced during 1934–37 with 4-cylinder 4850-cc (105 × 140 mm) petrol engine, developing 43 bhp at 1150 rpm. Max. speed 14.2 km/h. Model SH had 60-bhp engine, SHD had 66-bhp diesel engine. MH and MTH had 2760-cc 2-cyl. petrol engines.

Tractor, Full-Track, Artillery (Tatra V740) Produced as prototype in 1937. Overall dimensions 4.35 × 1.70 × 1.75 m. 5800 kg. Track centres 1.41 m. Powered by 4-cyl. water-cooled 7480-cc (115 × 180 mm) petrol engine, developing 65 bhp at 1200 rpm. 5F1R gearbox. Max. speed 15 km/h.

Tractor, Full-Track, Artillery (Praga T4) 4-cyl., 56 bhp, 5F1R, 4.14 × 1.67 × 2.18 m, 5070 kg. DBP 4500 kg. Payload 1000 kg or six men. Crew two. 5-ton winch. Praga N 6080-cc (110 × 160 mm) petrol engine. Track width 260 mm, centres 1.22 m. Max. speed 21 km/h. 1938.

Tractor, Full-Track, Artillery (Praga T6) 6-cyl., 75 bhp, 4F1R × 2, 4.80 × 1.80 × 2.42 (1.70) m, 6850 kg. DBP 6300 kg. Payload 800 kg or 8–10 men. 6-ton winch. Praga TN 7800-cc (105 × 150 mm) petrol engine. Track width 340 mm, centres 1.40 m. Max. speed 25 km/h. 1937.

Tractor, Full-Track, Artillery (Praga T9) Typical example of late-1930s Praga tractor with truck type cab and rear body. Praga 8V4 140-bhp 14½-litre V-8-cyl. petrol engine. Users included Peruvian Army (1938). The German *Wehrmacht* later also employed these and other, similar, Praga tractors.

ARMOURED VEHICLES

During the 1920s the armaments and heavy engineering firm of Skoda produced a series of four-wheel drive armoured cars of symmetrical design. They were known as the PA series (*Pancérový Automobil*). The first, designated PA 1, appeared about 1923. It could be driven at equal speeds in forward or rearward direction. The PA2 (1924) had a more rounded, turtle-shape, hull design with a fixed turret. The PA3 had faceted armour and a rotating turret. The PA4 and 5 were further modified and improved developments. Several armoured cars were built on commercial Tatra 6×4 chassis. Only relatively few types of Czech-designed tracked AFVs appeared but the latest, the TNHD (LT38), was used by the German *Wehrmacht* in considerable numbers. Most tracked AFVs were produced by CKD/Praga.

Armoured Car, 4×2 (Fiat/Skoda) Among the first AFVs of newly-formed Czechoslovakia in 1919 were some French Renault FT tanks, an Italian Lancia armoured car and this twin-turret vehicle built by Skoda on an Italian Fiat truck chassis.

Armoured Car, 4×4, PA1 (Skoda) 4-cyl., 80 bhp, wb 3.80 m, 6.00×2.10×2.50 m, 6500 kg. Duplicate steering controls at rear. Four-wheel steering. Max. speed 52 km/h in either direction. 9730-cc (135×170 mm) engine. Crew: five. Armament: two MGs. 1923.

Armoured Car, 4×4, PA2 (Skoda) 4-cyl., 80 bhp, wb 3.76 m, 6.20×2.20×2.60 m, 7000 kg. Solid rubber or bullet-proof (self-sealing) 40×8 tyres. Used by Czech and Yugoslav Armies and Vienna Police. Note symmetrical curved armour. Crew: five. Armament: two MGs. 1924.

Armoured Scout Car, 4×4 (Z) 4-cyl., 42 bhp, 4F1R×2, wb 2.53 m, approx. 4.00×1.60×1.80 m. Prototype, based on Z 4×4 field car chassis, by Czeskoslovenska Zbrojovka AS at Brno (manufacturer also of the well-known Ogar motorcycles and Zetor tractors). 1936.

Armoured Car, 6×4, VZOR 34 (Tatra) 6-cyl., 40–50 bhp, 4F1R×2, wb 5.20 m (overall), 7.60×1.86×3.10 m, 9000 kg. 2310-cc (70×100 mm) petrol engine. Commercial tubular backbone chassis with swing axles. Hull built up from riveted armour plates. Rotating turret. 1929.

Armoured Car, 6×4, VZOR 30 (Tatra) 4-cyl., 32–40 bhp, 4F1R×2, 4.02×1.52×2.02 m, 3100 kg. 1910-cc (80×95 mm) air-cooled flat-four engine. Armament two MGs. Armour up to 11 mm. Max speed 45 km/h. Gradability 25°. Based on commercial 1½-ton T72 chassis.

Carrier, Full-Track, A/T Gun, LKMVP (Skoda) 6-cyl., 76 (later 105) bhp, Praga-Wilson transmission, 4.30×2.10×1.60 m, 6700 kg (w/o gun). Air-cooled American tank engine. Carried or towed 37-mm anti-tank gun. Average speed 40 km/h. One built, 1938. Taken over by *Wehrmacht* in 1941.

DENMARK

In 1923 the Danish Army Technical Corps organized a 10,000-km trial of a 2-ton Triangel truck which took 18 days and was entirely successful. The makers, Forenede Automobil Fabriker SA of Odense, had produced motor vehicles since 1918 and supplied various types of trucks, as well as half-track tractors, to the Danish armed forces. Triangel was Denmark's premier marque. Another Danish product which the forces used in considerable numbers was the Nimbus motorcycle. This was a heavy machine with four-in-line engine and was used both as a solo machine and with sidecar. In addition the Danes used imported cars and trucks, notably American Fords. Ford V8 cars and trucks of 1939 were, in fact, still in service for many years after World War II, when, initially, most of the Army's vehicles originated from Allied surplus stocks. In the early 1930s the British Thornycroft company supplied at least one 6 × 4 truck to the Danish Army.

Motorcycle, with Sidecar, 3 × 1 (Nimbus) Danish-made machine with 18-bhp 750-cc OHV 4-cyl. in-line engine and shaft drive. Sidecar combination shown, which was produced about 1937, carried a Madsen machine gun with a crew of two. Note the telescopic front forks and the pressed steel frame.

Truck, 3-ton, 4 × 2, Cargo (Triangel) One of a series of military trucks produced about 1939 for the Danish armed forces by the Forenede Automobil Fabriker AS of Odense, which were Denmark's leading truck manufacturers. The body style suggests that this truck was used as a searchlight carrier.

Truck, 4-ton, Half-Track, Prime Mover (Triangel/Kornbeck) Designed in 1928 for hauling 10.5- and 15-cm coastal guns. Rear bogie was designed by M. Kornbeck. Dimensions: 5.90 × 2.50 m, wb 4.10 m. 70-bhp 6-cyl. engine. 4F1R gearbox with double reduction and 5-ton winch. Tyres: front 38 × 7, rear 890 × 140 (solid).

FINLAND

Like so many countries, the Finnish Army at the outbreak of hostilities in 1939 was still largely dependent on horsedrawn transport. To mobilize her forces it was necessary to obtain quickly a large fleet of motor vehicles. The Ford Motor Company in the United States helped to solve this problem and in 1940 supplied 4400 trucks on special credit terms agreed by President Roosevelt. The terms were 10 per cent cash with the balance to be settled five years after the cessation of hostilities. This gesture was commemorated by the Finnish Army in 1973 when a restored specimen was set on a plinth at an Army Ordnance centre near Helsinki. Prior to 1940 limited numbers of American trucks had been purchased from General Motors, e.g. Chevrolet 1½-ton trucks in 1932 and 1936. French, German and Russian vehicles were used also. Finland's own major truck maker was the Sisu company which was founded in 1931; during World War II this company was nationalized and in conjunction with another firm formed the Yhteissisu company which also produced military trucks. In 1948 Sisu became independent again.

Truck, 3½-ton, 4×2, Cargo (Sisu SH1) First production model of Sisu, delivered in 1931. The truck weighed 2500 kg and was powered by a six-cylinder petrol engine which had an output of 85 bhp at 2800 rpm. The gearbox had four forward speeds and one reverse. It was of thoroughly conventional design.

Truck, 1½-ton, 4×2, Cargo (Chevrolet RC) During the 1930s the Finnish Army acquired several batches of American Chevrolet trucks. Illustrated is the 1936 Model RC which had a 206-cu. in displacement OHV Six engine, 4F1R gearbox, 157-in wheelbase and 20-in wheels, single rear. Cabs and bodies were Finnish.

Truck, 1½-ton, 4×2, Cargo (Ford V8) In 1940 4400 regular Ford 158-in wheelbase trucks were supplied by the United States. The bodies were made locally. Truck shown is now exhibited at an Ordnance establishment near Helsinki. During the war similar Finnish Fords appeared with track bogies replacing the rear wheels.

FRANCE

It is an interesting coincidence that what is generally accepted as having been the world's first self-propelled road vehicle was a military vehicle. This was, of course, the Frenchman Nicholas-Joseph Cugnot's *Voiture a feu*, which in 1769 appeared on the streets of Paris. Cugnot had designed his three-wheeled steam-driven contraption, which carried a huge globular boiler up front, for artillery towing. Cugnot died in 1804, aged 79, without having been able to perfect his machine, and it was not until much later during that century that further progress was made. During the 1860s the firm of Lotz in Nantes, produced some steam locomotives, one of which, in 1867, was used to tow a gun and limber. In the 1890s the first practical motor vehicles appeared and by 1900 the French Army commenced employing cars and trucks in their annual manoeuvres. Some of these had special bodywork for military use. Between then and the beginning of the Great War of 1914–18 many experiments were carried out with a variety of vehicles. When it became evident that in a future war the use of trucks would be a necessity, the French Government (in common with those of Germany, Great Britain and Austro-Hungary) instituted a vehicle subsidy or subvention scheme. The system of subvention was much less costly for an army than direct purchase. In the latter case, work would have to be found in peace time to enable experience to be obtained resulting in the production of improved types and the vehicles originally purchased would then have to be superseded at high cost. In the case of the French subvention system, owners of approved trucks received an amount of 8200 francs in return for which their vehicle(s) had to remain at the Government's disposal, if needed, during a four-year period. As it was, when war broke out in 1914, large numbers of vehicles of many types had to be requisitioned and in addition substantial orders were placed with the motor industry, both in France and abroad, particularly in the United States. During 1917 the Army commenced disposing of surplus and war-worn trucks. This continued after the war when only sufficient quantities of selected types were retained for military use. Many of these trucks, as well as special purpose vehicles, remained in service until well into the 1930s; some were still in use as late as 1939/40 when the Second World War broke out.

During the 1920s and 1930s enormous progress was made in the design and development of cross-country vehicles, both of the wheeled and tracked types. The Kégresse-Hinstin tracked rear bogies, popularized by André Citroën, are an outstanding example and had an international following. The six-wheeled chassis designs of Renault and later Laffly and Latil and others were equally impressive. When France eventually fell to the Germans the booty of vehicles of all types was gigantic. Many of the serviceable French vehicles were taken into service by the German *Wehrmacht*. Most French plants were subsequently forced to produce, for the *Wehrmacht*, either their own or German designs. Some pre-1940 military vehicles remained in service with French forces overseas (notably North Africa), supplemented by fresh supplies from the Allies, again mainly from the United States.

Cugnot's steam-powered tractor of 1769 (replica).

Berliet trucks in Army trials before World War I.

Somua tractors and AFVs of the late 1930s.

Who's Who in the French Automotive Industry

Note: owing to the large number of makes and the complexity of some of their histories, only the more important ones which were still operative during the late 1930s are listed.

Berliet	Automobiles M.Berliet SA, Vénissieux (Rhône).
Bernard	Camions Bernard, Arcueil (Seine).
Chenard-Walcker	Sté Anon. des Ans. Ets. Chenard & Walcker, Gennevilliers (Seine).
Citroën	SA André Citroën, Paris (XVe).
De Dion-Bouton	De Dion, Bouton & Cie, Puteaux (Seine).
Delahaye	Sté des Automobiles Delahaye, Paris (XIIIe).
Delaunay-Belleville	Ateliers et Chantiers de l'Ermitage, St. Denis (Seine).
Ford	SA Ford Française, Poissy (Seine-et-Oise) and Asnières (Seine) (previously Matford, qv).
Hotchkiss	SA des Ans. Ets. Hotchkiss et Cie, St. Denis (Seine).
Laffly	Ets. Laffly, Asnières (Seine).
Latil	Cie des Automobiles Industriels Latil SA, Suresnes (Seine).
Licorne	Cie Française des Automobiles La Licorne, Courbevoie (Seine).
Lorraine	Sté des Moteurs et Automobiles Lorraine, Argenteuil (Seine-et-Oise).
Matford	SA Mathis, Strassbourg (Alsace) (see Ford).
Panhard	SA des Ans. Ets. Panhard et Levassor, Paris (XIIIe).
Peugeot	SA des Automobiles Peugeot, Paris (VIIIe), etc.
Renault	SA des Usines Renault, Billancourt (Seine).
Rochet-Schneider	Ets. Rochet-Schneider SA, Lyon-Monplaisir (Rhône).
Saurer	Automobiles Industriels Saurer, Suresnes (Seine).
Somua	Sté d'Outillage Mécanique et Usinage d'Artillerie, St. Ouen (Seine).
Unic	SA des Automobiles Unic, Puteaux (Seine).
Willème	Ets. Willème, Nanterre (Seine).

FRANCE
MOTORCYCLES

During World War I the armed forces employed a variety of civilian types, both solo machines (*Motocyclette*) and sidecar combinations (*Side-car*). During the inter-war period many new machines were acquired, varying from lightweights (*Velomoteur*) to heavy sidecar combinations specially built for military use. Some examples of *Velomoteurs*: Motobécane B1V2, Peugeot P53; *Motocyclettes*; Motobécane S5C (500-cc, 1934), Terrot RDA and RGMA (500-cc, *c.* 1938–40); *Side-car*: Gnome-Rhône 750 *Armée* (750-cc. *c.* 1938–40) and AX2 (804-cc, *c.* 1938–40), Monet-Goyon L5A1 (486-cc, *c.* 1935/36), René Gillet (350- and 750-cc, from 1928), Terrot TD (500-cc, 1939). Certain types were fitted with a machine gun, signals equipment or armoured sidecar. Some were used again after 1945. During the war many French motorcycles, particularly the Gnome-Rhône AX2, were employed by the German *Wehrmacht*.

Motorcycle, with Sidecar, 3×1 Typical application of civilian type motorcycle in French Army service during World War I. Note cane basket sidecar body, used for carrying mail. Photo was taken in the winter of 1916 and the machine was of British origin.

Motorcycle, with Sidecar, 3×1 (Monet-Goyon L5A1) 1-cyl., 486 cc, 4F, wb 1.15 m, 2.25×1.67×1.02 m, 300 kg. Maximum speed 55–60 km/h. Tyres 3.50-26. Magneto ignition. 6-volt battery for lighting. Sidecar interior dimensions 2.00×0.62×0.65 m. *c.* 1935/36.

Motorcycle, with Sidecar, 3×1 (Terrot RD) 1-cyl., 498 cc, 4F, wb 1.42 m, 2.33×1.74×1.04 m, 284 kg. Maximum speed 85 km/h. Tyres 4.00-27. Magneto ignition. 6-volt battery for lighting. Sidecar interior dimensions 2.00×0.53×0.50 m. 1939.

Motorcycle, with Sidecar, 3×1, Signals (René Gillet)
During the late 1920s the French developed various types of special purpose sidecars, exemplified by this telegraph/telephone unit. Others featured machine gun and armoured shields for gunner and driver. 1928.

Motorcycle, with Sidecar, 3×1 (Gnome-Rhône 750 Armée) HO-2-cyl., 750 cc, 4F, wb 1.45 m, 2.20×1.68×1.13 (1.05) m, 320 kg. Shaft drive. Tyres 4.00-27. Max. speed 78 km/h. OHV engine. Model AX2 had 804-cc SV engine and sidecar wheel drive (3×2). Late 1930s.

Motorcycle, Half-Track (Mercier) Experimental *motochenille* with driven tracked bogie at front. Note cut-out in armour shield to clear bogie during vertical articulation. JAP 350-cc air-cooled engine. Weight 160 kg. Max. speed 60 km/h. Gradability 25°. *c.* 1936.

Motorcycle, Full-Track (Lehaitre) This experimental and unsuccessful machine had a full-length rubber track of the Kégresse type. The 1-cyl. 500-cc Chaise engine drove the track via the rear sprocket. Note cable-operated (B) steering/stabilizer wheels (C). Weight 414 kg. *c.* 1938.

FRANCE

CARS, 4×2

Passenger cars, mainly of civilian type, have been used by the French armed forces since the very early days of motorization.

Especially during World War I the variety was great and even Paris taxi-cabs were used to carry troops. Some cars were fitted with machine guns, others were specially equipped or designed for military use in the deserts of Africa and Asia. During the 1930s quantities of civilian type four-door staff cars were acquired from big firms like Citroën (*Traction Avant* and conventional types), Peugeot (202, 402, etc.) and Renault (Monaquatre, Vivaquatre, etc.). Many of these types were later used by the German *Wehrmacht*, and employed again by the French forces in the early post-war years. Only a few typical examples are shown here. For purpose-built cross-country types see following section.

Car, 4×2, Command (Cottin-Desgouttes) One of several manufacturers to supply high-performance staff and command cars during World War I were Cottin et Desgouttes of Lyon. This was a heavy and powerful six-cylinder open model. Note the long cantilever-type leaf springs at rear.

Car, 4×2, Taxi-cab (Renault) One of the famous Paris taxi-cabs (*taximetre automobile* Renault AG), some 600 of which served successfully as troop carriers in the River Marne battle in September 1914. Engine was an 8 CV twin-cylinder, with three-speed gearbox. Top speed 35 mph.

Car, 4×2, Command (Berliet) Various Berliet cars and ambulances were used by the French and the British in WWI. A typical example (4-cyl. Model 12 or 15) is shown in a British ASC MT garage in 1918. Later Berliet military 4×2 cars included VIHCM and VURL four- and VRBM six-seaters.

Car, 4×2, Command (Renault) Model KZ of the 1920s, used as an Army command/staff car. The engine was a 4-cyl. 2120-cc (75×100 mm) with dash-mounted radiator. From 1933 militarized Renault Monaquatre cars were used in Morocco and the Sahara (*Voiture de Grand Raid*).

Car, 4×2, Desert Patrol (Hotchkiss) About 10 Hotchkiss AM80 cars were used by the French Army in the Syrian Desert during 1929–36. They had a 6-cyl. OHV 3-litre engine, 4F1R gearbox, 3.30-m wb and oversize tyres. During 1935–41 Chenard-Walcker cars were used for the same purpose.

Car, 4×2, Command (Citroën) Model C4-G liaison car for staff officers, based on modified 1931–32 C4 chassis with 4-cyl. 1767-cc 32-bhp engine, 4F1R gearbox, 2.78-m wb. Used in Morocco. Similar bodywork on Model C6 6-cyl. chassis. Known as *Voitures de Raid Saharien* or *de Grande Liaison*.

Car, 4×2, Command (Citroën) Model 15 CV command/staff car, 1932–34. Used by French Army in Algeria. 6-cyl. 2650-cc engine of 53–56 bhp with 4F1R gearbox. Wheelbase 2.91 m. Modified civilian *Berline* with double-skin roof, oversize tyres (160×40), special water tanks, towing hooks, etc.

Car, 4 × 2, Medium Sedan (Matford) Virtually standard civilian type car of the late 1930s. The Matford (Mathis/Ford) F82 was mechanically similar to the British Ford Model 62 2.2-litre. Model F81 had 3.6-litre engine. Both were V8s. Also with ambulance bodywork.

Car, 4 × 2, Medium Sedan (Peugeot) Several types of Peugeot cars were used by the French Forces. Shown is a Model 402B of 1938/39, officially designated *'Conduite Intérieure de moins de 15 CV, tôlée, 4 portes, 4 places'*. Engine was 2140-cc 4-cyl.

Car, 4 × 2, Medium Sedan (Renault) Model BDF1 (Primaquâtre) *Voiture de liaison de moins de 15 CV*. Model 603 4-cyl. 2383-cc engine. Wheelbase 2.71 m. Overall dimensions 4.10 × 1.54 × 1.58 m, weight 1200 kg. Tyres 5.75-16. 1938.

Car, 4 × 2, Heavy Sedan (Hotchkiss) This Hotchkiss Model 680 was known as a *Voiture de liaison de plus de 15 CV*. It had a 3-litre 6-cylinder engine. Wheelbase 3.25 m. Overall dimensions 4.90 × 1.80 × 1.62 m. Tyres 6.00-17. 1936–39.

FRANCE

FIELD CARS
4×4, 6×4 and 6×6

Makes and Models: Berliet VPB (6×6, 1926), VUR (4×4, 1928), VURB2, VURB2S (4×4, 1929/30), VUDB4 (4×4, 1932), VPDS (6×6, c. 1935), VPDS (6×4, 1938), etc. Hotchkiss R15R, W15T (6×6, 1938–39), etc. Laffly V10(C)M and V15R (4×4, 1938–40), S15R (6×6, 1937–39), etc. Latil M7, M7T1 (4×4, 1938–39), etc. Licorne V15R (4×4, 1938), etc. Lorraine 372 (6×4, 1933), etc. Renault MH (6×4, 1922–23), etc.

General Data: As can be seen in the following pages the French motor industry designed and produced some interesting cross-country cars during the twenties and thirties. They were known as VTT (*Voitures Tous Terrains*).

A common feature was the fitting of auxiliary bumper wheels, with pneumatic tyres, protecting the radiator and preventing the vehicle from digging its nose into the ground, whilst similar wheels mounted beneath the chassis protected the transmission and prevented the vehicle from 'bellying'. The earliest French six-wheeled car with articulating rear bogie was the Renault 10CV Model MH which made a name for itself in the first motor crossing of the Sahara. It featured dual tyres all round and a front-mounted winch and was Louis Renault's answer to André Citroën's Kégresse Autochenille half-track (*qv*). A specimen of the Renault was acquired by the British and evaluated by the Royal Army Service Corps in Aldershot. It initiated a successful line of British-made six-wheeler trucks and cars. The first Laffly VTT was powered by a Peugeot engine but production models had Hotchkiss engines. Hotchkiss also produced a considerable number of complete vehicles of the same type, between 1936 and 1940, under an agreement with Laffly before absorbing this company. After 1945 Hotchkiss produced several more. The Licorne company also produced some. Most of these cross-country vehicles, however, were produced as trucks and artillery prime movers (see relevant section) rather than field cars. A few appeared with armoured hulls. Berliet and Latil produced their own designs, both 4×4 and 6×6, and Lorraine licence-produced some Tatra designed cars, trucks and tractors (see section on 6×4 and 6×6 trucks and tractors).

For Citroën-Kégresse field cars see section on Half-Track Vehicles.

Vehicle shown (typical): Car, 4×4, Command and Reconnaissance (Latil M7T1)

Technical Data:
Engine: Latil 4-cylinder, I-L-W-F, 2720 cc (85×120 mm), 55 bhp at 2500 rpm.
Transmission: 4F1R × 2. Lockable diffs.
Brakes: hydraulic, power-assisted.
Tyres: 210 × 18 (puncture-proof type optional).
Wheelbase: 2.70 m. Track 1.45 m.
Overall l × w × h: 4.11 × 1.79 × 1.30 (min.) m.
Weight: (chassis) 1633 kg. GVW 3000 kg approx.
Note: IFS and IRS with coil springs. 12-volt electrics. Magneto ignition. 2.2-ton rear-mounted power winch. Canvas side curtains (some had four steel doors). 1939. Also employed as light artillery tractor (ammunition locker at rear).

72

Car, 4×4, Liaison (Laffly V10M) 4-cyl., 38 bhp, 4F1R×2, wb 2.05 m, 880 kg (gross: 1600 kg). Tyres 6.00-16. 1034-cc engine. Permanent four-wheel drive. One of several prototypes for a lightweight cross-country car, intended for field liaison and similar roles. Model V10CM had two-wheel steer. 1938–40.

Car, 4×4, Utility (Latil M7T1) 4-cyl., 55 bhp, 4F1R×2, wb 2.70 m. Basically same chassis as shown on page 71 but fitted with composite four-door utility body for Air Force. Designated *Break d'Aviation*. Produced in 1938/39. Similar *Break* was built by Laffly.

Car, 4×4, Command and Reconnaissance (Hotchkiss R15R) 4-cyl., 52 bhp, 4F1R×2, wb 2.14 m, 4.23×1.80 m, 2500 kg approx. Tyres 230×40. Rigid front axle, IRS with swing axles and quarter-elliptic springs. Twin prop. shafts fore and aft. Small auxiliary wheels with 420×150 aircraft tyres at front and centre of chassis. 1939.

Car, 4×4, Command and Reconnaissance (Laffly V15R) Hotchkiss 4-cyl., 52 bhp, 4F1R×2, wb 2.14 m, 4.23×1.80× 1.30 m, 2600 kg. Tyres 230×40. Similar to R15R but IFS with swing axles and coil springs, resulting in lower silhouette. Laffly V15T similar but with hub reduction gears for use as artillery prime mover.

73

Car, 4×4, Command (Berliet VUR) In 1928 Berliet produced an experimental four-wheel drive *Voiture de Liaison*. It had a rear-mounted engine, live axles and long cantilever-type leaf springs. Picture, taken on 15 Dec. 1928, shows revised prototype under test.

Car, 4×4, Command (Berliet VURB2) 6-cyl., 40 bhp, 4F1R×2, wb 3.00 m, 4.18×1.84 m, 2000 kg. GVW 3000 kg. 2½-litre engine. Tyres 36×8.25. 53 produced, 1929–30, incl. one for Belgian king. Two were modified for desert use in Algeria (VURB2S, S for Sahara).

Car, 4×4, Command (Berliet VUDB4) 4-cyl., 13 CV, 50 bhp approx., 4F1R×2, wb 3.00 m, 4.18×1.83 m. Lockable central differential with individual propeller shafts to each wheel (H-drive). Worm final drives. Spare wheels on idler hubs. Long cantilever springs. Live axles. 1932.

Chassis, 6×4, w/Test Body (Renault MH) 4-cyl., 10 CV, 3F1R, wb 3.02 (BC 0.90) m, 4.27×1.89 m, 1220 kg. Tyres 775×145, dual all round. Earliest light 6×4 with articulating bogie. Successful in 1924 Sahara crossings. Specimen shown tested by British RASC. French Army bought 22 in 1929.

Car, 6×6, Command and Reconnaissance (Berliet VPB)
16 CV *Voiture de Reconnaissance*, 1926. Long cantilever springs, front and rear, anchored at centre. Drive transmitted first to central differential in rear axle, thence via twin propeller shafts to centre and front wheels.

Car, 6×4, Liaison (Berliet VPDS) 4-cyl., 49 bhp, 4F1R×1, 5.26×1.82 m, 3000 kg. 2413-cc (80×120 mm) OHV engine. Tyres 230×40. Lockable diff. on rear axles. Mechanical brakes. One built in 1938 for Air Force. Note 'anti-bellying' wheels and long bogie wheelbase.

Car, 6×6, Personnel and Prime Mover (Laffly S15R)
Hotchkiss 4-cyl., 52 bhp, 4F1R×2, wb 2.83 m, 4.64×1.85×2.02 m, 4000 kg approx. Also with lower silhouette and IFS with swing axles instead of rigid front axle (Model W15R). 1937–39. Produced again during 1945–46 by Hotchkiss.

Car, 6×4, Command and Reconnaissance (Lorraine 72)
4-cyl., 30 bhp, 4F1R×2. This car was built under Czech Tatra licence and resembled the Tatra T71/72 (*qv*), featuring tubular backbone chassis with independent suspension all round but with additional bumper wheels at front.

FRANCE

BUSES AND TROOP CARRIERS
4×2

Like the British forces in World War I, the French employed large numbers of buses for the conveyance of troops to and from the battlefields. Some were requisitioned civilian types (incl. some 1100 Paris buses), others had special troop carrier bodywork with open sides. The chassis were the products of firms like De Dion-Bouton, Renault, Scemia and Schneider. Bus chassis were also used during the War with special bodywork, including ambulance, mobile operating theatre (*Voiture chirurgicale*), mobile pigeon loft (the French made great use of carrier pigeons), etc.

During the 1930s several types of commercial buses (coaches) were acquired, many on Citroën chassis. Some of these were used until after World War II.

Troop Carrier, 4×2 (De Dion-Bouton) Typical application during World War I of civilian type bus chassis with troop-carrying bodywork. Note circular radiator and driver's position above engine. Similar types produced on Schneider chassis.

Bus, 32-seater, 4×2 (Citroën P45) 6-cyl., 73 bhp, 4F1R, wb 3.62 m, 7.80×2.60×2.85 m, 4000 kg. Tyres 230×20. Max. speed 60 km/h. Hydraulic brakes. Model P38 4580-cc (84×110 mm) OHV petrol engine. Known as *Autocar, 32 places. c.* 1938.

Bus, 20-seater 4×2 (Unic) Used by the Free French forces in North Africa during World War II this was probably a requisitioned civilian bus. Note right-hand drive and semaphore-type direction indicator at rear (IWM photo E15134).

FRANCE

AMBULANCES

During the 1914–18 War an enormous variety of ambulances saw service with the French forces. Some were purpose-built, many others were conversions of passenger cars, taxis and buses. Even motorcycle/sidecar combinations appeared, fitted with two stretchers, one above the other. During the 1920s and 1930s new ambulances were purchased, mainly on car and light truck chassis, chiefly Citroëns, Peugeots and Renaults. A number of Matford V8 car-based ambulances with integral bodywork were acquired about 1938. After the outbreak of both the First and the Second World Wars, many ambulances were donated to the French Red Cross by civilian organizations overseas, including Britain and the USA. In turn, a number of Citroën and Renault ambulances went to the aid of Finland in the Russo-Finnish war under the auspices of the American and French Red Cross Societies.

Ambulance, Single-stretcher, 4×2 (Bedelia) The Bedelia was a cycle-car produced in Paris by Bourbeau & Devaux (1910–16). At least two types of ambulance conversions appeared shortly before and during World War I. Note belt-drive transmission and centre-pivot steering.

Ambulance, Three-stretcher, 4×2 This was a simple and quick conversion of a taxi-cab. Following removal of the bodywork a superstructure was installed made out of hollow tubing, in which three stretchers could be suspended. The modification was devised in 1914.

Vehicle, Medical Supply, 4×2 (Peugeot 502) Produced on 3.00-m wheelbase light truck chassis with 2212-cc Model FA four-cylinder engine, cone clutch, 4F1R gearbox and chain final drive. Horizontally-split hinged side panels. Speed 38 km/h. 1914.

Ambulance, Four-stretcher, 4×2 (Renault) Many private cars were converted into ambulances during World War I, both in Britain and in France. Renault car chassis of various types were commonly used and a typical example is shown. Note dash-mounted radiator. (IWM photo Q3392A).

Ambulance, Two-stretcher, 4×2 (Citroën B2) Typical light ambulance of the *Service de Santé Militaire* (Military Medical Service) in the mid-1920s. It was based on a Citroën 4-cyl. 10 CV Model B2 commercial chassis of 1925.

Ambulance, Three-stretcher, 4×2 (Renault AFB) This *Ambulance Légère* (light ambulance) of 1938 could accommodate three stretchers or six sitting patients. Overall dimensions: 4.85 × 1.82 × 2.18 m, wb 2.92 m, wt 2000 kg. Engine: Renault 85 2383-cc 4-cyl.

Ambulance, Heavy, Half-Track (Citroën-Kégresse) The Citroën-Kégresse appeared in many guises, including light and heavy ambulances. Shown is a *Voiture Sanitaire Lourde* of about 1936/37. Note the large bumper roller at front. See also section on half-track vehicles.

FRANCE

TRUCKS, 4×2
pre-1930

Light, medium and heavy trucks (*camions*) in service prior to 1930 were of a great variety of models and supplied by most of the French manufacturers, either directly or under a special Subvention Scheme. In the early days of WWI additional vehicles were requisitioned and more were ordered from the USA. For example, in 1914 alone the French ordered 50 Jeffery 1½-ton, 600 White 2-ton, 450 Packard 2- and 3-ton, 340 Kelly-Springfield 2-ton and 300 Pierce-Arrow 2-ton trucks. Many of the latter remained in service well into the 1930s. Among the requisitioned trucks was a fleet of Purrey steamers from a large sugar refinery. Britain and Italy also supplied a number of vehicles.

Under the pre-1914 Subvention Scheme owners of selected trucks were paid up to 8200 francs subsidy in return for which the vehicle had to remain at the Government's disposal for four years.

Truck, 3-ton, 4×2, Cargo (Ariès) GS load carrier with four-cyl. engine (90×150 mm), four-speed gearbox and chain drive During WWI SA Ariès of Courbevoie, Seine, supplied some 3000 trucks to the Government. The firm built trucks from 1906–1934.

Truck, 3½-ton, 4×2, Cargo (Berliet CAT) Prior to 1914 Berliet produced cab-over-engine trucks and vans of several types and some participated in French Army manoeuvres, from 1906. Shown is an entry in the 1911 events. Note iron-shod wooden rear wheels.

Truck, 3½-ton, 4×2, Cargo (Berliet CAT) 1913 Subsidy model, large numbers of which saw service in World War I. Four-cyl. engine (100×140 mm), four-speed gearbox, chain drive. Wheelbase 3.20 m. Handwheel by driver's side operated rear wheel brakes.

Truck, 4-ton, 4×2, Cargo (Berliet CBA) 4-cyl., 30 bhp, 4F1R, wb 4.00 m, 6.10×2.10×2.95 m, 3250 kg. 5300-cc (110×140 mm) four-cylinder 22 CV Model Z side-valve engine. Chain drive. Also short-wheelbase cab-over-engine version (Model CBD). 1914.

Truck, 4-ton, 4×2, Portee (Berliet CBA) Basically as truck shown on left but fitted with gas producer plant (*Gazogène*) and equipped for carrying of 75-mm gun and limber (*l'Artillerie de 75 portée*). Guns were also carried on Jeffery/Nash Quad, Latil, etc.

Truck, 3-ton, 4×2, Cargo (Cottin-Desgouttes) The firm of Cottin et Desgouttes of Lyon-Monplaisir produced trucks from 1905 until the mid-1930s. Shown are some of a fleet of shaft-drive trucks delivered to the Army in World War I. A larger model was also supplied.

Truck, 4-ton, 4×2, Workshop (Crochat) Typical French cab-over-engine chain-drive petrol-electric truck chassis produced by Ets. Henri Crochat in Dijon. Workshop body for Air Force (*Aviation Militaire*) was made by Ateliers de Construction Gabriel Perney of Paris. 1914/15. Also with ambulance bodywork.

Truck, 2-ton, 4×2, Cargo (De Dion-Bouton) One of several types of military trucks produced by the once famous and old-established firm of De Dion-Bouton et Cie of Puteaux, Seine. Used by the Army postal services this vehicle was of conventional design. 1914.

Truck, 2½-ton, 4×2, Cargo (Delahaye) Conventional type GS truck produced by Société des Automobiles Delahaye of Paris in 1914. Four-cyl. 3550-cc (90×140 mm) engine, developing 22 bhp at 1200 rpm. Four-speed transmission and shaft-drive. Weight 3150 kg.

Truck, 3½-ton, 4×2, Workshop (Delahaye) Chain-drive cab-over-engine chassis with comprehensively equipped workshop body 'for attending on a squad of service aeroplanes'. Supplied to French and British Government in 1913/14 at a cost of about £1800 each.

Truck, 5-ton, 4×2, Cargo (Delaunay-Belleville) Four-cyl. engine, four-speed gearbox and chain final drive, wooden wheels with solid rubber tyres. Built by SA des Automobiles Delaunay-Belleville of St. Denis. Bodywork by Gabriel Perney of Paris. 1914.

Truck, 7½-ton, 4×2, Tank Carrier (Dewald KL2) 4-cyl., 30 HP, 4F1R, wb 4.25 m, 6.25 × 2.38 m, 4586 kg. Max. GVW 12500 kg. *Camion Lourde* produced by Charles Dewald of Paris in 1921. Equipped with capstan power winch. Load: Renault light tank (6472 kg).

Truck, 1-ton, 4×2, Cargo (Panhard) One of the first self-propelled goods vehicles acquired and operated by the French Army was this *Camion à conduite avancée* (truck with forward control) produced by SA des Ans. Ets. Panhard et Levassor of Paris in 1905.

Truck, 1-ton, 4×2, Cargo (Peugeot 64) The Peugeot Type 64A and B were produced in Paris between 1906 and 1908 and had a two-cyl. 1817-cc (105 × 105 mm) engine of 7 HP with four-speed gearbox and chain final drive. One is shown here taking part in the *grandes manoeuvres du Sud-Ouest*.

Truck, 3-ton, 4×2, Cargo (Renault) 16CV (29-bhp) four-cylinder engine with four-speed gearbox and shaft drive. Solid tyres, size 1010 × 20. Wb 3.80 m. Length 5.57 m, width 2.02 m. GVW 6000 kg approx. Also ambulance, searchlight, etc. About 1000 produced, 1914–19.

F

Truck, 3-ton, 4×2, Signals (Renault) House-type signals van bodywork on shaft-drive Renault truck chassis. Like other contemporary Renault cars and commercial vehicles it had the radiator mounted behind the engine. T.S.F. stood for *Télégraphie sans fil* (wireless).

Truck, 7-ton, 4×2, Tank Carrier (Renault FU) 4-cyl., 38.8 HP, 4F1R, wb 4.57 m, 7.29×2.64 m, 5846 kg. GVW 12,500 kg. Prototype tank transporter, 1921. Epicyclic reduction gearing in rear wheel hubs. Inverted capstan power winch. Engine bore and stroke 125×160 mm.

Truck, 3-ton, 4×2, Cargo (Saurer) 4-cyl., 30 bhp, 4F1R. Weight 3475 kg. GVW 6515 kg. Shaft-drive transmission. 5300-cc (100×170 mm) side valve petrol engine, developing 30 bhp at 1000 rpm. Steel spoke wheels. 1914.

Truck, 3½-ton, 4×2, Cargo (Saurer B) Basically similar to truck on left but chain final drive and wooden spoke wheels. They were built in Arbon, Switzerland and by Automobiles Industriels Saurer in Suresnes, France. Some were converted to half-tracks (*qv*).

Truck, 1-ton, 4×2, Cargo (FIAT 2F) 4-cyl., 20 bhp, 4F1R, wb 2.84 m, length 4.08 m, weight 2500 kg. Pneumatic tyres, size 820×120, dual rear. Basically similar to 1911/12 FIAT 15ter but lower payload and specially built for French Army during WWI.

Truck, 5-ton, 4×2, Cargo (Pierce-Arrow R9) The Pierce-Arrow Motor Car Co. of Buffalo, NY, USA, supplied many trucks of 2- and 5-ton capacity to the French forces. Shown is a worm-drive 5-tonner, being loaded with a Renault light tank. 1921.

Truck, 6-ton, 4×2, Tractor (Knox) American tractor for semi-trailers. Hinged chassis with cantilever springs and semi-trailer turntable resting directly on unsprung rear axle. Widely used by French Army in WWI. 50-bhp four-cylinder engine with 3F1R gearbox.

Truck, 3-ton, 4×2, Cargo (White) One of a large number of imported American trucks of various types used by the French Army during World War I. Note the massive radiator guard, which was fitted in France, and the canvas 'apron' for the crew.

FRANCE

TRUCKS, 4×2
1930–1940

Makes and Models: Berliet VD Series (4½-ton, from 1936), GD Series (6-ton, from 1937, etc.). Chenard-Walcker U10 (tractor, c. 1930), etc. Citroën 11 UB (850-kg Van, 1937), 23 Series (1½- to 2-ton, from 1935), 32 Series (2- to 3-ton, from 1934), 45 Series (3½-ton, from 1934), etc. Delahaye 140 (1½-ton, 1938), etc. Dodge (US) VK62 (3-ton, 1939). Fiat (I) D9-11-20 (6-ton, 1938), 634 (9-ton, 1936). Ford (see Matford). GMC (US) AC-504 (3-ton, 1939). Laffly AB, AL, AR Series (4–6½-ton, from c. 1935), etc. Latil M2B1 (1½-ton, 1938), M2B4 (2½-ton, 1937), FB6 (3½-ton, 1936), etc. Matford V8-F81, -F817 (3-ton, 1938/39), V8-F81T, -F817T (4½-ton, 1938/39), V8-F917WS (5–6-ton, 1939), etc. Panhard K125 (5-ton, 1939), etc. Peugeot MK5 and DK5 Series (1.4-ton, from 1938), etc. Renault ADK (1½-ton, 1936), AGC (1½-ton, 1938), ADH (2½-ton, 1936), ADR (3½-ton, 1936), AGP (3-ton, 1937), AGR (4½-ton, 1938), AGK (6-ton, 1937), etc. Rochet-Schneider S420VL (5-ton, 1939), etc. Saurer 3CTID (7-ton, 1939), etc. Studebaker (US) K30 (3-ton, 1939), etc.

General Data: The majority of the above trucks were basically commercial models with cargo bodies and certain modifications to meet military requirements. The latter included the typical French ram's horn type front towing hooks and usually special bumpers. As at the beginning of World War I, the French Government in order to augment their Army motor transport fleet placed orders for substantial quantities of medium size 4×2 commercial type trucks in the United States. These orders went to Chrysler Corporation (for Dodge Model VK62 3-ton trucks), General Motors Corporation (for GMC AC504 3-ton trucks) and Studebaker (for Studebaker K30 3-ton trucks). Of the Dodges and the GMCs many were eventually delivered to Britain, after France had fallen to the Germans. The same also applied to numbers of 4×4, 6×4 and 6×6 trucks [for illustrations and details see the USA section of The Observer's Fighting Vehicles Directory—World War II (chapter on Vehicles supplied to other nations)].

Vehicle shown (typical): Truck, 1½-ton, 4×2, Cargo (Citroën 23U).

Technical Data:
Engine: Citroën 11 CV 4-cylinder, I-I-W-F, 1911 cc (79×100 mm), 50 bhp at 3800 rpm.
Transmission: 4F1R.
Brakes: mechanical.
Tyres: 15 or 16×50C.
Wheelbase: 3.38 m.
Overall l×w×h: 5.08×1.96×2.70 m (w/ tilt).
Weight: 2000 kg. GVW 3500 kg.
Note: commercial designation T23 or 1500 kg, works designation PUD. Produced 1935–41. Vehicles shown were used by Free French forces in North Africa, WWII. (IWM photo E887).

Truck, 3-ton, 4×2, Tractor (Chenard-Walcker U10) These tractors were equipped with an 'adhesion regulator' with which the special drawbar trailer (TTN) could be automatically turned into a semi-trailer by lifting the weight off the front wheels (the drawbar being fixed in horizontal position).

Truck, 2½-ton, 4×2, Cargo (Citroën 32U) 4-cyl., 48 bhp, 4F1R, wb 4.09 m, 6.00×2.12×2.95 m, 2850 kg. GVW 5200 kg. 3053-cc (94×110 mm) OHV engine. Tyres 170×20. Works designation P39. Produced 1934–39. Also available: diesel engine, shorter wheelbase and low chassis.

Truck, 2½-ton, 4×2, Road Sweeper (Citroën 32U) Special Air Force vehicle, equipped by Fernand Genève, for sweeping roads and runways. 2500-litre water tank. Nozzle on cab roof for antipoisonous gas washing. Non-standard cab with rear hinged doors and extra windows.

Truck, 3½-ton, 4×2, Cargo (Citroën 45U) 6-cyl., 73 bhp, 4F1R, wb 4.60 m, 7.12×2.35×3.14 (2.11) m, 3950 kg. GVW 7600 kg. Tyres 230×20. Works designation T45 or 3.5 Ton. Produced from 1934 until after World War II. Several body types, incl. house type vans.

Truck, 5–6-ton, 4×2, Cargo (Matford V8-F917WS) V-8-cyl., 85 bhp, 4F1R, wb 3.99 m, 6.88×2.20 m, 2500 kg. GVW 8500 kg. Tyres 230×20. French Ford, built in large numbers at Poissy, also as tankers for Air Force, during 1938–40; later for *Wehrmacht* (*Lkw. 5t*). Also used by French Red Cross.

Truck, 4-ton, 4×2, Cargo (Laffly AR35) One of a small batch of 'Colonial' trucks for desert service (hence special radiator and steel body), built in 1933. Other Laffly 4×2 trucks included AB2 Repair Shop (1936), AL Petrol Tanker (1938), BS Series (incl. fire trucks), etc.

Truck, 1½-ton, 4×2, Cargo (Latil M2B1) Latil M2 four-cylinder 4.08-litre (100×130 mm) engine, 4F1R gearbox. Mechanical brakes, 18×50 tyres, dual rear. Wheelbase 3.41 m, overall length 5.61 m. Some had Gohin-Poulenc gas producer. 1938–39.

Truck, 2½-ton, 4×2, Fuel Tanker (Latil M2B4) Engine as Model M2B1 on left. Various body types, incl. Cargo. Two wheelbase lengths: 3.96 m (M2B4N) and 4.34 m (M2B4L). GVW 5200 kg. 34×7 tyres. 1937–39. Some were supplied to Belgian Army with Gohin-Poulenc gas producer.

Truck, 3½-ton, 4×2, Cargo (Latil FB6) First introduced in 1934 the Latil FB6 was used by the French Army from about 1936 with various body types, incl. Cargo, Fuel Tanker and Artillery Workshop. GVW was 6–7 tons. Wheelbase was 4.10 m (shown) or 5.14 m, tyre size 42×9.

Truck, 3½-ton, 4×2, Van (Latil FB6) Found derelict by the Allies in 1944 this house-type van had been used by the German SS. Most of these Latils had a 110×160-mm four-cylinder petrol engine but diesel and compressed gas versions also existed.

Truck, 5-ton, 4×2, Cargo (Panhard K125) Panhard et Levassor supplied several types of trucks, including medical vans. This general service 5-tonner had a four-cylinder 105×140-mm petrol engine, 4.60-m wheelbase and measured 7.70×2.52×2.50 m. 1935–39.

Truck, 1.4-ton, 4×2, Cargo (Peugeot DK5/D5A) 4-cyl., 45 bhp, 3F1R, wb 3.38 m, GVW 3175 kg. *Camionnette basse bâchée Armée*, produced 1939–41, mainly for German *Wehrmacht*. French Army also used MK5 (1938–39), which had normal roof, and Radio Van. All had Peugeot 402 car type front end.

Truck, 1½–2-ton, 4×2, Cargo (Renault ADK) 4-cyl., 48 bhp, 4F1R, wb 3.34 m, 5.70×2.00×2.61 (2.04) m, 2200 kg. Used by French and later by the German forces, with various body types. Some had RHD and headlights mounted on the wings. From 1936.

Truck, 2½-ton, 4×2, Radio (Renault ADH) 4-cyl., 65 bhp, 4F1R, wb 3.47 m, 5.50×2.30×2.60 m, 4320 kg. Model 489 4-litre OHV engine. Interior dimensions 2.50×1.65×1.55 m. Tyres 210× 22. Max. speed 60 km/h at governed engine speed of 2200 rpm. From 1936.

Truck, 3½-ton, 4×2, Cargo (Renault ADR) 4-cyl, 65 bhp, 4F1R, wb 4.18 m, 6.80×2.25×2.15 m (cab), 4370 kg. Model 383 4-litre (100×129 mm) OHV engine with magneto ignition. Mechanical brakes with servo assistance. Tyres 19×50. Max. speed 62 km/h. From 1936.

Truck, 4½-ton, 4×2, Cargo (Renault AGR) 4-cyl., 65 bhp, 4F1R, wb 3.55 m, 6.33×2.32×2.55 m, 3000 kg (chassis). Also supplied with 2.55-m wheelbase. Model 383 4-litre OHV engine (or 4.7-litre diesel). 210×20/8.25-20 tyres. Max. speed 59 km/h. Servo brakes. 1938.

FRANCE

TRUCKS and TRACTORS
4×4

Makes and Models: Balachowsky et Caire (tractor, 1914). Berliet VUCT, VUD (truck/tractor, c. 1932), etc. Bernard TT4 (tractor, 1938–39). Châtillon-Panhard (tractor, 1912). Dodge (US) T203 (truck, 1½-ton, 1939). FWD (US) Model B (truck, 3-ton, 1915). GMC (US) ACK-353 (truck, 1½-ton, 1939). Jeffery (US) Quad (truck, 2-ton, 1915). Latil* LL (truck, 1913). TAR Series (tractors and trucks, from 1914), TH (truck, c. 1914). TP Series (tractors and trucks, from c. 1914), TL Series (tractors and trucks, from c. 1927), etc. Nash (US) Quad (truck, 2-ton, 1917). Panhard et Levassor K13, SK4 (tractors, 1914), etc. Pavesi (I) (tractor, 1919). Renault EG (tractor, 1914), etc. Schneider (tractor, 1914), etc.

* Early trucks known as Blum-Latil, tractors as Tourand-Latil.

General Data: With only a few exceptions the four-wheel drive vehicles used by the French armed forces prior to World War II were artillery tractors and the principal manufacturers of these were Latil, Panhard and Renault. Shortly before World War I the French Government began to test experimental four-wheel drive tractors, submitted by these and other manufacturers, coupled to artillery pieces and ammunition wagons. During the war improved versions of the Latil and Renault were used in large numbers (in November 1918 there were about 2000 Latils and just over 700 Renaults).

After the first World War further tests were held, the first in 1921. Modernized versions of the Latils were supplied, not only to the French but also to the Belgian Army. Some types were tested by the British as well. In addition the French experimented with the Italian Pavesi 4×4 articulated tractor and several semi- and full-tracked types of French and foreign origin. During World War II some types of 4×4 Latils were produced for the German armed forces, including a *Rail-Route* (road/rail) model.

Four-wheel drive trucks were mainly of US origin: FWD and Jeffery/Nash in World War I, Dodge and GMC at the beginning of World War II. Many if not most of the latter, however, came too late and these shipments were diverted to Great Britain.

For light 4×4 artillery prime movers see 'Field Cars'.

Vehicle shown (typical): Tractor, Light, Artillery, 4×4 (Latil TPEC)

Technical Data:
Engine: Latil 4-cylinder, I-I-W-F, 4849 cc (105×140 mm), 35 bhp at 1200 rpm.
Transmission: 5F1R. Spur wheel reduction.
Brakes: mechanical (foot brake on transmission, hand brake on rear wheels).
Tyres: solid, front 1000×140 single, rear 1000×130 dual.
Wheelbase: 2.10 m. Track 1.70 m.
Overall l×w×h: 5.33×2.08×1.77 m.
Weight: 3560 kg.
Note: winch and capstan at rear. Final drive by separate drive shafts to internal-tooth gear wheels at rear, external-tooth gear wheels at front. 3.50- and 3.75-m wb optional. 1921.

Tractor, Light, Artillery, 4×4 (Balachowsky et Caire) This tractor was produced in 1914 by Balachowsky et Caire of Paris, using a Belgian ACEC electric propulsion system with an electric motor in each wheel. The generator was driven by a conventional four-cylinder petrol engine.

Tractor, Heavy, Artillery, 4×4 (Latil TAR) Four-wheel steering tractor of which several thousand were built for use in World War I. 40/50-bhp four-cylinder engine with 5F1R gearbox. 12-ton towing capacity. Kerb weight 5800 kg. 1600×120 solid tyres, dual all round. Winch at rear.

Tractor, Heavy, Artillery, 4×4 (Latil TAR3) Post-war development of Latil TAR, with radiator in front. Four-cyl. engine (124 × 160 mm) with 5F1R gearbox and spur wheel reduction. 1160×120 tyres, dual front and rear. 5.36 × 2.31 × 3.00 m, wb 3.02 m. Mech. brakes (foot: trans., hand: rear wheels).

Tractor, Heavy, Artillery, 4×4 (Latil TAR5) A further development of the Latil TAR series of the late 1920s. Unlike the TAR3 and 4 it had a closed cab. The four-cyl. engine developed 61 bhp at 1750 rpm and the vehicle weighed 4780 kg. Shown fitted with Oriam anti-skid wheel chains.

91

Tractor, Light, Artillery, 4×4 (Châtillon-Panhard) One of the tractors which took part in the French Army tractor trials early in 1914. Produced by Panhard et Levassor it featured individual propeller shafts with worm drive to each wheel, four-wheel steering and a capstan at rear.

Tractor, Heavy, Artillery, 4×4 (Châtillon-Panhard) Heavy Panhard with engine-driven horizontal winch at front. The four-cylinder Knight sleeve-valve engine drove all wheels via individual propeller shafts. Shown is the 1914 trials machine; production models had an open-sided cab.

Tractor, Heavy, Artillery, 4×4 (Renault EG) Four-wheel steering tractor with horizontal power winch at rear. 45/60-bhp four-cylinder (130×160 mm) engine, driving both axles via 4F1R gearbox and enclosed propeller shafts. 1160×140 dual tyres. Wheelbase 3.63 m. Payload 5 tons; towed load 15 tons.

Tractor, Heavy, Artillery, 4×4 (Schneider) The Schneider tractors which took part in the 1914 Army trials differed from the others in several respects, notably the amidship position of the driver's seat. It had a worm-driven capstan at the rear. Note the circular radiator.

Tractor, Light, 4×4 (Latil TL) The TL was in production for many years in several versions for commercial and military applications. Shown is a basic model, fitted with dual-purpose wheels featuring spuds which, when not in use, were folded round the hub.

Tractor, Light, Artillery, 4×4 (Latil KTL) Four-wheel driven, steered and braked tractor with lockable differentials and six-speed gearbox. Produced during 1932–36 it had a four-cylinder (90× 130 mm) petrol engine. It was one of many variants in the Latil TL Series.

Tractor, Heavy, Artillery, 4×4 (Latil TARH) In 1932 the TAR Series was continued with the TARH1 (with Model B5 105×160 mm engine) but this was soon superseded by the TARH2 with Model F 110×160 mm engine. It weighed 6600 kg and had a six-speed gearbox and 270×28 pneumatic tyres.

Tractor, Light, Artillery, 4×4 (Bernard TT4) Prototype, built in 1938/39 to the design of the British/Hungarian engineer Nicholas Straussler. Front and rear unit could roll to either side, keeping the (unsprung) wheels on the ground at all times. Five seats, including two in the rear body.

Truck, 3½-ton, 4×4, Artillery Portee (Latil TH) Tractor-cum-Portee for 75-mm gun. 25-bhp engine. Rear-mounted winch. Towing capacity 8 tons. *c.* 1914. First 4×4 Latils were designed during 1912–13 by Ets. Charles Blum at Levallois-Perret. The American Walter was of similar design.

Truck, 2-ton, 4×4, Artillery Portee (Jeffery Quad 4015) Imported US chassis, widely used for towing and carrying 75-mm artillery pieces. It had four-wheel steering and a (French) cab top with special provisions to accommodate the gun barrel. Engine was a 32-bhp four-cylinder Buda, with 4F1R gearbox.

Truck, 2-ton, 4×4, Photographic Van (Latil M2TL6) Truck/trailer combination of MAM (*Ministère de l'Aéronautique Militaire*). Truck was variant of Latil HIB6 but with 70-bhp Model M2 petrol engine instead of Gardner diesel. GVW 6/7 tons. Tyres 42×9. Body by Aerazur. 1939.

Truck, 1½-ton, 4×4, Cargo (GMC ACK-353) Some of the US trucks which did reach France in 1939 but many of which later fell to the Germans. Engine was GMC 248 4.07-litre 77-bhp OHV Six with 4F1R×2 transmission. Steel cargo body with fixed sides, detachable canvas tilt and hoops.

FRANCE

TRUCKS and TRACTORS
6×4, 6×6, 8×8

Makes and Models: Berliet VPDF, VPE, etc. (6×4 and 6×6, from *c.* 1926). Bernard DH6 (6×4, *c.* 1939), etc. GMC (US) AFWX-354 (3-ton, 6×4, 1939), ACKW-353 (3-ton, 6×6, 1939). Hotchkiss W15T, S20TL, etc. (6×6, from 1936). Laffly S15L (2½-ton, 6×4), S15TL and S20TL (3.2-ton, 6×6), S25TL (4-ton, 6×6), S35TL (5.2-ton, 6×6), S45TL (12-ton, 6×6), S15T and W15T (tractors, 6×6), S25T, S35T, S45T (tractors, 6×6), etc., 1934–40. Latil M7TZ (tractor, 6×6, 1939/40), M2TZ (tractor and truck, 6×6, 1939), M4TX (tractor, 8×8, 1939), etc. Lorraine 28 (tractor, 6×4, 1934), 24/58 (10-ton, 6×4, 1933). Renault (various types, 6×4, from mid-1920s). White (US) 920 (12-ton, 6×4, 1939). Willème DG12 (12-ton, 6×2, 1935), etc.

General Data: The French were among the first to produce six-wheeled vehicles. Some if not most of the earliest models (*c.* 1903) had equal axle spacing and steering front and rear wheels, rather than a tandem rear bogie. Berliet still used this configuration on some models in the 1930s.

Laffly was one of the pioneers of French high-mobility light and medium six-wheel drive cross-country vehicles (built in part to the Austrian ADG design of Austro-Daimler) and together with Hotchkiss produced an impressive range of models during the five years prior to World War II. Certain types were exported (Afghanistan, Greece, Persia, etc.). Some are shown in this section and in the chapter on Field Cars. Latil also built several types of all-wheel drive multi-wheelers, including a huge 8×8 with 140-bhp engine, and Lorraine produced some 6×4 and 6×6 types under licence from Tatra in Czechoslovakia. Conventional 6×2 and 6×4 trucks were supplied by manufacturers like Berliet, Bernard and Willème. In 1939 additional six-wheelers were ordered from the USA but it would appear that most of these arrived too late and these shipments were diverted to Great Britain.

Vehicle shown (typical): Truck, 2-ton, 6×6, Reconnaissance and Prime Mover (Hotchkiss/Laffly W15T)

Technical Data:
Engine: Hotchkiss 11CV four-cylinder, I-I-W-F, 2300 cc (86×99.5 mm), 52 bhp at 3200 rpm.
Transmission: 4F1R×2.
Brakes: Bendix mechanical.
Tyres: 2.30×40 (bumper wheels: 42×150).
Wheelbase: 2.83 m. Track 1.51 m.
Overall l×w×h: 4.64×1.90×1.96 m.
Weight: GVW 4500 kg.
Note: known as *Véhicule de liaison et de reconnaissance tous terrains,* used also for towing loads of up to 2000 kg, e.g. 47-mm AT gun. Produced by Laffly and Hotchkiss (under Laffly licence). IFS, IRS. Max. speed 50 km/h.

Tractor, Artillery, 6×6 (Laffly S15T) Hotchkiss-engined *'Auto-caisson'* with 2½-ton winch. Bumper wheels fore and aft of front axle. 5.55 × 1.75 × 2.55 m. GVW 5500 kg. Towed load 2500 kg. *c.* 1936. Also wrecker version. Model W15T (*qv*) was low-silhouette variant with IFS.

Truck, 2½-ton, 6×4, Ambulance (Laffly S15L) Known as *Voiture Sanitaire Légère Tous Terrains*, this model had bumper wheels (with aero-type tyres) at the front only. Overall dimensions 5.00 × 1.85 × 2.50 m approx. First S15 tractor, in 1934, had Peugeot Model 601 engine.

Tractor, Artillery, 6×6 (Laffly S20TL) Known as *Voiture de Dragons Portés*, used for carrying personnel and towing light artillery pieces. Produced with and without doors. 5.42 × 2.00 × 2.30 m. GVW 5400 kg. Tyres 230 × 18. Max. speed 50 km/h. Later used by German *Wehrmacht*.

Truck, 3-ton, 6×6, Radio (Laffly S20TL) Known as *Voiture de Commandement* this was one of several vehicle types based on this chassis. All wheels were driven by individual propeller shafts. Rear suspension was independent with swinging half axles. Also with tanker body.

Tractor, Artillery, 6×6 (Laffly S35) One of the early prototypes of the heavier Laffly tractors on test. Note the relatively large bumper wheels, single at front. A similar test vehicle appeared with dual front bumper wheels and full-length touring car type body. *c.* 1935.

Tractor, Artillery, 6×6 (Laffly S25T) Prime mover for long 105-mm gun. 4.80 × 2.10 × 2.45 m approx. GVW 6550 kg. Tyres 230 × 20. Max. speed 45 km/h. Hub reduction gears. Laffly engine (like S35T and S45T; all others had Hotchkiss engines). 4-ton winch.

Tractor, Artillery, 6×6 (Laffly S35T) The S35T *Tracteur à 6 Roues Motrices* was introduced in 1935 for towing 155-mm guns. It had a towing capacity of 12 tons and a 6-ton winch with ground anchor. 5.50 × 2.10 × 2.68 m. GVW 9250 kg. Tyres 270 × 22. Max. speed 50 km/h.

Tractor, Recovery, 6×6 (Laffly S45T) Laffly 6230-cc (115 × 150 mm) 100-bhp 4-cyl. petrol engine with 8F2R gearbox built in unit with the lockable central differential. Tyre size 13.50-20. Air or vacuum brakes. 6½-ton power winch. Wheelbase 2.10 + 1.40 m. Overall length 5.87 m. 1938.

97

Tractor, Artillery, 6×6 (Latil M7TZ) Low-silhouette light artillery tractor produced shortly before the outbreak of World War II. Latil M7 2724-cc 50-bhp 4-cyl. engine with 4F1R×1 transmission. Tyre size 230×18. Basically 6×6 version of M7T1 field car (*qv*).

Tractor, Artillery, 6×6 (Latil M2TZ) Latil M2 4084-cc (100× 130 mm) 70-bhp 4-cyl. engine with 4F1R×2 transmission. Vacuum-servo brakes. Wheelbase 1.98+1.24 m. Track, front 1.82 m, rear 1.59 m. Tyres 270×22. GVW 7200 kg. Towed load 8 tons. 1939. Used by German *Wehrmacht* as *Schw. Radschlepper*.

Tractor, Recovery, 6×6 (Latil M2TZ) Alternative body style on M2TZ chassis with block and tackle hoist arrangement. The chassis was also available with long wheelbase (3.00+1.24 m). All three axle differentials were lockable simultaneously with a single lever. 4- or 5-ton winch was optional.

Tractor, Artillery, 8×8 (Latil M4TX) Latil M4 140-bhp 6-cyl. 11.2-litre (125×152 mm) SV petrol engine with 4F1R×2 transmission. Lockable diffs. Steering on outer axles. Wb 1.44+1.44+ 1.44 m. Tyres 12.75-24. Track 1.93 m. Dim. 6.45×2.38 m. 8700 kg approx. 10-ton winch. Prototype, 1939.

G

Truck, 3-ton, 6×2, Cargo (Lorraine-Dietrich) This *Fourgon* participated in manoeuvres in 1907. It was built by the Sté. Lorraine des Ans. Ets. de Dietrich of Luneville, Lorraine. Centre wheels were chain-driven; outer wheels steered. In 1906 a Borderal ambulance of similar configuration was tested.

Tractor, Artillery, 6×4 (Lorraine 28) *Voiture de Dragons-Portés* on chassis produced under Tatra licence. Lorraine 55-bhp 4-cyl. 4717-cc (100×150 mm) engine. 4F1R×2 trans. 250×20 tyres. Kerb weight 6500 kg. Payload 4–5 tons. Tubular backbone chassis with pivoting half axles at rear and IFS. 1934–35.

Truck, 10-ton, 6×4, Fuel Tanker (Lorraine 24/58) *Camion-Citerne* for aircraft fuel and oil transport. Capacity 9,000 litres. Tatra-licence chassis with 6-cyl. 110-bhp 11,220-cc (115×180 mm) petrol engine and 4F1R×2 trans. Wb 4.02+1.25 m. Track 1.80 m. Tyres 10.50-20. Speed 40 km/h. 1933.

Truck, 10-ton, 6×4, Cargo (Renault) During the early 1920s Renault introduced the light six-wheelers (see 'Field Cars') which were successful in the Sahara crossings in 1924. This heavier 'Colonial' truck, with a suspension system similar to that of the British Scammell Pioneer, appeared in 1931.

Tractor, Artillery, 6×4 (Berliet VPDF) Prototype tractor, fitted with personnel body, 1929. Another prototype with smaller body was tested for towing the 155-mm gun on four-wheel carriage. In 1926 three vehicles of similar type (VPD) crossed the Sahara from Algeria to Timbuktu (*Mission Sahara Niger*).

Truck, 12-ton, 6×4, Tank Transporter (Berliet) During 1933–34 Berliet built a number of 30-ton *Porte Char* trucks. Shown is one of a later series of 30, produced about 1939. Both had a 6-cyl. diesel engine. 12-ton hoist was operated by engine-driven screw spindles in body edges.

Truck, 12-ton, 6×2, Cargo (Bernard DH6) Introduced in 1935 this heavy duty truck had a 150-bhp Bernard-Gardner 6-cyl. diesel engine (built under licence). Following service in the German *Wehrmacht* this particular vehicle was used by the US Fifth Army in Italy (PWB mobile unit) in 1944.

Truck, 12-ton, 6×4, Fuel Tanker (White 920) In 1939 the French ordered a number of heavy six-wheeled commercial trucks from White in the USA, mainly for use as tank carriers. The fuel tanker shown was based on the same chassis. The tank carriers were eventually used by the British.

FRANCE

COMBAT VEHICLES, WHEELED

This section comprises a random selection of vehicles produced specifically for combat purposes. Some saw active service, others remained in the experimental stage, like several other designs of the inter-war period. Many vehicles which were used in 1939/40 were taken over by the Germans. It is interesting to note that some solid-tyred self-propelled De Dion-Bouton AA guns, which had been used in France and Britain during World War I, were still in service in 1940.

Also worth recording is that in 1941/42 the French set out to secretly produce a number of armoured cars on modified American GMC 1939 ACK-353 4×4 truck chassis. Over 200 chassis were converted but only one complete vehicle had been finished when the project was terminated and most of the material destroyed.

Car, 4×2, Machine Gun (Panhard 24CV/Genty) Chain-drive 1904 Panhard et Levassor car, converted into *Auto-Mitrailleuse* by a Captain Genty in 1906. Hotchkiss Puteaux machine gun with mounts in centre and at rear (shown). Rotating passenger/gunner seat; two rear seats.

Carriage, 4×2, Anti-Aircraft Gun (De Dion-Bouton 35CV) *Auto-Canon*, developed during 1910–12 (Capt. Houberdon), standardized in 1913. Still in service in 1940. Gun was famous *Soixante-Quinze* (75-mm). Used against aircraft and Zeppelins. Also used in Great Britain (Royal Navy, 1915).

Carrier, 4×2, Anti-Aircraft Ammunition (De Dion-Bouton 35CV) This *Auto-Caisson* carried the ammunition and the crew for the *Auto-Canon* shown on the left. It was based on the same 1913 De Dion-Bouton chassis which had a 35 CV V-8-cylinder engine, shaft drive and armour plate front end.

Truck, 4×2, Searchlight (Kriéger) Petrol-electric vehicle, produced by the Cie. Parisienne des Voitures Electriques (Système Kriéger) and tested by the Army in 1905. A four-cylinder petrol engine drove a 100-Amp. dynamo, providing current for the searchlight and the rear wheel motors.

Truck, 4×2, Searchlight (Ariès) Produced by SA Ariès of Courbevoie, Seine, this vehicle carried a searchlight mounted on a four-wheeled undercarriage which could be removed by means of ramps. Electricity was provided by a dynamo driven by the vehicle's engine. 1913/14.

Truck, 4×2, Searchlight (Renault) One of several types of searchlight vehicles produced by Renault. The *Projecteur* was mounted on a pedestal. Others (on the same chassis and on the solid-tyred 3-ton truck chassis) carried the searchlight on a four-wheeled undercarriage. 1917.

Truck, 4×2, Balloon Winch, M1918 (Delahaye) Observation balloon winch on Delahaye 78-35 chassis (also on Latil 4×4). The engine-driven winch (Saconney or Caquot) was at the rear and featured a sophisticated braking and winding system. Crew seats over rear wings. (IWM photo Q69726).

Armoured Car, 4 × 2 (Charron) As early as 1902 the Puteaux firm of CGT (Charron, Girardot et Voigt) exhibited a car with machine gun in circular armour plate rear tonneau. This fully-armoured type followed in 1904. In 1908 ten were ordered by the Russian Government.

Armoured Car, 4 × 2 (Peugeot 153) Armoured body on modified 1914 Model 153 car chassis. Machine gun on pedestal. Later models were heavier and had dual rear tyres; in 1918 there were 28 of these left, most of which were subsequently supplied to Poland.

Armoured Truck, 4 × 2 (Renault) Renault 3-ton truck chassis with armoured cab and front end, carrying 47-mm gun (*Auto-Canon Blindé*). Used by *Fusiliers Marins, c.* 1916. (IWM photo Q69717). Renault also produced lighter type armoured cars, as well as large numbers of light tanks.

Armoured Car, 4 × 2 (White/Laffly 50AM) During WWI the French built several hundred armoured cars on imported US White truck chassis. Many remained in service afterwards and during 1932–34 99 hulls were transferred to more modern Laffly LC2 chassis with 50-bhp engine and 4F1R × 2 transmission.

Armoured Car, 4×4 (Berliet VUDB) Prototype AMD (*Auto-Mitrailleuse de Découverte*) of 1929/30. Based on Berliet VUDB field car (*qv*). Eventual production, which took place at Monplaisir: 50 for French Army, 12 for Belgian Army. Vehicle carried a crew of three.

Armoured Car, 4×4 (Berliet UM) Another Berliet prototype, produced in 1934. This model weighed 7 tons and carried a crew of 4. It had a 76-bhp 6-cylinder petrol engine and the overall dimensions were about 4.85 × 2.25 × 2.60 m. Note the dual rear tyres.

Armoured Car, 4×2 (Panhard 165/175) *Auto-Mitrailleuse de Découverte* Panhard of 1926. Later models had larger tyres, reshaped turret and other improvements. Engine was a 20 CV (105 × 140 mm) 4-cyl., good for 60 km/h. Armament: one MG and one 37-mm gun. Weight 6350 kg.

Armoured Car, 4×4 (Panhard 178B) During the late 1930s Panhard built large numbers of AMDs and many fell to the Germans in 1939/40. They had a 115-bhp engine and weighed about 8 tons (there were several variants). Speed was about 75 km/h, length and width 4.77 × 2.00 m.

Armoured Car, 6×6 (Berliet UDB4) Produced in 1934/35 this armoured car had three driving axles, the centre one with dual tyres and the outer ones steering. It had front and rear driving positions and the armour thickness was 7–9 mm.

Carriage, 6×6, Anti-Aircraft Gun (Berliet VPR2) Twin 13.2-mm Hotchkiss machine guns on chassis similar to that shown on left. Wheelbase 1.82+1.75 m. Overall length 4.80 m, width 1.94 m. 51-bhp 2.74-litre six-cylinder engine with 4F1R×2 transmission. 1932.

Tractor, Armoured, 4×4 (Bernard) This twin-unit vehicle was built under Straussler licence and was basically similar to the tractor shown in the section on 4×4 trucks and tractors except that it had an armoured cab and midship engine location. *c.* 1939.

Carriage, 6×6, Anti-Tank Gun (Laffly W15T) Several types of combat vehicles were based on the ubiquitous Laffly 6×6 chassis. In 1939/40 some 70 tank hunters (with 47-mm gun) were hurriedly built but they were almost entirely open, unlike this full-armoured prototype.

FRANCE

HALF-TRACK VEHICLES

In 1915 Delahaye developed a track bogie to replace the driven wheels of a conventional vehicle. A number of Latil tractors and Delahaye and Saurer trucks were converted but the results were not satisfactory. The real breakthrough was brought about by Frenchman Adolph Kégresse who, as technical manager of the garages of the Russian Czar Nicholas II, had invented a bogie with an endless rubber band. After the Revolution he returned to France and in conjunction with the industrialist M. Hinstin produced a bogie which was immediately adopted by André Citroën. The first Citroën-Kégresse-Hinstin *Autochenille* performed extremely well during trials in 1921 and from then until World War II the system was further perfected and used on a large number of military and other vehicles in France and several other countries. A few of the numerous types are shown here. (See also HALF-TRACKS, Olyslager Auto Library/Warne).

Truck, 3½-ton, Half-Track (Saurer) In 1915/16 six Latil TAR 4×4 tractors (*qv*) were fitted with *Chenille Delahaye* attachments (one for each wheel). In 1917 147 sets were ordered for Saurer Model B (shown) and Delahaye trucks, strictly for off-road use. Performance was poor.

Car, 5-seater, Half-Track (Citroën-Kégresse) Early application of Kégresse-Hinstin bogies on 10 CV Citroën B2 car-cum-tractor of 1922. Driving sprockets were on fixed rear axle; forward end of bogie could move up and down. Average speed was 20 km/h, track life 3–4000 km.

Armoured Car, Half-Track (Peugeot/Kégresse) Another early Kégresse half-track conversion was carried out on this late-type World War I Peugeot armoured car. (for early type see section on Combat Vehicles, Wheeled). The trench crossing device at front was an additional modification. *c.* 1922.

Tractor, Half-Track (Citroën-Kégresse) Following several modifications to the original design (Models K1, P4T, etc), a new bogie design (P7) was introduced as seen here on a gun tractor. Visible in the rear body is a four-wheeled *Train Rouleur* (undercarriage for towing 75-mm gun).

Car, 5-seater, Half-Track (Citroën-Kégresse) Command car, fitted with the revised rear bogie (P7 *bis*) which was in production with subsequent detail improvements from 1927 until the late 1930s. Most French half-tracks now featured the large roller at the front, as shown here.

Tractor, Half-Track, Cavalry (Citroën-Kégresse) Known as *Voiture de Dragons-porté* (literally: vehicle for carrying dragoons) this type of vehicle could carry a machine gun crew and/or tow a gun. Note the front roller, which prevented the vehicle from digging in.

Vehicle, Half-Track, Telephone (Citroën-Kégresse) Another example with the later P7 *bis* bogies. The driving axle was now at the front and fitted with sprockets (instead of friction drive from the rear as on pre-1927 models). The tracks had metal cross-pieces and the suspension was redesigned.

Tractor, Half-Track, Engineers (Citroën-Kégresse) This Model C6P19 *Voiture de Sapeur-mineur de Génie* of the early 1930s carried a squad of Engineers (sappers) and their equipment. Note the two gear levers, one for the 4F1R main gearbox, the other for the 2-speed auxiliary box.

Carriage, Half-Track, Anti-Aircraft (Citroën-Kégresse) Model C6P14 chassis with twin 13.2-mm Hotchkiss machine guns on Type R4 mount. Skid plate instead of the front roller. Citroën C6 (1928–33) had 6-cyl. 42-bhp 2442-cc (72 × 100 mm) engine. C6P14 also appeared as *Tracteur de Dépannage* (recovery tractor).

Tractor, Half-Track, Artillery (Citroën-Kégresse) Produced during the mid-1930s this half-track (P75) was fitted with the same type of steel cab as the Citroën Models 23, 32 and 45 trucks, modified to accommodate the rear wings. There were two additional crew seats behind the cab.

Tractor, Half-Track, Engineers (Citroën-Kégresse) One of the last types of Citroën-Kégresse half-tracks was the P107 of the late 1930s. With the bodywork shown it was employed by the Engineers to haul bridging pontoon trailers. Similar vehicles were produced by Unic.

Truck, Half-Track, Fire (Delahaye 119 PSM) Delahaye chassis with Kégresse P16T bogies. Six-cyl. engine with 3F1R × 2 transmission. Weight 6900 (gross 9500) kg. Applevage 1200-kg crane (collapsible). Carried 800 litres of water and 20 litres of foam compound. Only one made (1935).

Tractor, Half-Track, Artillery (Panhard) Panhard also produced quantities of half-tracks using the Kégresse bogies. They were similar in appearance to the corresponding Citroëns but had a sleeve-valve engine. Unit shown was under test by the Polish Army in the late 1930s.

Tractor, Half-Track, Artillery (Somua MCG5) 4-cyl., 60 bhp, 5F1R, 5.40 × 1.97 × 2.89 m, 5300 kg. Towed 155-mm howitzer, tank recovery trailers, etc. The first of these Somuas (MCG) appeared in the early 1930s. They were produced during 1935–38 and later also used by the Germans (some with armoured body).

Tractor, Half-Track, Artillery (Somua MCG 11) This was an exp. derivation of the Somua MCG5 (*qv*) for towing the 155-mm Schneider howitzer M1917 in semi-trailer fashion (on wheels or on road-dolly). For sharp turns the steering gear actuated brakes on the bogie idler wheels. 1934.

Tractor, Half-Track, Artillery (Somua MCL5) Early Somua heavy type with special coupling for 155GPF gun. More powerful edition of the Somua MCG5, with 80–85-bhp 6.5-litre engine and a towing capacity of 5000 kg. It was 5.30 m long. *c.* 1933.

Tractor, Half-Track, Recovery (Somua MCL5) This heavy recovery vehicle was in service with cavalry units and featured a powerful engine-driven winch. The chain hoist was used for lifting tank turrets, engines, etc. The davit was collapsible.

Tractor, Half-Track, Artillery (Somua MSCL5) Designated Model MSCL5 this heavy Somua tractor was powered by a 105-bhp six-cylinder engine, hence the longer bonnet. It was designed and used for hauling four-wheeled 155-mm gun carriages. Kégresse rear bogies were used.

Tractor, Half-Track, Artillery (Somua MCJ5) Low-silhouette gun tractor of 1937/8. 160 were ordered for towing 47-mm M1937 AT gun. Production was to commence in 1940 but order was cancelled. Prototype shown still existed in 1948. Note the newly-designed bogie with larger wheels.

Tractor, Half-Track, Artillery (Unic P107) Produced by Unic and Citroën. Four-cylinder 55-bhp engine with 5F1R transmission. Towed 75-mm gun. After the fall of France they were used by the German *Wehrmacht* as *le. Zgkw. U (f) Typ P107 or Zgkw. U 304 (f)*.

Tractor, Half-Track, Artillery (Unic TU1) Produced during 1939–43, mainly for German *Wehrmacht* as *Zgkw. U 305 (f)*. 50-bhp 4-cyl. 2150-cc engine with 4F1R transmission. Overall dimensions 4.20 × 1.50 × 1.31 m. GVW 2910 kg. Tyres 5.25-18. Also appeared with crane as light wrecker.

Armoured Car, Half-Track (Citroën-Kégresse) Soon after the advent of the Citroën-Kégresse vehicle, some were fitted with armoured bodywork. 1926/27 model shown proved underpowered. Armament was 37-mm gun and one MG. Weight 2500 kg. 3.70 × 1.45 × 2.25 m. Crew 3.

Armoured Car, Half-Track (Citroën-Kégresse) Powered by a 66-bhp Panhard sleeve-valve engine this 1929 model could travel at 55 km/h. It had six speeds forward and reverse, weighed 6000 kg and measured 4.75 × 1.78 × 2.46 m. 20- or 37-mm gun plus MG. Crew 3.

Carrier, Half-Track (Citroën-Kégresse) Model N infantry supply carrier, produced in 1931 but rejected by the Army. It was then fitted with a Schneider turret (at rear) with 7.5-mm MG, of which 50 were ordered. The engine was to the right of the driver. Vehicle had cable-operated track brakes.

Carrier, Half-Track, Personnel (Citroën-Kégresse) Another example of the various armoured vehicles based on the ubiquitous Citroën-Kégresse chassis was this advanced design of an APC. Note the two electric extractor fans to the rear of the front doors. Vehicle had front roller.

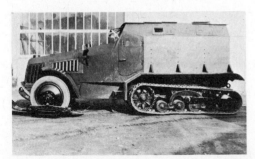

Carrier, Half-Track, Personnel (Citroën-Kégresse) Designated *Vehicule Blindé Transport Troupe* this APC was based on a Citroën-Kégresse Model P26A in 1932. To facilitate steering over snow it had detachable ski devices under the front wheels. Note the long rear bogies.

Vehicle, Half-Track, Bridge Layer (Somua/Coder) Armoured 22-ton *Poseur de Pont* with Coder hydraulic lift (F) and 8.3-metre ramp (H). Conical prong (E) engaged in aperture (G). Hydraulically-operated support legs at rear. Produced in 1939 on Somua MSCL5 chassis with Panhard engine.

FRANCE

FULL-TRACK VEHICLES

France was among the first nations to produce and employ tanks and several derivatives, including self-propelled guns and gun portees, which were built on tank chassis. Schneider also produced truck-bodied full-track artillery tractors, using components of their Model CA tank. In addition, imported American Holt tractors were employed as artillery tractors.

A number of Renault FT tanks were converted to tractors (Model HI), mainly for agricultural use, from 1918 until 1926. A variant of these (Model GP) took part in Army trials after the Armistice, together with track-laying tractors by Peugeot (T3), Pidwell (Neverslip) and Cleveland (Cletrac). Relatively few full-track tractors were used by the French Army, however, wheeled and half-track models finding more favour from the authorities.

Gun Portee, Wheel-cum-Track (Saint Chamond) The Compagnie des Forges d'Homecourt of St. Chamond designed several vehicles which could run either on tracks or on wheels. Shown is an 80 CV gun portee with the wheels lowered. Fully-armoured variants were also produced (1921–26).

Gun Portee, Full-Track (Renault) Known as *Caterpillar Porteur* this 14-ton vehicle could carry 8 tons and had a 110-bhp engine with four-speed gearbox. 350 were ordered in late 1916 but a lower number was delivered (120 in 1917). The tracks were patterned on the American Holt.

Gun Portee, Full-Track (Schneider) The Schneider gun portee was based on the same chassis as the Schneider Model CA assault tank. It had a front-mounted 60-bhp engine, driving the rear sprockets. Picture shows 155-mm gun M1917 being loaded.

Tractor, Full-Track (Schneider CD) Known as *Caterpillar Schneider CD* this truck-bodied artillery tractor had a 60-bhp (at 1000 rpm) engine with 4F1R gearbox. Payload rating was 3 tons, towing capacity in bottom gear 5.4 tons. Max. speed 8.2 km/h. 110 were in service by 1918. (IWM photo Q56453).

Tractor, Full-Track (ARA/Lorraine-Dietrich) 4-cyl., 30 bhp, 3F1R, 2.86 × 0.79 × 1.17 (rad.) m. Very narrow light tractor, tested by French Army for gun towing in 1922. Engine cubic capacity 1924 cc (70 × 130 mm). 400-kg capstan winch in centre. Width of tracks 0.15 m.

Tractor, Full-Track (Peugeot T3) 4-cyl., 25 bhp, 3F1R, 3.35 × 1.72 × 1.80 m, 3200 kg. 4714-cc (100 × 150 mm) engine. Maximum speed about 12 km/h. Diff.-brake steering. Crew 7. Took part in 1921 Subsidy trials, after having been used by the Army just before the Armistice in 1918.

Tractor, Full-Track (Renault GP) 4-cyl., 26/27-bhp, 4F1R, 3.90 × 1.90 × 2.30 m, 3250 kg. Agricultural type tractor, tested by the Army with and without four crew seats on either side. Engine bore and stroke 95 × 160 mm. Note transverse front spring, tiller steering and inclined radiator. 1919.

Tractor, Full-Track (Renault Yl) Truck-type artillery tractor, utilizing most of the chassis components of the Renault AMR tank of the mid-1930s. It was powered by an 80-bhp Renault 19 CV four-cylinder engine and weighed about 4500 kg. Produced in 1933.

Tractor, Full-Track (Renault YK) Basically similar to Model Yl (shown on left) but with more powerful six-cylinder engine (40 CV Model 380) and longer chassis. The suspension system of both types used hydraulic units rather than conventional springs. Both were private ventures by Renault. 1933.

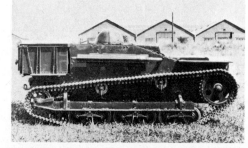

Carrier, Full-Track (Renault) Armoured personnel carrier based on Renault Model Yl artillery tractor. Note lever-type hydraulic shock absorbers at ends of piston rods of horizontal hydraulic suspension units. The front sprockets were driven. c. 1933/4.

Carrier, Full-Track (Renault UE/AMX) 4-cyl. (Renault 85), 35 bhp, 3F1R, 2.70 × 1.70 × 1.03 m, 2100 kg. Infantry supply carrier (*Chenillette de Ravitaillement d'Infantery, Mod. 1931R*), often used with tracked trailer. Some 6000 produced, first by Renault, then from 1936 until 1939 by AMX.

GERMANY

Motorization in Germany commenced in 1885 when Karl Benz in Mannheim launched the first practicable motor car and Gottlieb Daimler in Cannstatt near Stuttgart the first motorcycle.

In 1898 the motor vehicle made its debut in the German Army when a borrowed Daimler truck was used. One year later, in the 1899 *Kaisermanöver*, some motorcycles (NSU, Triumph) and cars (Benz, Cudell, Daimler, Eisenach, Marienfelde) were used and on 1 October the *Inspektion der Verkehrstruppen* was founded, in order to study and test motor vehicles. One of their first jobs was to test the first two trucks ordered by the Army from Daimler, one with a carrying capacity of 400 kg, the other a two-ton vehicle. More Daimler vehicles, as well as a Benz, were bought in 1900/01, followed by a French De Dion-Bouton and Serpollet steamer.

In 1905 some British Fowler steam traction engines were ordered from John Fowler & Co., Magdeburg (the Fowler was first introduced into Germany about the time of the Franco-German struggle of 1870 and had not been forgotten) and in the same year the German Volunteer Motor Corps (DFAC) was formed. In 1908 a Subvention (subsidy) scheme was introduced for the *leichter Armeelastzug*, i.e. a four-ton truck with two-ton four-wheeled trailer, built in compliance with certain standardized specifications laid down by the Government. In 1914 some 500 of these combinations were readily available and another 12,000 were built during the war.

Of the total of 64,000 motor vehicles existing in Germany in 1914, about three-quarters passed under the control of the military authorities. In 1918 the German Army possessed about 5400 motor cycles, 12,000 cars, 3200 ambulances, 25,000 trucks and 1600 trailers. After the war 5000 trucks had to be turned over to the Allied powers and most if not all of the combat and special vehicles had to be destroyed, in accordance with the Treaty of Versailles.

Most of the other transport equipment had already been surrendered under the terms of the Armistice.

Shortly after the war a small number of armoured cars was produced for police purposes and during the Twenties the *Reichswehr* commenced to re-equip itself, mostly under a screen of secrecy. Some new types of military vehicles were tested in disguise in Russia. During the early 1930s the process of re-arming was accelerated and many new developments appeared, culminating in a fleet of advanced combat and support vehicles for the '*Blitzkrieg*' of 1939/40.

Note: Most of those vehicle types shown in the following pages which were produced during the mid- and late 1930s were used in World War II and this section therefore overlaps the German section in *The Observer's Fighting Vehicles Directory—World War II*; in turn, the latter book contains pictures and information of certain vehicles which were produced prior to 1940 and which have not been repeated here.

The first two German military trucks, built by Daimler in 1898/99.

Who's Who in the German Automotive Industry

Note: Only those makes which still existed during the mid-1930s are listed.

Adler	Adlerwerke vorm. Heinr. Kleyer AG, Frankfurt/Main.
Audi	Auto Union AG, Werk Audi, Zwickau i.Sa.
Auto Union	Auto Union AG, Chemnitz.
BMW	Bayerische Motoren Werke AG, München.
Borgward	(see Hansa-Lloyd).
Büssing-NAG	Büssing-NAG Vereinigte Nutzkraftwagen AG, Braunschweig.
Daimler-Benz	(see Mercedes-Benz).

Demag	Demag AG, Wetter/Ruhr.
DKW	Auto Union AG, Werk DKW, Zschopau i.Sa.
Famo	Fahrzeug- und Motorenwerke GmbH, Breslau.
Faun	Faun-Werke GmbH, Nürnberg.
Ford	Ford Motor Comp. AG, Köln-Niehl.
Framo	Framo-Werke GmbH, Hainichen i.Sa.
Hanomag	Hannoversche Maschinenbau AG vorm. Georg Egestorff, Hannover-Linden.
Hansa-Lloyd	Hansa-Lloyd-Goliath Werke Carl F. W. Borgward, Bremen.
Henschel	Henschel & Sohn AG, Kassel.
Horch	Auto Union AG, Werk Horch, Zschopau & Zwickau i.Sa.
Kaelble	Carl Kaelble GmbH, Backnang.
Klöckner/ Humboldt-Deutz	Humboldt-Deutzmotoren AG, Köln-Deutz.
Krauss-Maffei	Lokomotivfabrik Krauss & Comp.—J. A. Maffei AG, München.
Krupp	Friedrich Krupp AG, Essen.
Lanz	Heinrich Lanz AG, Mannheim.
Magirus	C. D. Magirus AG, Ulm/Donau (1937: Humboldt-Deutzmotoren AG, Magirus-Werke, Ulm/Donau).
MAN	Maschinenfabrik Augsburg-Nürnberg AG, Nürnberg.
Maybach	Maybach-Motorenbau GmbH, Friedrichshafen/Bodensee.
Mercedes-Benz	Daimler-Benz AG, Stuttgart-Untertürkheim, Gaggenau i.B., Berlin-Marienfelde.
NSU	NSU-D-Rad Vereinigte Fahrzeugwerke AG, Neckarsulm.
Opel	Adam Opel AG, Rüsselsheim/Main.
Phänomen	Phänomen-Werke Gustav Hiller AG, Zittau.
Stoewer	Stoewer-Werke AG vorm. Gebr. Stoewer, Stettin-Neutorney.
Tempo	Vidal & Sohn, Tempo-Werke, Harburg-Wilhelmsburg.
Vomag	Vomag-Betriebs AG, Plauen i.Vo.
Wanderer	Auto Union AG, Werk Wanderer, Siegmar i.Sa.
Zündapp	Zündapp GmbH, Nürnberg.

Typical German motor transport of World War I: 1. 40 HP 6-seater car (Benz), 2. Semi-enclosed car with wire-catcher (Daimler), 3. Small car (Wanderer), 4. Subsidy-truck (Benz), 5. Light truck (NAG), 6. Army supply road train.

Büssing-NAG ZRW (10×10) prototype amphibious armoured car chassis in disguise; developed during 1927–29.

116

GERMANY
MOTORCYCLES

Pre-1940 motorcycles used by the *Reichswehr* and *Wehrmacht* were civilian types, usually 'militarized' in respect of paintwork and the fitment of standard type pannier-bags. Some had a pillion seat. Sidecars were used in conjunction with certain models. One of the main suppliers was the Bayerische Motoren-Werke AG (BMW) of Munich. This firm supplied single- and twin-cylinder shaft-drive machines from 1928, starting with R52 and R62 twins. Others were acquired from DKW, NSU, Triumph, Victoria, Zündapp, etc.

The *Wehrmacht* instituted three main classes: light (up to 350 cc), medium (up to 500 cc) and heavy (over 500 cc), officially known as *leichtes, mittleres* und *schweres Kraftrad* (*o*) or *l., m.* and *s. Krad* (*o*) respectively. The '(*o*)' suffix indicated *handelsüblich*, i.e. commercially available.

Motorcycles, Solo (Triumph, NSU) Motorcycles were first employed by the German Army in the *Kaisermanövern* (Imperial Manoeuvres) of 1899. Shown are two machines used by dispatch riders about 1904. They are a Triumph (left) and an NSU (right). NSUs were then known as Neckarsulm.

Motorcyle, with Sidecar (NSU 7 PS) V-twin-cyl. engine with foot-starter, 3F gearbox and chain-drive. Rear-wheel suspension with coil spring. Platform-type sidecar with machine gun on removable mount. 1915. Most German military motorcycles used in World War I were NSUs.

Motorcycle, Solo (Victoria KR VI) HO-twin-cyl., 18 bhp, 3F, wb 1.50 m, 2.32 × 0.85 × 1.02 m, 170 kg approx. 598-cc (77 × 64 mm) OHV engine. Produced during 1927–32 and used by *Reichswehr*, solo and with sidecar. Also appeared with two rear wheels in tandem (exp.).

Motorcycle, Light, Solo (NSU 251 OS) Single-cyl., 10.5 bhp, 4F, wb 1.34 m, 2.04 × 0.78 × 0.95 m, 144 kg. 241-cc (64 × 75 mm) OHV engine. Max. speed 100 km/h. Tyres 3.00-19. 1938–40. *Wehrmacht* 'light' requirements specified engine size under 350 cc, weight under 150 kg, etc.

Motorcycle, Medium, Solo (Victoria KR35 SN/WH) Single-cyl., 18 bhp, 4F, wb 1.40 m, 2.16 × 0.78 × 1.00 m, 154 kg. 342-cc (69 × 91.5 mm) OHV engine. Max. speed 100 km/h. Tyres 3.25-19. 1938. *Wehrmacht* 'medium' requirements specified engine size under 500 cc, weight under 190 kg, etc.

Motorcycle, Heavy, Solo (BMW R11) 1935 *Wehrmacht* 'heavy' requirements specified: engine: over 500 cc; weight: max. 200 kg; GVW: 400 kg; payload on/off roads: 180–200/120–150 kg; max. overall dim.: 2.30 × 0.90 × 1.15 m; clearance 110 mm; gradability: 20°. Shown: BMW 750 twin, 1929–34.

Motorcycle, Heavy, with Sidecar (BMW R12) HO-2-cyl., 18 bhp, 4F, wb 1.40 m, 2.52 × 1.61 × 1.00 m, GVW 560 kg. 746-cc (78 × 78 mm) SV engine. Shaft-drive. Max. speed 85 km/h. Produced 1935–41 for use as solo machine or with commercial or *Einheits* (standardized) sidecar. The latter is shown.

GERMANY
CARS
1900–1918

It was in 1898/9 that the Prussian Army first used motor vehicles. The first Army-owned passenger car was a 10 HP Daimler 'Phaeton', followed in 1901 by a Benz and another Daimler. Other early cars included French De Dion-Bouton and Serpollet (steamer).

These and others were used at manoeuvres and for other activities. During World War I a large number of cars was employed, mainly of civilian-type. Largest quantity was produced by Opel; others included Adler, Audi, Benz, Bergmann, Daimler (Mercedes), Dürkopp, Fafnir, Hansa, Horch, Komnick, Nacke, NAG, Protos, Stoewer, Wanderer, etc.

By 1918 some 12,000 cars of many different types were in service. A random selection of typical models is shown.

Car, 6-seater, 4 × 2 (Daimler) This was the first German military passenger car, ordered in Nov. 1899, delivered in early 1900. The 'Phaeton' had a four-cylinder 10-hp engine with four-speed gearbox, chain-drive and solid tyres. Maximum speed was 40–45 km/h.

Car, 4-seater, 3 × 1 (Phänomobil) During WWI the Germans had two types of *Dreiradwagen*: the Cyklonette (made by Cyklon, Berlin) and the Phänomobil (Phänomen-Werke Gustav Hiller, Zittau). One of the latter is shown, in Dinant, Belgium, 1914. Note steering lever, passing under windscreen.

Car, 2-seater, 4 × 2 (Wanderer) Popularly known as *Puppchen* (little doll) this tandem-bodied patrol car had a 4-cyl. 1150-cc 5/12 HP engine with cone clutch and 3F1R gearbox. The steering wheel was mounted centrally. Max. speed 55 km/h. ¾-elliptic rear springs. 1914.

Car, 2-seater, 4×2 (Stoewer) Known as *leichter zweisitziger Kraftwagen* (light two-seater car), this model was based on the Stettin-built Stoewer Model C2 chassis with 10/28-hp four-cylinder 2412-cc engine and 4F1R gearbox. Civilian models had wire-spoke wheels.

Car, 4-seater, 4×2 (Dürkopp) Typical *Doppelphaeton*, used as *Offizierswagen* (officers' car). Example shown was produced by the Dürkoppwerke AG of Bielefeld in 1913/14. Note the two windscreens, the rifle holders between the doors and the usual German crest.

Car, 4/5-seater, 4×2 (Protos) By 1915 German rubber supplies had all but dried up. From 1916 many cars, trucks and ambulances were shod with sprung wheels; some incorporated rubber blocks or small semi-elliptic springs, others coil springs, as shown.

Car, 4/5-seater, 4×2 (Opel) Adam Opel of Rüsselsheim was one of the main suppliers of cars (as well as trucks and aero engines) for the German war effort during 1914–18. This is a typical model of the period. All had 4-cyl. engine, 4F1R gearbox and shaft-drive.

Car, 6-seater, 4×2 (Benz) Benz & Cie of Mannheim supplied heavy 16/40 hp cars with open (shown) and closed bodywork. They had a 3945-cc 4-cyl. L-head engine, leather-cone clutch, 4F1R gearbox and shaft-drive. Wheelbase 3.25 m. Maximum speed 85 km/h.

Car, 6-seater, 4×2 (NAG) The Nationale Automobil-Gesellschaft of Berlin produced heavy open and closed staff cars and light and heavy trucks. The steel rails over the top were a common feature; they served to cut or lift wires which were frequently stretched across roads.

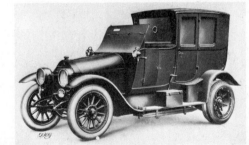

Car, 6-seater, 4×4 (Mercedes) Four-wheel drive car with 45-bhp engine and four-wheel steering, produced in 1907 for the *Reichskolonialamt*. Special large radiator, augmented by horseshoe-shaped radiator at the scuttle. It was used satisfactorily in German South West Africa in 1908.

Car, Armoured, 4×2 (Mercedes) The Daimler Motoren Gesellschaft of Stuttgart-Untertürkheim supplied cars under the Mercedes name. This chain-drive *Landaulet* was a special job with armour plate bodywork. Note the dual rear tyres to cope with the extra weight. *c.* 1914.

GERMANY

CARS, LIGHT
4×2 and 4×4

Makes and Models: *Note: 4×2 unless indicated otherwise* BMW 3/15 (1929–31), 303 (1933–34), 309 (1934–35), 315 (1934–36), *E.Pkw.* 325 (4×4, 1937–40). Dixi 3/15 (1928). DKW (various, from *c.* 1931). Framo MW (1936–37). Hanomag 3/16 (1928–29), 4/20 (1930–31), 4/23 (1932–33), Garant (1934–37), Kurier (1934–38), Rekord (1934–38), *E.Pkw.* 20B (4×4, 1937–40). Mercedes-Benz 130H (1933). Opel P4 (1936–38), Olympia (1938–40). Röhr Junior (1934). Stoewer Greif Junior (1936), *E.Pkw.* R180 *Spezial* (4×4, 1936–38) and R200 *Spezial* (4×4, 1938–40). Tempo G1200 (4×4, 1936–39), etc.

General Data: Among the first light cars used by the *Reichswehr* after World War I was the Dixi 3/15 PS, which was the British Austin Seven, produced under licence by Dixi, of Eisenach. In 1929 Dixi was absorbed by BMW and the 3/15 PS was continued under the BMW name until 1931. It was used as a small two-seater scout car and also as the basis for dummy armoured cars (the real thing being forbidden under the Treaty of Versailles). In the early 1930s it was decided that cars with an engine of under 1500-cc capacity were to be termed 'light cars' (officially *leichter Personenkraftwagen*, or *l.Pkw.* for short). Until 1936 they were standard or modified civilian models, indicated by the letter (o). In 1936/7 the first military pattern light car chassis appeared, produced by BMW, Hanomag and Stoewer to a common *Wehrmacht* specification. This was known as the *Einheitsfahrgestell I für l.Pkw.* and was fitted mainly with four-seater soft-top bodywork. The engine cubic capacity was just under 2 litres. Meanwhile, and especially in 1939/40, many civilian small cars were used and some manufacturers had designed special cross-country cars for military, police and export purposes, as private ventures.

Civilian type chassis with special military bodywork, usually open two- to four-seaters were classed as *leichter geländegängiger Personenkraftwagen mit Fahrgestell des l.Pkw.* (o). The passenger-carrying version was known as *Kfz.1*, the two-seater signals version as *Kfz.2*, etc.

Vehicle shown (typical): Car, Light, 4×2, Signals, Kfz. 2 (Hanomag Garant) (*Kleiner Fernsprechkraftwagen* (*Kfz.* 2) m. Fahrgestell des l. Pkw. (o))

Technical Data:
Engine: Hanomag 4-cylinder, I-L-W-F, 1089 cc (63×88 mm), 23 bhp at 3500 rpm.
Transmission: 4F1R.
Brakes: hydraulic.
Tyres: 4.75-17.
Wheelbase: 2.47 m.
Overall l×w×h: 3.70×1.50×1.55 m approx.
Weight: 1000 kg approx.
Note: cable reel brackets on rear deck. Special two-seat body with rear locker, also on BMW chassis. *c.* 1935. Later models, on Mercedes-Benz 170V and *I.E.Pkw.* chassis, had three seats.

Car, Light, 4×2 (Dixi/BMW 3/15 PS) 4-cyl., 15 bhp, 3F1R, wb 1.90 m. Usually with two-seater soft-top bodywork. Shown as basic vehicle for dummy armoured car (wooden superstructure removed). Transversal leaf spring at front, quarter-elliptic springs at rear. *c.* 1929.

Car, Light, 4×2 (DKW Front F2 600) 2-cyl., 18 bhp, 3F1R, wb 2.40 m. Front-wheel drive. *c.* 1932. Example of *l.Pkw* (*o*). 1935 *Wehrmacht* 'light' requirement specified engine under 1500 cc and 30 bhp, overall dim. 3.6–4.1 × 1.4–1.5 × 1.5–1.6 m, wt 650–1000 kg, GVW 950–1300 kg, etc.

Car, Light, 4×2 (Framo MW 'Sachsen') DKW 2-cyl., 18 bhp, 3F1R × 2, wb 2.20 m, 575 kg. Tyres 4.00-19. 584-cc (74 × 68 mm) DKW 2-stroke engine mounted amidships. Chain drive to solid rear axle. IFS. Central driver's seat; two rear seats. Prototypes only, 1936/37.

Car, Light, 4×2 (DKW F7 Meisterklasse) 2-cyl., 20 bhp, 3F1R, wb 2.61 m. Civilian DKW car chassis with some modifications (lower gear ratio; no freewheel). 3-seat open body with rear locker. Transverse 684-cc 2-stroke engine, driving front wheels. Built for police use, 1937.

Car, Light, 4×2, Kfz. 1 (Röhr Junior) HO-4-cyl., 30 bhp, 4F1R, wb 2.66 m. *l.gl.Pkw* on Röhr backbone chassis, produced under Tatra licence (1932–35). Four bucket-type seats, hence known as *'Kübelsitzer'* or *'Kübelwagen'*. (Archiv-Photo v.Fersen). Stoewer Greif Junior was similar.

Car, Light, 4×2, Signals, Kfz. 2 (BMW 303) 6-cyl., 30 bhp, 4F1R, wb 2.40 m, 4.10×1.50×1.50 m approx., 1000 kg approx. Tyres 5.25-16. Shown with top and side curtains installed. 1173-cc (56×80 mm) OHV engine. BMW 309 similar but 4-cyl. 845-cc engine. Also with *Kfz. 1* bodywork. 1934.

Car, Light, 4×4 (Tempo G1200) Two ILO 2-cyl. 19-bhp 596-cc engines, one at each end and each driving through 4F1R gearbox. IFS, IRS. Wb 2.83 m. Length and width 4.00×1.68 m. Tyres 5.00-17. Designed by Ing. Otto Daus for Vidal & Sohn. Private venture. Demonstrated and sold, mainly for military use, in some 40 countries. 1936–39.

Car, Light, 4×4, Kfz. 1 (Stoewer R200 Spezial) 4-cyl., 50 bhp, 5F1R×1, wb 2.40 m, 3.85×1.69×1.90 m, 1700 kg. Four-wheel steering. IFS and IRS with coil springs. Standardized *Wehrmacht* design (*Einheits-Pkw.*), produced also by BMW (6-cyl.) and Hanomag. 1997-cc OHV engine. Tyres 6.00-18. 1938–40. Alternative roles: *Kfz. 2, 2/40, 3* and *4*.

GERMANY

CARS, MEDIUM
4 × 2 and 4 × 4

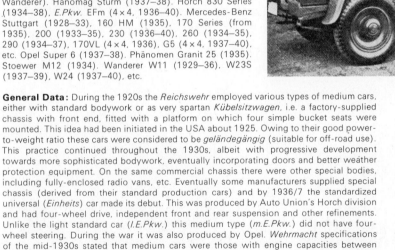

Makes and Models: *Note: 4 × 2 unless indicated otherwise* Adler Favorit (1925–33), Standard 6 (1932–35), Diplomat 3 Gd (1936–40). Auto Union (see Horch and Wanderer). Hanomag Sturm (1937–38). Horch 830 Series (1934–38), *E.Pkw.* EFm (4 × 4, 1936–40). Mercedes-Benz Stuttgart (1928–33), 160 HM (1935), 170 Series (from 1935), 200 (1933–35), 230 (1936–40), 260 (1934–35), 290 (1934–37), 170VL (4 × 4, 1936), G5 (4 × 4, 1937–40), etc. Opel Super 6 (1937–38). Phänomen Granit 25 (1935). Stoewer M12 (1934). Wanderer W11 (1929–36), W23S (1937–39), W24 (1937–40), etc.

General Data: During the 1920s the *Reichswehr* employed various types of medium cars, either with standard bodywork or as very spartan *Kübelsitzwagen*, i.e. a factory-supplied chassis with front end, fitted with a platform on which four simple bucket seats were mounted. This idea had been initiated in the USA about 1925. Owing to their good power-to-weight ratio these cars were considered to be *geländegängig* (suitable for off-road use). This practice continued throughout the 1930s, albeit with progressive development towards more sophisticated bodywork, eventually incorporating doors and better weather protection equipment. On the same commercial chassis there were other special bodies, including fully-enclosed radio vans, etc. Eventually some manufacturers supplied special chassis (derived from their standard production cars) and by 1936/7 the standardized universal (*Einheits*) car made its debut. This was produced by Auto Union's Horch division and had four-wheel drive, independent front and rear suspension and other refinements. Unlike the light standard car (*l.E.Pkw.*) this medium type (*m.E.Pkw.*) did not have four-wheel steering. During the war it was also produced by Opel. *Wehrmacht* specifications of the mid-1930s stated that medium cars were those with engine capacities between 1500 and 3000 cc but the *m.E.Pkw.* had a 3.5-litre V8 engine. On the other hand, the 1.7-litre Mercedes-Benz 170V was often used for applications within the *l.Pkw.* (light car) classification.

Vehicle shown (typical): Car, Medium, 4 × 2, Kfz. 12 (Wanderer W11) (*mittlerer geländegängiger Personenkraftwagen mit Zugvorrichtung (Kfz. 12) mit Fahrgestell des m. Pkw.* (o))

Technical Data:
Engine: Wanderer 6-cylinder, I-I-W-F, 2970 cc (75 × 113 mm), 65 bhp at 3000 rpm.
Transmission: 4F1R.
Brakes: hydraulic.
Tyres: 6.00-20.
Wheelbase: 3.10 m. Track 1.42 m.
Overall l × w × h: 4.50 × 1.70 × 1.85 m approx.
Weight: 1700 kg. GVW 2000 kg.
Note: typical example of *Kfz. 12*. Also on other chassis makes and without towing hook (*Kfz. 11*). Over 2700 produced by Wanderer, 1933–36. Earlier models had 50-bhp 2.5-litre engine.

Car, Medium, 4×2, Kfz. 11 (Adler Favorit) 4-cyl., 27 bhp, 4F1R, wb 2.80 m, 3.80×1.50×1.80 m approx., 1100 kg approx. 1550-cc (67×110 mm) L-head engine. Chassis produced during 1925–29 by Adlerwerke vorm. Heinrich Kleyer AG, Frankfurt/Main. 1929–33 models had 35-bhp 1840-cc engine.

Car, Medium, 4×2, Kfz. 4 (Mercedes-Benz Stuttgart) 6-cyl., 50 bhp, 3F1R (later with OD), wb 2.81 m, 4.06×1.68 m. Rare application of twin anti-aircraft machine guns on early-1930s *Kfz. 12*, known as *le. Truppen-Luftschutz-Kraftwagen*. Superseded by *Kfz. 4* on *l.E.Pkw*.

Car, Medium, 4×2, Kfz. 18 (Mercedes-Benz 200) Officially known as *Gefechtskraftwagen* (combat car) the *Kfz. 18* of the 1930s had open 2-seat bodywork with large equipment locker and towing hook. It was produced during 1933–36, measured 4.40× 1.70×1.82 m and weighed 1610 kg.

Car, Medium, 4×2, Kfz. 15 (Horch 830) V-8-cyl., 70 bhp, 4F1R (OD top), wb 3.20 m, 4.80×1.80×1.85 m, 1950 kg. *Nachrichten-Fernsprech-* or *Funk-Kraftwagen* (communications/signals car). Locker and towing hook. Mainly on Horch, Mercedes-Benz and Wanderer chassis, from *c.* 1933.

Car, Medium, 4×2, Kfz. 11 (Stoewer M12) The Stoewer M12 was used in limited numbers with *Kfz. 11/12 'Kübel'* and *Kfz. 15* bodywork. One of the former is shown, taking part in the 1934 3-day Harz-trials. Engine was a 2963-cc 60-bhp 8-in-line. Wheelbase 3.10 m. Tyres 5.50-18.

Car, Medium, 4×2, Kfz. 12 (Wanderer W14) The Wanderer W14 was developed from the W11 (*qv*) and was produced during 1933–36. The engine was a 2970-cc 65-bhp 6-cyl. Shown is the late model bodywork with four detachable doors and integral locker. Wheelbase 3.10 m. Tyres 6.00-20.

Car, Medium, 4×2, Kfz. 12 (Adler 3Gd) 6-cyl., 60 bhp, 4F1R, wb 3.35 m, 4.80×1.80×2.00 m, 2210 kg (gross). Nearly 4300 produced during the late 1930s and widely used in the *Blitzkrieg*. IFS with two transversal leaf springs. Tyres 6.50-20. Max. speed 80 km/h.

Car, Medium, 4×2, Kfz. 12 (Mercedes-Benz 320) 6-cyl., 78 bhp, 4F1R, wb 2.88 m, 4.80×1.80×2.00 m, 2200 kg (gross). Special military bodywork with exception of grille and bonnet (like Adler 3Gd and Wanderer W14, *qv*). Note truck-type wheels and wire cutters under front door.

Car, Medium, 4×2, Kfz. 11 (Mercedes-Benz 170V) 4-cyl., 38 bhp, 4F1R, wb 2.84 m, 4.11×1.58×1.80 m approx. Daimler-Benz 4-door *Kübelsitzer* bodywork with folding top and windscreen. 1938. Over 19,000 170Vs were supplied for military purposes (various body styles).

Car, Medium, 4×4 (Mercedes-Benz G5/W152) 4-cyl., 45 bhp, 5F1R×1 (OD top), wb 2.53 m, 3.99×1.68×1.90 m, 1880 kg. Four-wheel steering. Commercially available. Over 300 produced during 1937–41, with various body styles. DB M149 2006-cc (82×95 mm) SV engine. Few used by *Wehrmacht*.

Car, Medium, 4×4, Kfz. 15 (Auto-Union/Horch) V-8-cyl., 80 bhp, 4F1R×2, wb 3.10 m, 4.70×1.86×2.07 m, 2600 kg. *Nachrichtenkraftwagen* on early universal *m.E.Pkw* chassis. 3517-cc (78×92 mm) SV engine. Lockable differentials in transfer case and rear axle. Four seats. 1936/37.

Car, Medium, 4×4, Kfz. 17 (Auto-Union/Horch) V-8-cyl., 80 bhp, 4F1R×2, wb 3.10 m, 4.80×1.80×1.85 m, 3525 kg (laden). *Kabelmesskraftwagen* on later model *m.E.Pkw* chassis. Similar vans used for other purposes, incl. radio, telephone exchange, etc. Produced until 1943 by Horch and Opel.

GERMANY

CARS, HEAVY
4×2, 4×4, 6×4, 6×6

Makes and Models: Auto Union (see Horch). BMW 335 (4×2, 1939–40). Ford V8 (4×2, 1937–39), *E.Pkw.* EGa/b/d (4×4, 1939–40). Horch 8 Geländewagen (Horch/Argus, 6×4, 1926–28), 600 V12 (4×2, 1931–33), 830 BL (4×2, *c.* 1935), 850 (4×2, 1932–36), 5-Litre (Horch/Argus, 4×4, 1933–34), *E.Pkw.* 1a, 1b (4×4, 1937–40), etc. Krupp L2H143 (6×4, 1939–40). Maybach V12 (4×2, 1934), etc. Mercedes-Benz G1 (6×4, 1926–28), Nürburg 460 (4×2, 1928–33), G4 (6×4 and 6×6, 1933–39), 320 (4×2, 1938–40), 340 (4×2, 1938–39), 540K and 770 (4×2, 1938), etc. Opel Admiral (4×2, 1938–39). Phänomen Granit (4×2, *c.* 1935–37). Selve/Voran M (6×6, 1928–29), etc.

General Data: The *Wehrmacht* specification for a 'heavy car' (*schwerer Personenkraftwagen*) in 1935/36 stipulated an engine cubic capacity of over 3000 cc and a power output of over 60 bhp. However, several cars in the 'medium' category had engines of this size and power, e.g. the Horch 830 models with their 70-bhp 3249-cc (and later 75- and 82-bhp 3517-cc) V8 engines. The specification also listed the overall dimensions as 5.10–5.60 metres long, 1.80–1.85 metres wide and 1.70–1.85 metres high. The standardized universal heavy car, which was introduced in the late 1930s, had a 3.8-litre Auto-Union/Horch or 3.6-litre Ford V8 engine and measured 4.85 × 2.00 × 2.04 metres and to confuse the matter even further the 6-seater *Kfz.21 s.gl. Pkw* was based on the medium *E.Pkw.* chassis.

Covered in this section are some typical examples of cars with engines of over 3000-cc cubic capacity as well as 6-wheeled cars. The latter first appeared in the late 1920s when Auto-Union (Horch), Daimler-Benz (Mercedes-Benz) and Selve-Voran introduced 3-axle cross-country cars, as private ventures. The Selva-Voran featured all-wheel drive, the others had tandem rear axle drive only, except for 15 of the 72 Mercedes-Benz G5 cars, which had a driven front axle.

Some pre-1940 models, like the Ford V8 and Opel Admiral, were not or seldom used until after the war had started; many pre-war cars, particularly captured and impressed ones, were then taken into service with either the original or special bodywork.

Vehicle shown (typical): Car, Heavy, 4×4, *Kfz.* 21 (Horch/Argus) (*schwerer geländegängiger Personenkraftwagen* (*Kfz.* 21) mit Fahrgestell des s. gl. Pkw. (o))

Technical Data:
Engine: Horch 8-cylinder, I-I-W-F, 4911 cc (87 × 104 mm), 100 bhp at 3600 rpm.
Transmission: 4F1R × 2.
Brakes: hyd., vacuum-assisted.
Tyres: 6.00-18.
Wheelbase: 3.47 m. Track, front/rear 1.45/1.93 m.
Overall l × w × h: 5.10 × 2.10 × 2.15 m.
Weight: 2800 kg. GVW 3400 kg.
Note: one of a small batch of staff cars built for the *Wehrmacht* during 1933–34. Note dual rear tyres.

J

Car, Heavy, 4×2, Armoured (Mercedes-Benz Nürburg 460) 8-cyl., 80 bhp, 4F1R, wb 3.67 m. Partly-armoured *Panzerlimousine* on 4.6-litre 18/80 PS chassis and similar in appearance to the standard 1929 sedan. 8-in-line SV engine had 4592-cc (80×115 mm) capacity. Note RHD.

Car, Heavy, 4×2, Convertible (Mercedes-Benz 540K) 8-cyl., 115 bhp (with supercharger 180 bhp), 4F1R, wb 3.29 m, 2300 kg. This 1938/39 car was used by Göring and is shown after being found by the British Army in 1945. 5401-cc (88×111 mm) OHV engine. IFS and IRS (IWM photo BU 3525).

Car, Heavy, 4×2, Sedan (Mercedes-Benz 770) 8-cyl., 150 bhp (with supercharger 200 bhp), 3F1R, wb 3.75 m. One of several cars used by Hitler, this edition of *'Der Grosse Mercedes'* featured a good deal of special equipment. The 8-in-line engine's capacity was 7655 cc (95×135 mm).

Car, Heavy, 4×2, Convertible (Opel Admiral) 6-cyl., 75 bhp, 3F1R, wb 3.15 m, 5.26×1.80×1.63 m, 1600 kg. This was Opel's largest car (1938–39) and was frequently used by the military with either original or special bodywork, e.g. van and ambulance. Engine: 3626-cc OHV Six.

131

Car, Heavy, 4×2, Kfz. 21 (Phänomen Granit) Basically a light truck chassis the Granit appeared in small numbers with six-seat bodywork. Some were exported. Later models had spare wheels behind front wings and body with detachable doors. Engine was air-cooled 2.5- or 3-litre.

Car, Heavy, 4×4, Limousine (Auto Union/Horch) There were several special 'heavy' body styles on the *m.E.Pkw.* chassis, including this unusual all-steel *Limousine*, seen in front of the *Deutsche Botschaft* (German Embassy) in Moscow. From 1954 this was the Embassy of the DDR.

Car, Heavy, 4×4, Kfz. 23 (Auto Union/Horch 1a, 1b) V-8-cyl., 81 bhp, 5F1R×1, wb 3.00 m, 4.85×2.00×2.04 m, 3000 kg. Standardized heavy car chassis (*s.E.Pkw.II*), here as Kfz. 23 *Fernsprechkraftwagen*. (communications vehicle). On model 1a the rear wheels could be made to steer if required. 1939.

Car, Heavy, 4×4, Kfz. 24 (Auto Union/Horch 1a, 1b) The *s.E.Pkw.II* chassis was also used for the mounting of closed bodywork like this *Verstärkerkraftwagen*. The *s.E.Pkw.I* chassis had the engine (3823-cc Horch V8) at the rear and was used for armoured cars. Vehicle shown measured 4.85×2.00×2.76 m.

Car, Heavy, 6×4, Kfz. 21 (Horch/Argus 71) 8-cyl., 80 bhp, 4F1R×2, wb 2.73 (BC 0.87) m, 4.90×2.00×2.10 m, 2680 kg. 4180-cc (73×118 mm) 8-in-line OHV engine. 5 produced in 1926, in conjunction with Argus of Berlin. Tyres 32×6 (shown with overall tracks on rear bogie). Six seats.

Car, Heavy, 6×6, Kfz. 21 (Selve/Voran M) 6-cyl., 53 bhp, 4F1R×2, wb 3.01 (BC 0.95) m, 4.70×1.86 m, 2664 kg. 3096-cc (74×120 mm) SV engine. Produced about 1928 as 6–8-seat staff car. IFS and IRS. Built by Selve of Hameln; front wheel drive design by Voran (front-drive car makers) of Berlin.

Car, Heavy, 6×4, Kfz. 21 (Mercedes-Benz G1) 6-cyl., 50 bhp, 4F1R, wb 2.84 (BC 0.95) m, 2400 kg. Basically similar to the makers' Model G3 light 6×4 truck chassis. Rigid axles with mechanical brakes and semi-elliptical springs (inverted at rear). PTO-driven capstan winch at rear. 1926–28.

Car, Heavy, 6×4, Convertible (Mercedes-Benz G4/W31) 8-cyl., 100 bhp, 4F1R×2, wb 3.57 (BC 0.95) m, 5.34×1.87×1.90 m, 3500 kg. 57 produced, 1933–39, for Army and Party officials. 5018-cc 8-in-line engine. Also with 115-bhp 5252-cc engine and six-wheel drive (G4/W131).

GERMANY
AMBULANCES

Most ambulances of the *Reichswehr* and *Wehrmacht* were based on heavy car (*s.Pkw.* and *s.E.Pkw.*) and light truck (*l.Lkw.*) chassis. In the *Wehrmacht* all ambulances, regardless of the chassis make and type, were designated *Kfz. 31* and officially known as *Krankenkraftwagen* or *Sanitätskraftwagen*. The latter name led to the widely used abbreviation '*Sanka*'. Most common ambulance chassis during the 1930s was the Phänomen Granit with air-cooled engine. Others were based on Adler, Horch, Mercedes-Benz, Opel and other chassis. A few typical examples are shown here. In 1939/40 a number of captured ambulances (British, French, etc.) were taken into service.

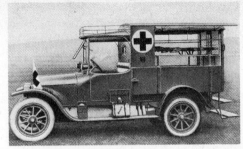

Ambulance, 4×2 (Mercedes 12/32 PS) 4-cyl., 35 bhp, 4F1R, wb 3.24 m. 3175-cc (84×140 mm) SV engine. Shaft drive. Typical World War I four-stretcher ambulance. There were also two- and six-stretcher versions on similar chassis. Some featured four full-elliptic leaf springs suspending each stretcher.

Ambulance Trailers, Two-Wheeled For mass-transportation of casualties these ambulance trains appeared in 1915. Each train consisted of a car towing three light two-wheeled canvas-covered ambulance trailers. The trailers were well-sprung and contained two or three stretchers.

Ambulance, 4×2 (Mannesmann-Mulag) Known as *schwerer Krankenkraftwagen* (heavy ambulance) this bus-type model of *c.* 1914 was based on a forward-control chain-drive truck chassis by Mannesmann-Mulag AG of Aachen. It was probably used as a mobile first aid post in battle zones.

Ambulance, 4×2 (Protos C1) 4-cyl., 45 bhp, 4F1R, wb 3.30 m. Basically a 1924–26 car chassis, with 2596-cc (80×130 mm) OHV engine. Protos was acquired by NAG in 1926 and production of this chassis was continued for vans and light trucks.

Ambulance, 4×2 (Adler K-3 I) 6-cyl., 65 bhp, 4F1R, wb 3.35 m, 4.90×1.74×2.14 m, 1750 kg. Tyres 7.00-17. 2916-cc (75×110 mm) Model 3G side-valve engine. Commercial chassis with IFS, derived from Adler Diplomat car, 1938. Used for civilian and military purposes.

Ambulance, 4×2, Kfz. 31 (Phänomen Granit 25) 4-cyl., 37–40 bhp, 4F1R, wb 3.25 m. The Granit 25 (2497-cc, 85×110 mm) appeared with various types of ambulance bodywork. Model shown has folding cab top and windscreen; others had hardtop cab with doors or side curtains. c. 1935.

Ambulance, 4×2, Kfz. 31 (Opel Blitz 2.5–32) 6-cyl., 55 bhp, 4F1R, wb 3.25 m. Introduced in 1937/38 and used by the *Wehrmacht* mainly with truck and ambulance bodywork. Tyres 5.50-18, dual rear. 2473-cc (80×82 mm) OHV engine (as in Opel Super 6 and Kapitän cars). Shown in *Luftwaffe* service.

GERMANY
BUSES

Chassis for buses were supplied by several truck manufacturers, notably Daimler-Benz (Mercedes-Benz), Faun, Henschel, MAN and Vomag. From 1939 the Opel Blitz Model 3.6-47 *Wehrmacht-Omnibus* was widely used, for personnel carrying as well as mobile command posts, offices, ambulances, etc. In 1939/40 Gräf & Stift of Austria supplied a fleet of buses for command and communications purposes. Known as *Kraftomnibus* the *Wehrmacht* knew three classes: *l.Kom.*(o) (light), *m.Kom.*(o) (medium) and *s.Kom.*(o) (heavy). The medium type was also used as *Befehls-Kw.* (command), *Funkauswerte-Kw.* (radio intelligence) and *Laboratoriums-Kw.* (laboratory vehicle). During the war large numbers of civilian buses were placed into service.

Bus, Light, 4×2 (Mercedes-Benz LO2000) Light buses were those with up to 15 seats and a max. GVW of 4000 kg. Vehicle shown was made during 1932–36 with 4-cyl. 3770-cc (100× 120 mm) OM59 diesel or M60 petrol engine and 4F1R gearbox. Both engines developed 55 bhp.

Bus, Medium, 4×2 (Vomag 3LR443) Medium buses had up to 30 seats and a max. GVW of 7300 kg. They were also used for command posts and other special purposes. This Vomag of the mid-1930s had a 4-cyl. 6080-cc (110×160 mm) diesel engine of 68 bhp at 1900 rpm.

Bus, Heavy, 4×2 (Henschel 4J5) Heavy buses had accommodation for over 30 passengers and a GVW of about 12,000 kg. This mid-1930s Henschel bus chassis was intended for up to 50 seats. The bodywork was military pattern. Engine was an 11.8-litre 6-cyl. 125-bhp diesel with 5F1R overdrive (*Schnellgang*) gearbox.

GERMANY

TRUCKS, LIGHT

Light trucks existed in several chassis configurations and with a variety of body types, including conventional load carriers, house-type vans, ambulances (*qv*), etc. They were almost exclusively of the 4 × 2 type, until about 1930 when the 6 × 4 type, as developed in France and Britain, was adopted for road and cross-country work; relatively large numbers of these were subsequently made by Büssing-NAG, Daimler-Benz, Krupp and Magirus. In 1937 the standard military *l.E.Lkw.* (6 × 6, diesel) appeared. Some 7500 were produced by several manufacturers, until 1940. These appeared with open and closed bodywork and also as snow plough and wrecker. Light 4 × 4 trucks became common after WWII had commenced.

Truck, Light, 4 × 2, Cargo (Daimler Canstatt) 4-cyl., 14 bhp, 4F1R, wb 3.17 m, length 5.40 m, height 2.55 m, net weight 2340 kg, GVW 4590 kg. Wheelsize, front 1.15 m, rear 1.45 m. Daimler M934 10/14 PS engine. Max. speed about 10 km/h. Produced during 1900–02.

Truck, Light, 4 × 2, Aircraft (Benz) Typical example of light truck used in World War I. Most had pneumatic tyres (later also solid tyres on sprung wheels). Shown is an aircraft carrier-cum-maintenance/repair vehicle. Note additional tyres/rims attached to single rear wheels.

Truck, Light, 4 × 4, Chassis (Daimler 70 PS) Designed as a fast machine gun carrier (twin MGs) this vehicle was powered by a 70-horsepower four-cylinder aero engine. Transmission was by individual drive shafts to each wheel. (H-drive). Max. speed 52 km/h. Two were completed, 1914.

Truck, Light, 4×2, Cargo (Phänomen Granit 30) 4-cyl. 55 bhp, 4F1R, wb 3.90 m, 6.00×2.20×2.60 m, 2500 kg. Air-cooled 3054-cc engine. Tyres 6.50-20, dual rear. Known as *l.Lkw. m. geschl. Aufbau (o)*. Also produced with canvas tilt and soft-top cab. Mid-1930s.

Truck, Light, 4×2, Cargo (Opel Blitz 2,0-12) 6-cyl., 36 bhp, 4F1R, wb 2.85 m, 4.64×1.99×2.28 m, 1325 kg approx. Two-litre (1932-cc; 67.5×90 mm) side-valve engine. Commercial 1-tonner (*l.Lkw. (o)*), also with van body, 1938. Photo (IWM BU6185) shows captured vehicle.

Truck, Light, 6×4, Signals, Kfz. 77 (Büssing-NAG G31) 4-cyl., 65 bhp, 3F1R×2, wb 3.19 (BC 0.95) m, 5.75×2.22×2.35 m, 3330 kg. Known as *Fernsprechkraftwagen* or *Leichter Fernsprech-Bautrupp (mot)*. 2300 chassis built and fitted with various open and closed body types, 1931–35.

Truck, Light, 6×4, Chassis (Krupp L2H) About 1933 Friedrich Krupp AG of Essen designed a cross-country chassis of which large numbers were made during 1934–42. This is a prototype with front bumper wheels. Note coil-sprung IRS. Engine was air-cooled 'boxer' type.

Truck, Light, 6×4, Prime Mover, Kfz. 69 (Krupp L2H43) HO-4-cyl., 60 bhp, 4F1R×2, wb 2.90 (BC 0.86) m, 5.00×1.90× 2.00 m, 2700 kg. Produced 1934–36, then superseded by improved L2H143. Both had 3308-cc (90×130 mm) air-cooled engine. Numerous body types.

Truck, Light, 6×4, Cargo (Magirus M206) 6-cyl., 70 bhp, 4F1R×2, wb 3.19 (BC 0.95) m, 5.35×2.25 m, 3750 kg. 4562-cc SV engine. 1153 chassis produced, 1934–37, for military and other users. Also with soft-top cab. Payload on/off roads 1650/1100 kg.

Truck, Light, 6×4, Engineers (Mercedes-Benz G3a) 6-cyl., 68 bhp, 3F1R×2, wb 3.47 (BC 0.95) m, 5.75×2.10×2.35 m, 3300 kg. 3689-cc SV engine. Over 2000 produced (various body types) during 1929–35, following 3460-cc Model G3 of 1928, of which 89 were made.

Truck, Light, 6×6, Chassis ('Einheitsdiesel') Standardized chassis with IFS/IRS, 80-bhp 6234-cc diesel engine, 4F1R×2 transmission and air brakes. Produced 1937–40, mainly by Büssing-NAG, Henschel, Magirus (shown) and MAN. Open and closed body types. Tyres 210-18. Wb 3.65 (BC 1.10) m.

GERMANY

TRUCKS, MEDIUM

During World War I most German transport vehicles were in the 3–4-ton payload class, made by Adler, Benz, Bergmann, Büssing, DAAG, Daimler, Dinos, Dixi, Dürkopp, Dux, Ehrhardt, Faun, Hansa-Lloyd, Horch, LUC, Magirus, Mannesmann-Mulag, Nacke, NAG, NSU, Opel, Podeus, Richard & Hering, Stoewer, Vomag, etc.

By the mid-1930s a *m.Lkw* (*mittlerer Lastkraftwagen*) was by definition one which carried 3500 kg or somewhat less if the bodywork was heavy (e.g. special van bodies). At this time there were many 6×4 types (*m.gl.Lkw.*); later the 4×4 became more popular. Most numerous during all these periods, however, was the commercial type 4×2 truck, often with modifications such as soft-top cab, etc. Some typical examples of all types are shown here.

Truck, Medium, 4×2, Cargo (Adler) 4-cyl., 4F1R, wb 4.15 m, 6.15×2.04×1.98 m. One of several types of vehicles supplied by Adlerwerke (vorm. Heinrich Kleyer) AG of Frankfurt/Main. This specimen, of conventional design, was captured and taken to England for examination in 1918.

Truck, Medium, 4×2, Cargo (Benz Gaggenau) 4-cyl., 4F1R, wb 4.77 m, 6.55×1.90 m. Many 3-ton trucks were produced by the Gaggenau works of Benz & Cie during WWI. Specimen shown was fitted experimentally with Benz-Bräuer half-track attachments, about 1917.

Truck, Medium, 4×2, Cargo (Büssing) Büssing of Braunschweig was another supplier of a variety of vehicles in WWI. Illustrated is a shaft-drive *Kavallerie-Lkw.* (cavalry truck). Note the coil-type helper springs at front. Similar chassis also with house-type operating-room body.

Truck, Medium, 4×2, Cargo (Daimler Marienfelde) In 1902 Daimler-Motoren-Gesellschaft of Cannstatt took over Motorfahrzeug- und Motorenfabrik Berlin AG of Berlin-Marienfelde. This became Daimler's truck factory. Shown is a 3-*t-Lkw mit Ritzelantrieb* (pinion and internal gear drive). *c.* 1910.

Truck, Medium, 4×2, Cargo (Daimler Marienfelde) Typical 3-ton chain-drive German Army truck of World War I, built by Daimler at Marienfelde. Lockers under rear body contained (left to right) chains, sand and spare parts. Note hard-top cab with glass screens.

Truck, Medium, 4×2, Cargo (Dürkopp L60) 4-cyl., 38–40 bhp, 4F1R, wb 4.37 m, 6.20×1.93×2.13 m, 3910 kg. 5401-cc (115×130 mm) side-valve engine. Internal cone-type clutch. Chain final drive. Dürkoppwerke of Bielefeld also produced cars (*qv*) and light trucks.

Truck, Medium, 4×2, Cargo (Dux LD) 4-cyl., 4F1R, wb 4.11 m, 6.40×2.00×2.53 m. Wheel diameter, front 30 inches, rear $33\frac{1}{2}$ inches, solid rubber tyres, dual rear. Dux cars and trucks were made by Dux-Automobil-Werke-AG, Leipzig-Wahren.

Truck, Medium, 4 × 2, Cargo (Ehrhardt) Heinrich Ehrhardt AG of Düsseldorf and Zella-Mehlis were well known for their heavy artillery tractors (*qv*), four-wheel drive AA gun carriages (*qv*), etc., but they also produced military cargo trucks. Later production models had a hard-top cab.

Truck, Medium, 4 × 2, Cargo (Hansa-Lloyd) Damaged example of World War I Hansa-Lloyd truck. Engine was four-cylinder (cast in pairs) 50-bhp with side valves and cone clutch, transmission 4F1R with chain final drive. Foot brake acted on transmission, hand brake on rear wheels.

Truck, Medium, 4 × 2, Cargo (Horch) The famous firm of Horch in Zwickau produced cars, trucks and heavy tractors for the German war effort in World War I. Shown is a four-cylinder shaft-drive 3-ton truck, captured by the British Army in France (Rouen, 1918).

Truck, Medium, 4 × 2, Cargo (LUC) 22/40 HP engine with four cylinders cast in pairs, overhead inlet and side exhaust valves. Pallas carburettor. Cone clutch. 4F1R gearbox. Foot brake on front of diff. housing. Chain drive. Made by Loeb & Co. GmbH of Berlin-Charlottenburg.

Truck, Medium, 4×2, Cargo (MAN) The old-established Maschinenfabrik Augsburg-Nürnberg (MAN) commenced production of trucks under Saurer licence in 1915. They had a 37-bhp 4-cyl. engine and chain drive. Note supplementary front springs which acted as shock dampers.

Truck, Medium, 4×2, Cargo (Nacke) Until the late 1920s E. Nacke of Coswig-Sachsen (Saxony) was a well-known German truck maker. This model was built for the Germany Army. Earlier models had soft-top cab. Before 1914 Subsidy-type trucks and trailers were produced.

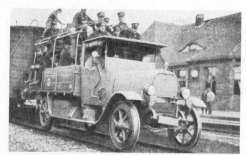

Truck, Medium, 4×2, Cargo (NAG) The Nationale Automobil-Gesellschaft AG (NAG) of Berlin-Oberschöneweide (which later merged with Büssing) produced many military cars and trucks. This truck was converted for use on railroads about 1917.

Truck, Medium, 4×2, Cargo (Podeus) 4-cyl., 45 bhp, 4F1R, wb 4.01 m. Imported and exhibited in Britain in the summer of 1914 this 3–4-ton commercial Subsidy model was produced by Maschinenfabrik Podeus AG of Wismar, Mecklenburg. It featured a Bosch self-starter.

Truck, Medium, 4×2, Cargo (Stoewer) Stoewer-Werke AG (vormals Gebrüder Stoewer) of Stettin manufactured trucks until the mid-1920s, including (from *c.* 1909) several light, medium (shown) and heavy military types. Aircraft engines were also produced.

Truck, Medium, 4×2, Cargo (Vomag P30z) 4-cyl., 35/40 bhp, 4F1R, wb 4.00 m, 6.25×1.95 m. Shaft-drive (Model P40zK had chain-drive and 4.20-m wb). 110×160-mm engine. Tyres, front 930×120, rear 1010×120 (dual). Max. speed 30 km/h. 1918.

Truck, Medium, 4×2, Cargo (Büssing) Typical general service load carrier with dropside body as used by the *Reichswehr* during the late 1920s. Registration numbers were carried at both sides of the body at the rear. Basically a commercial truck (*m.Lkw.*(*o*)).

Truck, Medium, 4×2, Fuel Servicing (DAAG ACO) Produced during 1928–30 this commercial truck had a DAAG Model C3 4-cyl. 6080-cc (110×160 mm) engine, developing 50 bhp at 1200 rpm. Officially known as *mittlerer Betriebstoffkesselkraftwagen* (*o*) on *m. Lkw.* (*o*) chassis.

Truck, Medium, 4×2, Medical (Ford V8-51) Mobile operating room on Ford truck chassis. German Ford trucks were basically similar to US production but continued longer. For example, the 1936 (US) Model 51 and earlier Model BB were still made in Cologne in 1937/38.

Truck, Medium, 4×2, Cargo (Ford G917T St III a,b) American-pattern 1939 Ford truck was produced in Cologne until 1941, first with 3.6-, later with 3.9-litre V8 engine. (1940 US pattern was made during 1941—45). Soft-top *Einheits*-cab was used on several commercial trucks.

Truck, Medium, 4×2, Signals (Hansa-Lloyd Merkur) Humboldt-Deutz F6M313 diesel, 6-cyl., 75 bhp, 4F1R, wb 5.20 m. Used by *Luftwaffe* (Air Force) as *Lkw für Telegrafen-Bautrupp*, by *Heer* (Army) as *Fernsprechbaukraftwagen*. Basically commercial chassis, c. 1935.

Truck, Medium, 4×2, Signals (Krupp L3,5M242) 4-cyl., 75 bhp, 4F1R, wb 4.00 m. House-type van body, designated *Kfz. 42*. Used for Signals workshop (*Nachr. Werkst. Kw.*) and other purposes. Chassis (commercial) was also produced with 2-cyl. Junkers diesel engine. 1936/37.

Truck, Medium, 4×2, Signals (Magirus) Commercial Magirus truck chassis of the mid-1920s (chassis 1 CV-100 or 2CI-V110) with military pattern bodywork as mobile radio station. Officially known as *leichte Funkstation*, or, abbreviated, as *l.Funkst*. Note swivel-type aerial.

Truck, Medium, 4×2, Cargo/Signals (MAN) Commercial chassis with general service truck body (*mittlerer Lastkraftwagen, offen*(o)). It was employed also as *Telegrafenbaukraftwagen* (*Tel. Baukw.*) for telegraph line construction. Payload capacity was about 3500 kg.

Truck, Medium, 4×2, Cargo/Engineers (Mercedes-Benz L1) This was one of the first trucks produced after the merger in 1926 of Daimler and Benz. Introduced in 1927 it had a 4-cyl. 45-bhp M14 engine, replaced in 1928 by the 6-cyl. 50-bhp M15. Shown is a *Pionier Kw. II* of 1928/9.

Truck, Medium, 4×2, Cargo (Mercedes-Benz L3000) From 1930/1 Daimler-Benz AG's Gaggenau works used sales model designations which indicated the vehicle payload. The L3000 (internally known as L57) was a 3-tonner. It had a 70-bhp 6-cyl. M56 petrol engine. O3000 (N58) was bus variant.

K

Truck, Medium, 4×2, Cargo (Mercedes-Benz L4000) This L4000 of the *Reichswehr* was powered by a 7.07-litre 6-cyl. M26 petrol engine, developing 70 bhp at 1500 rpm. Weight was 4900 kg; official (military) carrying capacity 3150 kg or up to 30 persons. *c.* 1931.

Truck, Medium, 4×2, Personnel (Mercedes-Benz L3500) Daimler-Benz M68 6-cyl., 95 bhp (LO3500: OM67 6-cyl. diesel), 4F1R, wb 4.60 m. Open cab with full-length canvas top extending over rear body which had seven bench seats and double dropsides. Spare tyre at rear. *c.* 1935.

Truck, Medium, 4×2, Signals (NAG Z) Officially known as *leichter Fernsprechbaukraftwagen* this vehicle, which could be based on a commercial 2–2½- or 3-ton chassis, was used for telephone line construction. Shown is a NAG Model Z, delivered to the *Reichswehr* about 1928.

Truck, Medium, 4×2, Cargo (Opel Blitz 3,6-36S) 6-cyl., 68 bhp, 5F1R, wb 3.60 m, 6.02 × 2.26 × 2.60 m, 2500 kg approx. Most common truck of the *Wehrmacht*. Some 70,000 chassis were supplied by Opel's Brandenburg/Havel truck plant during 1937–44. From 1940 also 4×4 variant.

Truck, Medium, 4×2, Van, Kfz. 305 (Opel Blitz 3,6-36S) 6-cyl., 68 bhp, 5F1R, wb 3.60 m, 6.10 × 2.15 × 2.85 m. Standard body, used for numerous purposes (shown: *Kfz. 305/94 Wasseraufbereitungs Kw.II*, water purification plant). Also on Mercedes-Benz chassis. Note military cab.

Truck, Medium, 4×2, Workshop (Opel Blitz 3,6-47) 6-cyl., 68 bhp, 5F1R, wb 4.65 m, chassis : 7.36 × 1.97 × 1.91 m. Low-frame chassis for bus and special bodies. Being a GM product, Opel Blitz was similar in design to Chevrolet and Bedford. Main exception was 5-speed gearbox.

Truck, Medium, 4×4, Cargo (Mercedes-Benz LG65/2) 4-cyl. diesel (OM65; or petrol M66), 70 bhp, 4F1R × 2, wb 3.80 m. Chassis wt 2800 kg. IFS and IRS. Commercial venture of mid-1930s. Also as 6-cyl. 6×6 (LG65/3) and 8×8 (LG65/4) with many components in common.

Truck, Medium, 6×4, Cargo (Büssing-NAG III GL6) 6-cyl., 90 bhp, 5F1R, wb 4.25 (BC 1.25) m, 7.65 × 2.30 × 2.80 m, 6500 kg approx. 300 chassis produced during 1931–38, mainly for Army and export (Turkey). Open and closed body types. 9350-cc OHV engine. Knorr air brakes.

147

Truck, Medium, 6×4, Cargo/Personnel (Henschel 33D1) 6-cyl., 100 bhp, 5F1R, wb 4.30 (BC 1.10) m, 7.10×2.50 m, 6100 kg. approx. About 22,000 produced during 1933–42 (from 1937 also with diesel engine, Model 33G1). Various open and closed body types, including fire fighters.

Truck, Medium, 6×4, Signals (Krupp L3H163) 6-cyl., 110 bhp, 4F1R×2, wb 4.20 (BC 1.10) m, 7.40×2.50 m, 5800 kg approx. 7542-cc OHV engine. Produced 1936–38, superseding 90-bhp 6107-cc engined Model L3H63 (1931–35). Various bodies; shown as signals equipment carrier.

Truck, Medium, 6×4, Signals, Kfz.72 (Magirus 33G1) Deutz 6-cyl. diesel, 100 and 125 bhp, wb 4.20 (BC 1.10) m. 3815 chassis made during 1938–41 under Henschel licence. Shown as *Fernschreibvermittlungs-Kw.* with *Sd.Ah.24 Anhänger für schw. Maschinensatz* (generator trailer).

Truck, Medium, 6×4, Cargo/Signals (Mercedes-Benz LG3000/LG63) 6-cyl. diesel, 95 bhp, 5F1R×2, wb 4.42 (BC 1.05) m. 7434 chassis produced, 1935–38, as *m.gl.Lkw.* (*o*). Various body types, some with closed cab. Truck shown was used as FFK equipment carrier for telephone construction troops.

GERMANY

TRUCKS, HEAVY

Heavy trucks were generally those with a payload of 4 tons and over. During World War I these included Subsidy-type vehicles, in conjunction with four-wheeled trailers; these combinations were known as (*staatlich subventionierte*) *Lastzuge*. They were supplied by firms like Benz (Gaggenau), Büssing, Daimler (Marienfelde), Dürkopp, Hansa-Lloyd, Mannesmann-Mulag, Nacke, NAG, etc. During the 1920s and 1930s heavy trucks were mainly commercial 4×2 types with certain military modifications. Some 9-ton 6×4 types appeared in 1937 (Büssing-NAG, Faun) and during World War II the majority of new heavy trucks were 4–5-ton 4×2 and 4×4 models. Unlike the light and medium types, the heavy trucks usually had an open load-carrying body.

Truck, Heavy, 4×2, Cargo (Daimler-Loutzky/MMB Marienfelde) 2-cyl., 12.9 bhp, 4F1R, wb 3.97 m, 5.80×2.03×2.32 m, 3400 kg. GVW 8400 kg. Ordered in 1901 by Russian Government through Borus Loutzky. Delivered in 1903; shown in Decembrists' Square, St. Petersburg (Leningrad).

Truck, Heavy, 4×2, Cargo (Büssing IIIA) 4-cyl., 36 bhp, 4F1R. Heavy chain-drive *Heeres Lastwagen* (Army truck) of 1914. From 1909 Büssing had produced *Subventionswagen* (Subsidy trucks) of similar appearance. Note shock-damping coil springs at both ends of front springs.

Truck, Heavy, 4×2, Cargo (Büssing) Known as *Schienen-Lastkraftwagen* this Büssing truck was converted during World War I for running on railway tracks. Similar rail/road conversions appeared again in World War II (Henschel, Magirus, Mercedes-Benz).

Truck, Heavy, 4×2, Cargo (Daimler Marienfelde) Iron-wheeled chain-drive military truck, built in 1907 by Daimler's branch at Berlin-Marienfelde (formerly the Motorfahrzeug- und Motorenfabrik Berlin AG, taken over by Daimler of Cannstatt in 1902).

Truck, Heavy, 4×2, Cargo (Daimler Marienfelde) Heavy Subsidy-type truck, produced before World War I. Final drive was by shaft, pinions and internally-toothed gearwheels on the rear wheels (*Ritzelantrieb*). Lockers under body contained chains, sand, spare parts, etc.

Truck, Heavy, 4×2, Cargo (Daimler Marienfelde) Subsidy-type truck, produced in 1913/14. It had a 4-cyl. overhead-valve engine, cone clutch, 4.19-m wheelbase, and measured 6.12 × 1.96 × 2.15 m. Note hard-top cab with 5-piece glass windscreen. Payload was 4500 kg.

Truck, Heavy, 4×4, Cargo (Daimler Marienfelde) This four-wheel drive vehicle with 60-bhp engine was produced in 1908/09 and supplied to Portuguese West Africa. It was later converted into a half-track, after adding extra wheels fore and aft of the rear wheels.

Truck, Heavy, 4×2, Cargo (Dürkopp) Typical example of Subsidy-type truck with four-wheeled drawbar trailer as produced prior to and during the first World War. It was also Dürkopp's first truck with cardan shaft-drive. Note cabin for trailer attendant.

Truck, Heavy, 4×2, Gas Supply (NAG) One of the many types supplied by NAG of Berlin-Oberschöneweide in 1914 was this 45-bhp 5-ton chain-drive chassis fitted out with gas cylinders for use by Zeppelin airships. Radiator shape was typical of heavy NAGs.

Truck, Heavy, 4×2, Cargo (SAG) 36-bhp 6-tonner, produced in 1907/08 for the *Heeresverwaltung* by the Süddeutsche Automobilfabrik GmbH of Gaggenau (taken over about that time by Benz). Cab-over-engine layout was more common in France and the USA.

Truck, Heavy, 4×2, Cargo (Stoewer) 5-ton truck produced by Stoewer of Stettin in 1909 and supplied, like the products of several other manufacturers, to the *Versuchs-Abteilung der Verkehrstruppen* (transport research department). Note pioneer tools and location of fuel tank.

Truck, Heavy, 4×2, Fuel Servicing (Henschel 6J3) 6-cyl., 125 bhp, 5F1R, wb 5.60 m, 8.50×2.50×2.60 m, 8000 kg approx. 11,781-cc (125×160 mm) OHV engine. Known as *schwerer Betriebstoffkesselkraftwagen* on chassis of *s.Lkw.(o)*. Tank capacity about 6500 litres. Mid-1930s.

Truck, Heavy, 4×2, Oxygen Tank (Krupp L5N62) 6-cyl., 110 bhp, 5F1R, wb 5.20 m. Krupp M12 7540-cc OHV petrol engine. Known officially as *Sauerstoffkesselkraftwagen* on chassis of *s.Lkw.(o)*. Similar chassis available with 3- or 4-cyl. Krupp-Junkers diesel engine, 1935.

Truck, Heavy, 4×2, Cargo (Magirus) Heavy diesel truck of the *Luftwaffe* in difficulty on the only main supply route north of the Polar circle in Norway (World War II). Hinged triangle on cab roof was erected to indicate trailer being towed. Truck dates from late 1930s.

Truck, Heavy, 4×2, Fire Fighting (Magirus) Deutz F6M516 6-cyl. diesel, 125 bhp, 4F1R with overdrive, wb 4.57 m, 8.68×2.20×2.70 m, 7800 kg. Known as *Kraftfahrspritze (o)* or *K.S.(o)* this was a universal (*Einheits*) type with 2500-l/min. centrifugal pump. Crew 8+1. 1938–40.

Truck, Heavy, 4×2, Personnel (MAN D1) 6-cyl., diesel, 80 bhp, 4 or 5F1R, wb 5.70 m. 5-ton payload chassis. Bodywork: *Mannschafts-Transportwagen*, with double dropsides and transverse troop seats. 7274-cc engine. 8.25-20 tyres. Vacuum-servo brakes. 1934/35.

Truck, Heavy, 4×2, Cargo (MAN F2H6) 6-cyl. diesel, 100 bhp, 4F1R, wb 5.25 m, 7.65×2.50×2.75 m approx. Chassis carrying capacity 7500 kg. MAN D2086 12,214-cc engine. MAN/Bosch mech. brakes with vacuum or air assistance. 38×9 tyres. 1935.

Truck, Heavy, 4×2, Cargo (Mercedes-Benz L5000) 5-ton truck produced by Daimler-Benz at Gaggenau for the *Deutsche Reichswehr*. Standard engine was the 95-bhp OM5S diesel but a petrol engine was available for export and special purposes. Note solid rubber tyres.

Truck, Heavy, 6×6, Cargo (Mercedes-Benz LG4000/LG68) 6-cyl. diesel, 95 bhp, 4F4R×2, wb 4.40 (BC 1.20) m, 6.92× 2.19×2.03 m (chassis), 6300 kg. IFS/IRS. Fitted with self-locking differentials and winch. Also with petrol engine, dual rear tyres and no bumper wheels. 1937/38.

Truck, Heavy, 6×4, Tank Transporter (Büssing-NAG 900)
6-cyl. diesel, 130 bhp, 4F1R, wb 6.17 (BC 1.45) m, 9.90×2.50× 2.60 m, 8900 kg approx. Used mainly to carry light tanks (*Pz.II*) and tow low-loader tank transporter trailer. 13,538-cc engine. 13.50-20 tyres (single rear). 1937–39.

Truck, Heavy, 6×4, Tank Transporter (Faun L900D567)
Deutz F6M517 6-cyl. diesel, 4F1R, wb 5.52 (BC 1.40) m, 9.80× 2.50×2.60 m, 9200 kg. Bosch air brakes. 13,538-cc engine. 12.75 or 13.50-20 tyres, single rear. Used with trailer. About 80 supplied during 1937–39.

Truck, Heavy, 8×8, Cargo (Mercedes-Benz LG65/4) 6-cyl. diesel, 100 bhp, 4F4R. IFS and IRS with double wishbones, coil springs and balancer beams. Hyd. brakes and shock absorbers on all wheels. Self-locking differentials. Max. speed 60 km/h. Also as 6×6 and 4×4. Mid-1930s.

Truck, Heavy, 8×8, Chassis (MAN) Derived from the 6×6 '*Einheitsdiesel*' was this 8×8 truck with 120-bhp V-8-cyl. 4942-cc diesel engine. Payload on/off roads was 5000/4000 kg. Four were built in 1937/38 but later converted to amphibians with 150-bhp 6-cyl. engine.

GERMANY

TRACTORS, WHEELED

Makes and Models: Austro-Daimler (see Austria). Benz (4×2, WWI), VRZ (4×4, 1920). Büssing, various models (4×2 and 4×4, WWI). Daimler 25 PS and 45 PS (4×2, 1907), 60 PS (4×4, 1907), 100 PS (1912), KDI-46/100 PS (4×4, from 1917*), V8-92/70 PS (4×4, 1918*), etc. Deutz 100 PS (4×2, WWI). Dürkopp 80 PS (4×2, WWI). Ehrhardt 36/80 PS (4×2 with oscillating rear axles, WWI). Faun ZR (4×2, 1939). Hanomag SS100, etc. (4×2, from c. 1936). Horch (4×2, WWI). Kaelble Z6GN125 (4×2, 1935). Lanz ZF 100 PS (4×2, WWI), Bulldog (4×2, from c. 1937). Magirus 70 PS (4×4, WWI). Müller (petrol-electric, 1910). NAG, various models (4×2, from 1903). Podeus (4×2, WWI). Pohl 100 PS (4×2, WWI). Siemens-Schuckert (petrol-electric, from 1903), etc.

* In conjunction with Krupp.

General Data: Most interesting of the above tractors were the *Artillerie-Zugmaschinen* (artillery prime movers) of World War I. Unlike the Americans and the British, the Austrians (*qv*) and the Germans had developed a number of large and very powerful special tractors. They were fitted with efficient winches, generally with fairleads of such a type that the lead could be pulled in any direction. Some had four-wheel drive and generally the rear wheels were of large size, the proportions of front and rear wheels following traction-engine practice (the German Army had used a number of British Fowler steam traction-engines in 1905 and earlier). Actual brake horsepower generally varied from about 75 to 135. Power-to-weight ratio resulted in maximum speeds of up to 20 mph. The vehicles were very well engineered and some appeared over-complicated, frame side members having bends in every direction. Great attention was paid to spuds on the rear wheels, usually arranged so that they could either be withdrawn quickly inside the rim or alternatively so arranged as to be easily removable. The extensive use of steel pressings for various parts of the vehicles, including bodywork, was noteworthy at the time. Winches were generally of substantial construction, set either on a horizontal spindle across the frame or on a vertical spindle and arranged much like a steam plough engine. Some had an automatic coiling gear. During the 1930s the Germans developed semi-tracked vehicles for artillery towing and wheeled prime movers for fast haulage on their *Autobahnen*.

Vehicle shown (typical): Tractor, Heavy, 4×4, Artillery (Daimler KDI).

Technical Data:
Engine: Daimler M1574 4-cylinder, I-F-W-F, 12,020 cc (150×170 mm), 100 bhp at 1200 rpm. CR 4.5:1.
Transmission: 4F1R × 2.
Brakes: mech. (foot on trans., hand on rear wheels).
Tyres: steel (later solid rubber).
Wheelbase: 3.57 m. Track 1.67 m.
Overall l × w × h: 6.70 × 2.20 × 3.00 m.
Weight: 5200 kg. Payload 2000 kg.
Note: also known as Krupp-Daimler and Mercedes. Winch amidships with automatic coiling mechanism. 63 built in 1918; further batch of 49 for *Reichswehr* during 1920s. Later known as *Sd.Kfz. 2*. Chassis variant used as SP gun carriage (*Sd.Kfz.1*).

Tractor, Heavy, 3×2, Artillery Heavy prime mover of unknown make. Transmission was located in front of engine, driving rear wheels via roller chains. Note large drum above single front wheel. Reputedly used in Poland during World War I.

Tractor, Heavy, 4×2, Artillery (NAG) The Neue (later: Nationale) Automobil-Gesellschaft m.b.H. of Berlin-Oberschöneweide built artillery tractors as early as 1903. This model was powered by a NAG Model VM IV 4-cyl. engine of 40 bhp at 650 rpm.

Tractor, Heavy, 4×4, Artillery (Daimler 60PS) In 1907 Daimler built their first all-wheel drive *Zugwagen*. It had internal tooth gear wheel drive, front and rear. At the same time they produced smaller but similar-looking 25 and 45 PS (hp) 4×2 models.

Tractor, Heavy, 4×4, Artillery (Daimler 100 PS) This tractor was built in 1912 and delivered to Spain. It was powered by a 100-bhp airship engine with special tropical cooling system and during preliminary tests hauled four siege guns on 4-wheeled carriages.

Tractor, Heavy, 4×2, Artillery (Benz) 4-cyl., 4F1R, wb 3.95 m, 6.40×2.40 m, 10,820 kg approx. 165×220-mm engine. Worm-driven 3F1R vertical-spindle winch with 200-m cable. 6-ft dia. rear wheels with retractable spuds. Lockable diff. Shown under test in England, *c.* 1920.

Tractor, Heavy, 4×2, Artillery (Büssing KZW 1800) 4-cyl., 80 bhp, 4F1R, wb 3.65. Also produced with equal size wheels. Gun crew bench seat at rear of cab. Straight frame with front winch. Retractable rear wheel spuds. Some had small rear body for ammunition. 1916/17.

Tractor, Heavy, 4×2, Artillery (Büssing) Basically similar to Model KZW 1800 but used principally for ammunition haulage, hence different rear body. Also produced with 6-cyl. engine (cast in three pairs); these models had a taller radiator.

Tractor, Heavy, 4×4, Artillery (Büssing) 6-cyl. tractor, introduced in 1918 to supersede the earlier Büssing types. It had four-wheel drive and looked very modern for its time. In accordance with the Treaty of Versailles all this equipment was destroyed in 1920.

Tractor, Heavy, 4×4, Artillery (Daimler 92/170 PS) This *Zugwagen* was designed by Daimler in conjunction with Krupp in 1918. Powered by a 200-bhp V8 engine it was intended to haul loads of up to 25 tons (5 tons on tractor, 20 tons on trailer or gun of similar weight). No quantity production.

Tractor, Heavy, 4×2, Artillery (Deutz 100 PS) 4-cyl., 100 bhp, 3F1R×2, wb 3.50 m, 6.40×2.89×2.96 m, 12,500 kg. 10-ton vertical-spindle winch amidships. Drawbar pull 8000 kg. Rear wheels with non-skid bars and retractable spade-type spuds. 160×200-mm Kämper F-head engine.

Tractor, Heavy, 4×2, Artillery (Dürkopp 80 PS) 4-cyl., 80 bhp, 4F1R×2, wb 3.90 m, 6.42×2.28×2.70 m, 7600 kg. 91 built, some with taller (1.72-m) radiator. 165×178-mm SV engine. 4F1R gear-driven winch on vertical post under gearbox, with guide rollers front and rear.

Tractor, Medium, 4×2, Artillery (Ehrhardt 36/80 PS) 4-cyl., 80 bhp, 4F1R×2, wb 4.35 m, 7.32×2.23×2.99 m. 156×160-mm T-head engine. Winch amidships. Four rear wheels, in line, each pair in a swivelling frame. Drive chain sprockets on outer wheels. Load platform 3.75×2.00 m.

Tractor, Heavy, 4×2, Artillery (Horch) 4-cyl., 55 bhp, 4F1R, wb 3.60 m, 6.10×2.18×2.75 m, 7500 kg approx. Double-cranked frame with centre-mounted winch. Tractor shown was one of about 50 captured vehicles taken to England for evaluation by the RASC at Aldershot.

Tractor, Heavy, 4×2, Artillery (Lanz ZF) 6-cyl., 100 bhp, 3F1R×2, wb 3.70 m, 6.65×2.20×2.61 m, 8575 kg approx. Canopy removed. Winding drum under rear axle. L-head engine with cylinders cast in pairs. Bore and stroke 150×160 mm. Unsprung centrally-pivoted front axle.

Tractor, Heavy, 4×4, Artillery (Magirus 70 PS) Four-wheel drive *Kraftprotze* with equal size front and rear wheels, produced about 1918 by C.D. Magirus in Ulm/Donau to the design of Heinrich Buschmann. At the rear there were back-to-back bench seats for the gun crew.

Tractor, Heavy, 4×2, Artillery (Podeus) Heinrich Kämper 4-cyl. F-head (inlet-over-exhaust) 160×200-mm engine with 4F4R transmission and spur gear final drive to 2.50-m diameter steel rear wheels. Produced by Podeus of Wismar, Mecklenburg, originally as an agricultural tractor.

Tractor, Heavy, 4×2, Road/Rail (Faun ZRS) Deutz F6M517 diesel 6-cyl., 150 bhp, 4F1R with overdrive (between engine and gearbox), 6.45 (without buffers) × 2.44 × 2.55 m, 10,000 kg approx. Shown with railway wheels installed. Known as *Schienenzepp*. From 1939.

Tractor, Heavy, 4×2, Road Haulage (Hanomag Gigant) During the late 1930s Hanomag of Hanover offered a range of *Schnelltransporter*, of which this is the 100 PS model with wood-burning Imbert gas producer. On level roads it could haul two 7-ton trailers at speeds up to 45 km/h.

Tractor, Heavy, 4×2, Road Haulage (Hanomag SS100) 6-cyl. diesel, 100 bhp, 4F1R, wb 3.00 m, 5.54 × 2.46 × 2.42 m, 6540 kg. Hanomag D85 8553-cc engine. These tractors, with double cab, were widely used to tow special drawbar trailers of *Luftwaffe*, Engineers, etc.

Tractor, Heavy, 4×2, Road Haulage (Kaelble Z6GN125) 6-cyl. diesel, 135 bhp, 5F1R, wb 2.88 m, 5.64 × 2.32 × 2.33 m, 6100 kg. Kaelble GN125S 13,253-cc engine. Photo shows *Luftwaffe* personnel hauling bomb on special sledge in the battle for Stalingrad. Note covered crew seats.

GERMANY

COMBAT VEHICLES, WHEELED

Most pre-1914 German combat vehicles were so-called *BAK-Wagen*. Produced by Daimler and Ehrhardt these consisted of a *Plattformwagen* (mostly with all-wheel drive) mounting an anti-aircraft (*Flak*) gun.

Aircraft included observation and dirigible balloons or airships, hence the common designation *BAK* for *Ballon-Abwehr-Kanone* (Balloon Defence Gun). In 1915 these two manufacturers, as well as Büssing, introduced 4×4 armoured cars. After the Armistice all this material was destroyed. In the early 1920s some armoured vehicles were built for police use and about 1930 the first 6×4 armoured cars appeared, based on commercial cross-country chassis.

These were later superseded by new military 4×4 and 8×8 designs, executed by Auto Union and Büssing-NAG.

Armoured Car, 4×2, Command (Opel 40PS) Semi-armoured *Kriegswagen für höhere Truppenführer*, designed and produced in 1906 by Ing. Emil August Schmidt, technical manager of the Berlin Opel branch. Armament: 2 Mauser MGs and 2 Mauser pistols. Engine: 40-bhp 6880-cc T-head Four.

Carriage, 4×2, 5-cm BAK (Ehrhardt/Rheinmetall) In 1906 Heinrich Ehrhardt had produced an armoured vehicle with Rheinmetall 5-cm *Flak* L/30 gun in a fixed turret. In 1908 this open version appeared. They were intended for anti-aircraft and anti-balloon/airship purposes.

Carriage, 4×4, 6.5-cm BAK (Ehrhardt/Rheinmetall) Chassis by Heinrich Ehrhardt Automobilwerke AG with 6.5-cm *Flak* L/35 gun by Rheinmetall, from 1910. *BAK* stood for *Ballon-Abwehr-Kanone* (balloon defence gun). The carriage was referred to as *Rad-Selbstfahr lafette* or *Plattformwagen*.

L

Carriage, 4×4, 7.7-cm BAK (Daimler/Krupp) From 1908/09 Daimler built several *BAK-Wagen*. Some models had varying types of armour-plate superstructures, others were entirely open. Similar bodywork and gun were later mounted on the popular KDI art. tractor chassis (KW19).

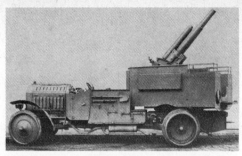

Carriage, 4×4, 7.7-cm BAK (Daimler/Krupp) 4-cyl., 54 bhp, 4F1R×2, wb 3.84 m, length 5.57 m, chassis wt 3500 kg. *Plattformwagen* by Daimler (1911), gun by Krupp. 920-mm diameter solid tyres. 9850-cc L-head engine. Gun mounted on turntable.

Carriage, 4×4, 7.7-cm K-Flak (Daimler/Krupp) 4-cyl., 60–80 bhp, 4F1R×2, wb 3.84 m, length 6.27 m, GVW 8000 kg. This model was used during World War I. A similar model was produced by Ehrhardt (Model E V/4). Both had ammunition lockers and crew seats for up to 10 men.

Carrier, 4×4, Field Gun (Daimler/Krupp) *BAK-Wagen* chassis were also used as field gun portees. Again the superstructure was made by Fried. Krupp AG of Essen. A winch was used to haul the gun aboard and the hinged ramps doubled up to secure the gun wheels in position. 1910.

Combat Vehicle, 3×2 (Hansa-Lloyd 'Treff-As') This huge machine with rear wheels of about 3 metres (10 ft) in diameter was produced experimentally in 1917 and known as *'Treff-As'* (Ace of Clubs). A similar machine was made in 1942 by Lauster of Stuttgart.

Carrier, Armoured, Personnel, 4×2 (Mannesmann-Mulag) Armoured personnel carrier, produced on conventional 4×2 truck chassis by Mannesmann-Mulag of Aachen. Engine was also covered by armour plating. Powerful spotlights were fitted at front and rear. *c.* 1916.

Armoured Car, 4×4 (Büssing A5P) 6-cyl., 90 bhp, 5F5R, 9.50×2.10 m, combat weight 10,250 kg. Crew 9. Armament 3 heavy MGs. Armour 5–7 mm. Max. speed 35 km/h. Steering controls front and rear; four-wheel steering. 1915. Picture taken in Rumania, 1917.

Armoured Car, 4×4 (Daimler M1915) 4-cyl., 80 bhp, 4F4R, wb 3.84 m, 5.61×2.05 m, combat, weight 9000 kg. Crew, armament and armour as Büssing A5P (*qv*). Steering controls front and rear; front-wheel steering. Chassis same as *BAK-Wagen*, with 9850-cc engine. 1915.

Armoured Car, 4×4 (Ehrhardt E-V/4) 4-cyl., 70 bhp, 4F4R (later 6F6R), wb 4.70 m, 5.60 (later 5.30) × 2.00 × 3.00 m, combat wt 9000 (later 7750) kg. Front wheel steering with dual controls (front and rear). Some 33 built during 1915–18. (IWM photo Q23751).

Armoured Car, 4×4 (Benz VP21) 4-cyl., 100 bhp, 6F6R, wb 3.60 m, 5.95 × 2.60 × 3.32 m, GVW 11,000 kg. Known as *Schupo-Sonderwagen* a number of these vehicles were built by Benz at their Gaggenau plant in 1921. Crew 5. Armament 2 heavy MGs. Speed 55 km/h.

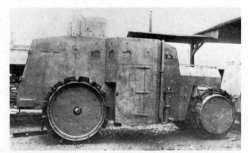

Armoured Car, 4×4 (Daimler DZVR) 4-cyl., 100 bhp, 5F5R, wb 3.75 m, 5.58 × 2.10 × 3.10 m, GVW 10,500 kg. *Schupo-Sonderwagen* (special vehicle for *Schutzpolizei*), produced during 1919–22. Front and rear steering controls. 1921–22 models had equal-size wheels.

Armoured Car, 4×4 (Daimler DZVR) This model was used by the *Reichswehr* as *Sd.Kfz. 3* until well into the 1930s as a turret-less armoured personnel carrier. Daimler M1574 12-litre 100-bhp engine. Overall dim. 5.95 × 2.20 × 2.72 m, wt 10,600 kg. Crew 6.

Dummy Armoured Vehicle, 4×2 (Opel P4) From about 1928 until well into WWII the German Army used dummy 'tanks' based on 4×2 car chassis (Dixi 1928, Adler 1930–32, Opel P4 (shown) 1937–38 and finally VW *'Kübel'*). Known as *Panzer-Attrappe* or *-Nachbildung*.

Armoured Car, Medium, 4×2, Kfz.14 (Adler Standard 6) 6-cyl., 60 bhp, 4F1R, wb 2.84 m, 4.20×1.70×1.50 m, 1900 kg. Lightly armoured radio car, crew 3. Also as Kfz.13 *MG-Kraftwagen* with one machine gun and crew of 2. Produced 1932–34.

Armoured Car, Medium, 4×4, Sd.Kfz.223 (Auto Union) Horch V-8-cyl., 75 (later 90) bhp, 5F1R, wb 2.80 m, 4.80×1.95× 1.75 m, 3950 kg. Radio car, based on rear-engined *Einheitsfahrgestell I für s.Pkw.* (heavy universal car chassis) of the late 1930s. Four-wheel steering. IFS/IRS.

Car, Armoured, Command, 4×4, Sd.Kfz.247 (Auto Union) Horch V-8-cyl., 81 bhp, 5F1R, wb 3.00 m, 5.00×2.00×1.80 m, 3700 kg. Relatively rare vehicle, used by high-ranking staff officers, produced in 1939 on the *Einheitsfahrgestell II für s.Pkw*, which had the engine mounted at the front.

Armoured Car, Heavy, 6×4 (Mercedes-Benz G3a(p)) About 1930/31 a new type armoured car was developed on a 6×4 truck chassis with dual controls. The hulls, made by Deutsche Edelstahl-Werke, Krefeld, were mounted on the chassis by Deutsche Werke AG, Kiel. Prototype shown.

Armoured Car, Heavy, 6×4, Radio, Sd.Kfz.263 (Magirus M206(p)) 6-cyl., 70 bhp, 4F1R×2, wb 2.95 (BC 0.90) m, 5.57 × 1.82 × 2.93 m, 5250 kg. Similar vehicles were built on Büssing-NAG G31(p) and Mercedes-Benz G3a(p) chassis. Only the Magirus (shown) had 'belly support wheels'. 1934–36.

Car, Armoured, Command, 6×4, Sd.Kfz.247 (Krupp L2H143) HO-4-cyl., 60 bhp, 4F1R×2, wb 3.35 (BC 0.91) m, 4.60×1.96× 1.70 m, 4600 kg. Produced on Krupp *l.gl.Lkw.* chassis, 1937–38. A number of different armoured cars on these chassis were exported to the Netherlands East Indies.

Armoured Car, Heavy, 8×8, Sd.Kfz.231 (Büssing-NAG GS) V-8-cyl., 150 (later 180) bhp, 3F3R×2, wb 1.35+1.40+1.35 m, 5.85×2.20×2.34 m. Crew 4. Max. speed 85 km/h. IFS/IRS. Dual controls, front and rear; all-wheel steering. 2-cm gun and co-axial MG in turret. Several variants.

GERMANY

SEMI-TRACK VEHICLES

The first German vehicles with track bogies instead of rear wheels appeared in World War I. They were the *Bremerwagen* (a Daimler conversion by Bremer) and the Benz-Bräuer *Kraftprotze*. The former was unsuccessful and superseded by Daimler's own *Marienwagen* of which several variants appeared (incl. four- and full-track). During the 1920s a few models were made by Dürkopp and Maffei, followed by the Daimler-Benz ZD5 in 1931. From then on progress was rapid and a wide range was developed in various weight classes. The models which were eventually standardized were built by about a dozen firms until well into World War II. There were 'soft-skin' and armoured types. (See also HALF-TRACKS, Olyslager Auto Library/Warne).

Truck, Medium, Semi-Track, Cargo (Nacke/Aquilon) During and after World War I several trucks were modified for over-snow operation. This Nacke ran on four skis and was propelled by a track bogie, chain-driven from sprockets on the rear wheel hubs.

Tractor, Medium, Semi-Track (Benz-Bräuer) One of several pilot models of the Benz-Bräuer *Kraftprotze*, developed in 1917. For road use the outer wheel/track bogies could be lifted clear. Production models (50 of 200 ordered) were scrapped in 1920.

Truck, Medium, Semi-Track (Daimler Marienwagen II) Daimler 4-ton truck, converted at the company's Marienfelde works. It was a continuation of experimental but unsuccessful work by Ing. Hugo E. Bremer (*Bremerwagen*) and also appeared in full-track form.

Tractor, Medium, Semi-Track (Maffei MS) One of at least two road tractors with auxiliary track bogies produced in 1928/29 by the old-established locomotive builders J. A. Maffei AG (later Krauss-Maffei). Weight 4200 kg. Payload 12 persons or 1000 kg. Towed load 5000 kg.

Tractor, Heavy, Semi-Track (Mercedes-Benz ZD5) Originally intended for the Russian Government this vehicle was built by Daimler-Benz at Marienfelde in 1931/2. Front tyres were solid rubber and wheels were fitted on cranked arms, inter-connected with rear bogie suspension.

Tractor, Light, Semi-Track (Demag D 11 2) Second prototype (1934) for the 1-ton Series, developed by Demag AG of Wetter/Ruhr. It was powered by a rear-mounted BMW 319 6-cyl. engine. Eventually this became the (front-engined) *leichter Zugkraftwagen 1 t (Sd.Kfz. 10)*.

Tractor, Light, Semi-Track (Hansa-Lloyd H Kl 3) One of the prototypes for the 3-ton Series, developed by Hansa-Lloyd-Goliath AG of Bremen and powered by one of their own engines. Resulted in standardized Maybach-engined *leichter Zugkraftwagen 3 t (Sd.Kfz. 11)*. Model shown appeared in 1936.

Tractor, Light, Semi-Track (Demag D6) Early model of *Sd.Kfz. 10* 1-ton Series. Version shown was *le.Eg.Kw. (Sd.Kfz. 10/2)* (decontamination vehicle), measuring 4.75 × 1.84 × 1.62 m. It was powered by a Maybach NL38 6-cyl. 90-bhp engine and produced during 1937–38.

Tractor, Light, Semi-Track (Demag D7) Maybach HL42 6-cyl., 100 bhp, 7F3R, 4.75 × 1.93 × 1.57 m, 3400 kg. Used as *Sd.Kfz. 10 (l. Zgkw. 1 t)* and *Sd.Kfz. 10/1 (Gasspü. Kw.*; gas detection vehicle). Produced 1939–44 by Demag (parent firm) and four co-producers, incl. Saurer in Vienna.

Tractor, Light, Semi-Track (Adler HK300/A3) Maybach HL25 4-cyl., 65 bhp, 6F1R. Concurrently with Demag, Adler designed a range of light semi-track vehicles, known as the HK300 Series. They were not adopted and during 1940–44 Adler co-produced the Demag D7 design.

Tractor, Light, Semi-Track (Borgward HL kl 6) Maybach NL38 6-cyl., 90 bhp (later HL42, 100 bhp), 4F1R × 2. Chassis of *l. Zgkw. 3 t (Sd.Kfz. 11)*, mounting ambulance bodywork for use by the German coast-guard. Several other body styles appeared on this widely-used chassis.

Tractor, Medium, Semi-Track (Büssing-NAG BN L 5) Maybach NL35 6-cyl., 90 bhp, 8F2R. Büssing-NAG were the parent firm of the *m.Zgkw. 5 t* (*Sd.Kfz. 6*); co-producers included Daimler-Benz. Shown is 1935 *Sd.Kfz. 6/1* gun tractor. 1938–43 models had longer track bogies.

Carriage, Medium, Semi-Track, AA Gun (Büssing-NAG BN L 8) Officially known as *Selbstfahrlafette* (*Sd.Kfz. 6/2*) this vehicle mounted a 3.7-mm *Flak 36* gun. It was based on the late type 5-ton *m.Zgkw.* chassis, powered by a Maybach NL38 100-bhp 6-cyl. engine. Produced in 1938.

Tractor, Medium, Semi-Track (Krauss-Maffei KM m 8) Maybach HL52 6-cyl., 115 bhp, 4F1R × 2. Krauss-Maffei were the parent firm of the *m.Zgkw. 8 t* (*Sd.Kfz. 7*) and produced 6129 units during 1934–45. This is the 1934/35 model. Co-producers included Büssing-NAG and Daimler-Benz.

Tractor, Medium, Semi-Track (Krauss-Maffei KM m 9) 1936 model of *m.Zgkw. 8 t* (*Sd.Kfz. 7*), which had 130-bhp Maybach HL57 engine. KM m 10 of 1936–37 was similar but had 140 bhp Maybach HL62 engine. KM m 11 of 1937–45 had longer track bogies with more wheels. Also SP variants.

Tractor, Heavy, Semi-Track (Mercedes-Benz DB s 7) Maybach DSO8 V-12-cyl., 150 bhp, 4F1R×2. First of the *s.Zgkw. 12 t* (*Sd.Kfz. 8*), developed by Daimler-Benz for towing heavy artillery (15-cm guns). 1934/35. Later models co-produced by Krauss-Maffei, Krupp and Skoda.

Tractor, Heavy, Semi-Track (Famo F2) Maybach HL 98 V-12-cyl., 230 bhp, 4F1R×2. Heaviest of German semi-tracks, known as *s.Zgkw. 18 t* (*Sd.Kfz.9*). Usually with two rows of seats and cargo body. Also basis for mobile cranes and SP mount for 8.8-cm *Flak* (AA gun).

Carrier, Armoured, Light, Semi-Track (Demag D7p) Based on the *Sd.Kfz. 10* light tractor but with shorter track bogies this was one of a wide range of AFVs in the *Sd.Kfz. 250* Series. Specimen shown (*Sd.Kfz. 250/11*) had 28-mm heavy AT rifle with sandwich-type armour shield.

Carrier, Armoured, Medium, Semi-Track (Hanomag H kl 6p) The *Sd.Kfz. 251* Series, based on the *Sd.Kfz. 11* 3-ton tractor, was used, with suitable modifications, for over 20 different roles. Shown is an early basic model (APC; *Sd.Kfz. 251/1*, *Ausf.A.*). Late models had simplified hull design.

GERMANY

FULL-TRACK VEHICLES

During World War I the Germans built only a few full-track vehicles, including the A7V tank which ran on lengthened American Holt pattern track bogies; it was produced by Daimler at Marienfelde and a carrier-cum-tractor variant was built by Heinrich Büssing at Braunschweig. Both were powered by twin four-cylinder Daimler engines, mounted amidships. Other full-track vehicles of the period included the exp. Dür-Wagen, Orion-Wagen and four-track Bremer-Wagen.

During the 1930s light, medium and heavy commercial crawler tractors were introduced; these were known as *leichter, mittlerer* and *schwerer Kettenschlepper* (o) and had drawbar towing capacities of 1500, 2500 and 3500 kg (later increased to 3500, 4600 and 9500 kg respectively).

Truck, Medium, Four-Track, Cargo (Daimler/Bremer) Ing. Hugo Bremer modified a 4½-ton Daimler truck to semi-track configuration in 1915 and in 1916 replaced the front wheels by (undriven) track bogies in order to improve cross-country and over-snow performance.

Truck, Medium, Four-Track, Armoured (Daimler/Bremer) Officially known as *Bremer-Überland-(Sturm) Panzerwagen* (Bremer overland (assault) armoured car), this experimental vehicle was designed in 1917 by Ing. Joseph Vollmer on the four-track *Bremer-Wagen* chassis.

Truck, Medium, Four-Track, Chassis (Daimler Marienwagen) After the *Bremer-Wagen* experiments were discontinued, Daimler at Marienfelde endeavoured to improve the design in semi-, four- (shown) and full-track form. The rear bogies were based on the American Holt design. 1918.

Carrier/Tractor, Heavy, Full-Track (Büssing) The *Überlandwagen* or, popularly, *'Munitions-Tank'*, was a 10-ton carrier (or tractor for 15-ton loads) based on the chassis of the A7V tank (which it outnumbered). It had two 100-bhp 4-cyl. Daimler engines and 3F3R transmission.

Tractor, Heavy, Full-Track (Lanz) Except for the track bogies this tractor was similar to the Lanz ZF four-wheeled tractor (*qv*). It was used during World War I for heavy artillery towing. Shown is a captured specimen, minus bonnet top and body sides, in England.

Tractor, Medium, Full-Track (Krupp LaS) In 1933/34 Krupp designed what was called a *Landwirtschaftlicher Schlepper* (*LaS*) (agricultural tractor). In reality it was the forerunner of the *Panzer I* light tank, popularly known as the 'Krupp Sport'. Engine: Krupp 60-bhp 'boxer'.

Tractor, Amphibious, Full-Track Maybach HL120 V-12-cyl., 300 bhp, 5F1R, 8.60 × 3.16 × 3.13 m, 13,000 kg. Known as *Land-Wasser-Schlepper* this tractor-cum-river tug was designed in 1935/36 by Rheinmetall-Borsig. Its performance in water was very much better than on land. Few were made.

GREAT BRITAIN

The earliest recorded use of mechanical transport (as it was then called) on active British military service was in the 1873–74 Ashanti War when the Royal Engineers used one of their two steam traction engines. Even earlier than that, the War Department had acquired a Boydell traction engine (1857), as well as a Bray and a Burrell. In the Second Boer War (1899–1902) both steam and internal-combustion-engined vehicles were employed successfully and as a result the War Office set up a Mechanical Transport Committee to benefit from this experience and to guide future developments. In 1901 the Committee organized a competitive trial for self-propelled vehicles and of eleven entrants five actually took part: one Foden, one Milnes-Daimler, one Straker and two Thornycrofts. Apart from the Milnes-Daimler all were steam-propelled. Although steamers won the prizes, it was felt that there was potential in 'internal-combustion lorries' and further trials were planned for these. In 1903 the first MT Company of the Army Service Corps was set up at Woolwich. More followed. Among the vehicles used were Brooke, Brush, Lanchester, Napier and Wolseley cars, Fowler traction engines and Thornycroft steam wagons. In 1905 the ASC had designed and produced by Hornsby of Lincoln a paraffin-engined tractor, which in 1907 was converted into a track-laying tractor. In 1909 there were trials for paraffin-engined tractors at Aldershot (again won by a Thornycroft) and the 'Hastings Run' (see page 188).

In 1911 the Secretary of War made a revolutionary decision. He told an astounded House of Commons that he was contemplating the replacement of horses by MT on a large scale. Following tests in the South of England it was decided that petrol-driven vehicles of about 3-ton capacity were most suitable and a Subvention or Subsidy scheme was authorized whereby certain sums were paid to owners of approved makes and types of vehicles. Similar schemes were also operated in France, Germany and Austria. In 1912 and 1913 further Subsidy Trials were held and by the end of 1913 there were about 1000 Subsidy-type vehicles in use by commercial operators, who, in return for keeping their vehicle ready for Government use in time of national emergency obtained a subsidy of £110, £30 being paid when the owner enrolled his vehicle and £80 payable in six half-yearly instalments. There were two classes:

Class A (3-ton) and Class B (30-cwt). The main objects of the scheme in terms of vehicle specifications were stated by the War Office as follows: (1) to make the manipulation and control of all vehicles the same and (2) to minimize the number of different makes of vehicles of which the transport columns of the Army would be composed.

As it was, when war broke out in August 1914 the majority of vehicles used by the BEF were Subsidy types, supplemented by impressed vehicles (including many buses). During the war years the manufacturers of the Subsidy type vehicles continued producing these directly for the Government but supply did not catch up with demand, as a result of which many thousands of additional trucks were imported, mainly from the United States. On the other hand, a number of British trucks went to the Allied forces (816 to Belgium, 1171 to France, 1126 to Russia, 4307 to the US Army).

After the 1918 Armistice stock was taken of British vehicle holdings at home and in the various theatres of war, France, Italy, Mesopotamia, East Africa, etc. The grand total of vehicles in service was 165,128, sub-divided as follows: Motorcycles: 48,175, Cars and Ambulances: 43,187, Trucks: 66,352, Steam Wagons: 1,293, Tractors and Misc.: 6,121.

The majority of these were in France and Flanders. No records are available of wartime losses. After the war thousands of surplus vehicles were sold to 'Civvy Street', causing considerable problems for the motor industry which was left with increased productive capacity and few customers for new vehicles. The Army kept sufficient numbers of various types and during the 1920s purchased relatively few new vehicles.

Those new vehicles which were procured were of advanced designs, notably newly-designed 30-cwt trucks with pneumatic tyres (with a new Subsidy system in 1923) and six-wheeled cross-country trucks with military design articulated rear bogie.

The latter vehicles, in order to boost production, were again subsidized when bought by civilian operators. Half- and full-track vehicles were also experimented with. These and many other vehicles of different types were further developed throughout the 1930s and although there was no Subsidy scheme when the Second World War broke out in 1939, history repeated itself in that again large numbers of vehicles had to be impressed and substantial orders for additional vehicles were placed in the United States. That the American vehicles could be supplied was because, as in 1914, the US did not enter the war until a few years later, until when it was possible for these orders to be met through commercial channels. In addition the Canadian automotive industry supplied many vehicles for the British and Commonwealth forces.

Who's Who in the British Motor Industry

Note: listed are manufacturers of British military vehicles (excluding tanks and similar) which were, or still were, operating in the 1930s.

AEC	The Associated Equipment Co. Ltd, Southall, Middlesex.
Albion	Albion Motors Ltd, Scotstoun, Glasgow.
Alvis-Straussler	Alvis-Straussler Ltd, Coventry.
Ariel	Ariel Motors Ltd, Selly Oak, Birmingham.
Austin	Austin Motor Co. Ltd, Longbridge, Birmingham.
Bedford (GM)	Vauxhall Motors Ltd, Luton, Bedfordshire.
BSA	BSA Cycles Ltd, Birmingham.
Commer (Rootes)	Commer Cars Ltd, Luton, Bedfordshire.
Crossley	Crossley Motors Ltd, Gorton, Manchester.
Daimler	Daimler Motor Co. Ltd, Coventry.
Dennis	Dennis Bros Ltd, Guildford.
Dodge	Dodge Brothers (Britain) Ltd, Kew, Surrey.
Foden	Fodens Ltd, Sandbach, Cheshire.
Ford(son)	Ford Motor Co. Ltd, Dagenham, Essex.
Garner	Garner Motors Ltd, Tyseley, Birmingham, etc.
Guy	Guy Motors Ltd, Wolverhampton.
Hillman (Rootes)	The Hillman Motor Car Co. Ltd, Coventry.
Humber (Rootes)	Humber Ltd, Coventry.
Karrier (Rootes)	Karrier Motors Ltd, Luton, Beds.
Leyland	Leyland Motors Ltd, Leyland, Lancs (also Kingston-upon-Thames, Surrey).
Matchless	Matchless Motor Cycles, London SE18.
Maudslay	The Maudslay Motor Co. Ltd, Alcester, Warwickshire.
Morris (Nuffield)	Morris Motors Ltd, Cowley, Oxford.
Morris-Commercial (Nuffield)	Morris-Commercial Cars Ltd, Adderley Park, Birmingham.
Norton	Norton Motors Ltd, Birmingham.
Riley	Riley (Coventry) Ltd, Coventry.
Rolls-Royce	Rolls-Royce Ltd, Derby.
Rover	The Rover Co. Ltd, Solihull, Warwickshire.
Royal Enfield	The Enfield Cycle Co. Ltd, Redditch, Worcs.
Scammell	Scammell Lorries Ltd, Watford, Herts.
Standard	Standard Motor Co. Ltd, Coventry.
Straussler	Straussler Mechanisation Ltd, London SW1.
Sunbeam	Sunbeam Motor Car Co. Ltd, Wolverhampton, Staffs.
Talbot	Clement-Talbot Ltd, London W10.
Thornycroft	John I. Thornycroft & Co. Ltd, Basingstoke, Hants.
Tilling-Stevens	Tilling-Stevens Motors Ltd, Maidstone, Kent.
Triumph	Triumph Engineering Co. Ltd, Coventry.
Trojan	Trojan Ltd, Croydon, Surrey.
Vauxhall (GM)	Vauxhall Motors Ltd, Luton, Beds. (trucks named Bedford).
Velocette	Veloce Limited, Hall Green, Birmingham.
Vulcan	Tilling-Stevens Motors Ltd, Maidstone, Kent.
Wolseley (Nuffield)	Wolseley Motors Ltd, Birmingham.

Advanced thinking by the MWEE produced this Morris-Commercial 8×8 conversion *c.* 1929. (IWM photo STT 5777).

GB

MOTORCYCLES incl.-TRICYCLES and -QUADRICYCLES

In the Army manoeuvres of 1910 the first extensive use was made of motorcycles for military purposes, when despatch riders on numerous types were successfully employed. Since then continuous improvement in design made the motorcycle so reliable that it became indispensable for many purposes.

By the end of World War I the British had more than 48,000 motorcycles of 54 makes, the most numerous being BSA, Clyno, Douglas, Phelon & Moore (for RAF) and Triumph. The Douglas $2\frac{3}{4}$ and 4 HP and the Triumph 4 HP were standardized Army machines.

Motortricycles and -quadricycles were rare and appeared mainly in the very early days of army mechanization. Of particular interest was the 1915/16 Scott gun carrier, a light car with three wheels arranged in sidecar-combination fashion.

The first purpose-built motorcycles for the War Office were single-cylinder 4 HP machines supplied in 1906 by A. W. Hall of Guildford.

Motorcycle, Solo (BSA 4 HP) Typical solo machine of the first World War, when the 1914/15 4 HP model was the most numerous. Engine was single-cyl. 85 × 98 mm. In 1918 there were 537 of these, as well as 551 BSAs of other types, mainly $3\frac{1}{2}$ and $4\frac{1}{4}$ HP models of 1912–17.

Motorcycle, Solo (Douglas $2\frac{3}{4}$ HP) Douglas Bros. of Kingswood, Bristol, supplied large numbers of $2\frac{3}{4}$ and 4 HP HO-2-cyl.-engined models, the latter with sidecar. In 1918 there were 13,477 and 4816 resp. and a few of other types. Most were in France. Also supplied to Australian and Belgian forces.

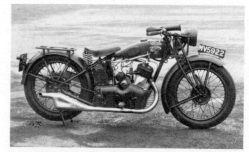

Motorcycles, Solo (Triumph) The British armed forces have always been large-scale users of Triumphs. In 1918 there were 17,998, incl. several hundred with sidecar. Just over 17,000 of these were standardized 4 HPs (85 × 97 mm) of 1915–18, known as the 'Trusty'. Machine shown was supplied in 1933.

Motorcycle, with Machine Gun Sidecar (Clyno/Vickers) 5–6 HP) Designed and built jointly by Clyno and Vickers this combination carried a .303" machine gun on a tripod. Based on similar machines were an ammunition carrier and a passenger sidecar. In 1918 there were just under 1800.

Motorcycle, Solo, 3-wheeled (JAP) In the mid-1920s, when 6 × 4 cross-country trucks became popular, some motorcycles were converted into tandem-drive three-wheelers. Tests proved, however, that this configuration was not advantageous. One, a Triumph, still exists.

Motortricycle, Machine Gun (AC Autocarrier) 1-cyl., 5.6 HP (90 × 102 mm), 2F, wb 5'0", 8'0" × 4'0" × 3'6". Payload 4–5 cwt. In 1910 the 25th County of London Cyclist Regt acquired these two Maxim MG carriers and two ammunition carriers. They had wooden frames and tiller steering with hinged steering arm.

Motorquadricycle, Machine Gun (Simms/Vickers) Designed by Frederick R. Simms for Vickers, Son & Maxim Ltd, this 1½ HP 'motor scout' made its public debut at Richmond, Surrey, in 1899. It carried an air-cooled Maxim MG with armour shield. In 1902 a larger 'Motor War Car' was demonstrated by the same designer.

M

GB

CARS, 4×2

During the period prior to 1940 the British armed forces used more makes and models of passenger cars than could possibly be listed in any detail. During World War I alone more than 160 makes were operated and the number of models ran into hundreds. Standardized by the Army were the Sunbeam and the Vauxhall; other widely-used makes included Austin, Clement Talbot, Crossley, Daimler, Ford, Napier, Siddeley Deasy, Studebaker and Wolseley. Altogether in 1918/19 there were 43,187 cars (including car-based ambulances); war-time losses are not known. Many cars had been impressed, others were of foreign origin or had been captured. The RAF had standardized on the Crossley. During the 1920s and 1930s small batches of new cars were bought at infrequent intervals.

Car, 4-seater, 4×2 (Wolseley 8 HP) Among the earliest cars used by the British forces was this Wolseley with 2-cyl. horizontal engine of 1904/05. It belonged to Admiral Scott and was fitted with a Maxim machine gun. Similar Wolseleys but with entirely open bodywork were also in service.

Car, 4-seater, 4×2 (Crossley 20/25 HP) 4-cyl., 50 bhp, 4F1R, wb 11'3", 16'0"×5'9"×5'8", 41 cwt. 4½-litre (102×140 mm) L-head engine. Cone clutch. Tyres 920×120 (or 895×135), dual rear. Used extensively by RFC/RAC as 'Squadron 4-seater' from WWI until about 1930. Also with closed bodywork.

Car, 4-seater, 4×2 (Daimler TR 20 HP) 4-cyl., 35 bhp, 4F1R, wb 10'11". This was the most numerous of the many types of Daimlers used in World War I. It had a 3308-cc (90×130 mm) Knight sleeve valve type engine and was also made with Staff Limousine and other body styles.

Car, 4-seater, 4×2 (Ford T 20 HP) 4-cyl., 20 bhp, 2F1R (epicyclic), wb 8'4", 11'6"×5'8"×6'0". Widely used, with various types of bodywork, varying from Tourer (shown) to Soup Kitchen. At the end of WWI there were a total of 18,984 in British service, mainly cars and light trucks (IWM photo Q3707).

Car, Staff Limousine, 4×2 (Rolls-Royce 40/50 HP) 6-cyl., 50 bhp, 4F1R, wb 11'11½". Several hundred of the famous 'Silver Ghost' cars were used by the armed forces in World War I. There were open and closed types, as well as light trucks (tenders) and armoured cars. 23 chassis types were listed.

Car, 4-seater, 4×2 (Sunbeam T 12/16 HP) 4-cyl., 33 bhp, 4F1R, wb 10'4", 14'4"×5'7"×6'6" (with hood up), 32 cwt. 3016-cc (80×150 mm) side-valve engine. Tyres 815×105. Standardized for Army use; built in large numbers by Sunbeam and the Rover Co. Also with ambulance body (Model S).

Car, Staff Limousine, 4×2 (Vauxhall D 25 HP) 4-cyl., 50 bhp, 4F1R, wb 10'10", 14'9"×5'10"×7'7" (Tourer 6'5"). 3969-cc (95×140 mm) side-valve engine. Tyres 880×120 (some dual rear). Built in large numbers for British Army during WWI, mainly with open bodywork. In 1918/19 there were 1855 (1570 overseas).

Car, 4-seater, 4×2 (Jowett 7 HP) HO-2-cyl., 16 bhp, 3F1R, wb 8'6", 11'6"×4'6"×5'10". 907-cc (75.4×101.5 mm) flat-twin engine. Tyres 27×4. Mech. brakes at rear only. Produced in 1928 for RASC. 6-cwt Van was supplied in 1929 on same chassis with 5.125 (in lieu of 4.66) axle ratio.

Car, 4-seater, 4×2 (Austin Hertford) 6-cyl., 38 bhp, 4F1R, wb 9'4", 14'0"×5'8"×5'10", 26½ cwt. 1935. Available as 16 HP or 18 HP, the latter with 44-bhp engine. During the 1930s Austin supplied several batches of open and closed small, medium and large cars for military use.

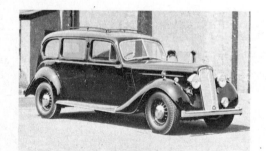

Car, 6-seater, 4×2 (Humber Pullman) 6-cyl., 85 bhp, 4F1R, wb 11'0", 16'0"×6'2". Tyres 7.50-16. 4086-cc (85×120 mm) side-valve engine, rated at 26.9 HP. Supplied in mid-1938. Similar to contemporary Humber Super Snipe but longer and more luxurious. Note roof luggage rail.

Car, 4/5-seater, 4×2 (Ford V8 30 HP) V-8-cyl., 85 bhp, 3F1R, wb 112". During the late 1930s the Ford Motor Co. of Dagenham delivered several types of American-type Saloon and Estate cars to the Army and RAF. Shown is a Model 81A Fordor of 1938, fitted with 9.00-13 tyres, supplied to the RAF.

GB

FIELD CARS
4×2 and 6×4

In addition to what were basically standard civilian type cars the British armed forces during the 1920s and 1930s used a number of cars which were specially adapted for use off the beaten track. Some were cars of conventional design but fitted with special or modified soft-top bodywork and oversize tyres. Others were based on light 6×4 truck chassis, notably by Morris-Commercial Cars Ltd. All-wheel drive cars were not made in quantity; prototypes were built by French's Motor Engineering Works of Balham (Holverter car, designed by Alex. Holle; 4-wh. steer, 1923), Straussler (Ford engined, 4-wh. steer, 1933), etc., but these remained experimental. In 1926 Vickers modified a Wolseley Tourer into a wheel-cum-track car, again experimentally.

Car, 2-seater, 4×2 (Austin Seven) 4-cyl., 10.5 bhp, 3F1R, wb 6'3", 9'2"×4'1½"×4'11", 9 cwt. Issued for Cavalry reconnaissance and for use by junior officers and NCOs of mechanized units. WD body design. 1928. Usually fitted with 'knobbly' tyres. Later models had civilian type body.

Car, 2-seater, 4×2 (Morris Eight) 4-cyl., 22.5 bhp, 3F1R, wb 7'6", 11'9"×4'6½"×5'1", 13½ cwt (basic). Modified civilian 1936 Series I Tourer with 918-cc (57×90 mm) SV engine. Fitted with radio equipment. Earlier there was a signals version of the 1933 Minor, with WD bodywork like the Austin Seven.

Car, 4-seater, 4×2 (Riley Nine) 4-cyl., 27 bhp, 4F1R, wb 8'10", 12'8"×4'10½"×5'8", 20 cwt. Known as Riley Touring Mk IV Plus Series. 1089-cc (60.3×95.2 mm) OHV engine. 5.25-21 tyres (pressure 29 lb/sq. in). Rear axle ratio 6.75:1. Speed, cruising 30–40, max. 65 mph. 77 supplied in 1930/31. Also with closed bodywork.

Car, 5-seater, 4×2 (Commer Raider) The RAF used a relatively large number of Commer 30-cwt trucks (*qv*) in the Middle East. This command car had a similar chassis but fitted with special Tourer type bodywork for the SSO, Palestine, about 1933. The tyres were size 9.00-22 on special wheels.

Car, 4-seater, 4×2 (Hillman Hawk) 6-cyl., 75 bhp, 4F1R, wb 9'0½", 14'6"×6'1"×6'0" approx. 3181-cc (75×120 mm) SV engine. Tyres 7.00–16. One of a batch of 21 HP Hawk Tourers supplied to the War Office in 1936. Bodywork was made by Mulliners of Birmingham (shown) and Coachcraft.

Car, 6-seater, 6×4 (Morris-Commercial D) This 'general purpose cross country and road car', issued to mechanized formations as a staff car, was based on the makers' 30-cwt 6×4 truck chassis (*qv*). Weight 52 cwt. Speed, road 30, cross country 15–20 mph. 1927–32.

Car, 6-seater, 6×4 (Morris-Commercial CD) Basically similar to Model D (shown on left) but based on later type chassis (30-cwt, 6×4, 1933–39). 55-bhp 4-cyl. engine with 5-speed gearbox. Worm-drive rear axles. Tyres 7.50-20. Full-length folding top.

GB

BUSES and COACHES

Probably the most famous buses in the history of military motor transport were the hundreds of standard London B-Types (LGOC/AEC and Daimler) which were shipped from England to Belgium and France during World War I. These were used to convey troops (initially in their original bright-coloured livery) but also in modified form as ambulances, field kitchens, pigeon lofts, etc. Many had their double-deck bodies replaced by GS cargo bodywork or special types. A number were later returned or lost in action but by war's end there were still some 1082 AECs and 113 Daimlers in the Government's inventory, the majority in France. Between the wars very few new types were taken into service.

Motor Bus, Double-Deck (AEC/Daimler/LGOC) Some of the many buses of the London General Omnibus Co. which went to war in 1914/15, with the Army and the Royal Marines. They had been produced by AEC and Daimler to the same general design but the latter had Silent Knight sleeve valve engines.

Coach, 12-seater, 4×2 (Talbot) Produced about 1923 on a Clement-Talbot 30-cwt Subsidy-type truck chassis for the RASC. Known as 'all-weather coach' this body style was a development of the earlier charabanc (derived from the French word 'char-à-bancs'). Note sliding roof.

Coach, 20-seater, 4×2 (Morris-Commercial Viceroy) During the early and late 1930s the British Army took delivery of some Morris-Commercial passenger vehicles. These were basically civilian types. Illustrated is a 1932 Viceroy, which had a 70-bhp six-cylinder side-valve engine.

GB

AMBULANCES

During the initial stages of World War I, ambulances were built on almost any available chassis. Large numbers were based on civilian touring car chassis, after removal of the rear body and many of these were donated by private citizens, institutions, etc. Heavy ambulances carried, in addition to the driver and one attendant, four stretcher cases or eight sitting cases. These were used for the transport of wounded and sick from the forward area dressing-stations to the hospitals, hospital trains, etc., and for inter-hospital work. Light ambulances (mainly Ford T), carrying crew plus four sitting and two lying patients were similarly used and proved more mobile over bad terrain. During the 1920s and 1930s batches of more modern types (4×2 and 6×4) were acquired.

Ambulance, Heavy, 4×2 (Wolseley-Siddeley) Prior to World War I the Army Service Corps (later Royal) operated a few 'motor ambulance vans', starting with a Straker-Squire in 1906 (with air-suspended body). This somewhat smaller shaft-drive Wolseley-Siddeley was acquired two years later.

Ambulance, Heavy, 4×2 (Siddeley-Deasy) Over 500 of these saw service in World War I. Other widely-used makes included Austin, Clement-Talbot, Crossley, Daimler, Ford, Napier, Star, Sunbeam, Vauxhall and Wolseley as well as American Buicks, Fords and GMCs, etc. (IWM photo Q2645).

Ambulance, Heavy, 4×2 (Sunbeam, Rover) The Sunbeam was the standard Army ambulance and was produced mainly by Rover. By 1918 over 1600 were in service. WD requirements included central passage, interior lining, dual rear wheels, special ventilators and Red Cross special heating system.

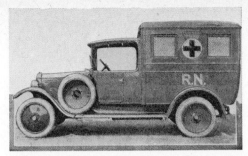

Ambulance, Light, 4×2 (Austin 20 HP) This light ambulance was one of a batch of six supplied to the Royal Navy in 1922. The all-enclosed composite rear body contained two stretchers, one above the other. Alternatively it could accommodate eight sitting patients.

Ambulance, Heavy, 4×2 (Albion AM463) Introduced in 1934 this 4-stretcher model was specially made for the Air Ministry. 1900 chassis were supplied, also for the mounting of other body types. It had a 4-cyl. 65.5-bhp engine with four-speed gearbox and the wheelbase was 12'0".

Ambulance, Heavy, 4×2 (Morris-Commercial CS11/30F) During 1935–39 Morris-Commercial supplied 30-cwt chassis of conventional type for GS bodies as well as a number of semi-forward control variants for the mounting of what became the standard WD heavy ambulance body. Vehicle weighed 61 cwt.

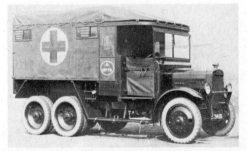

Ambulance, Heavy, 6×4 (Vulcan RSW) In 1927 the RASC acquired some ambulances on Morris and Vulcan light six-wheeler chassis. The standard body was designed 'as advised by the medical authorities'. In 1929 an improved version followed, again on the Morris and Vulcan chassis. The latter is shown.

GB

LIGHT VANS

Illustrated in this section are some typical examples of panel or box vans. During World War I motor vans were used extensively for the rapid transit of light stores and small loads urgently needed and the transport of which in heavy vehicles would be uneconomical. In Mesopotamia and Palestine the Model T Ford van and light truck were used and formed the principal form of MT over sandy terrain.

In Britain and France several makes were used, incl. Daimler, Ford and Studebaker. Like ambulances, most vans were based on car chassis. Between the wars the RAF acquired Trojans (5/7-cwt, from 1924) and later Singer 9 HP vans (5/7-cwt, from 1934) and the Army used several makes incl. Bedford, Ford, Morris, etc.

Note: in the British Army the word 'van' was used also for light trucks with open box-type body and tilt.

Van, 8-cwt, 4×2 (Ford T) 4-cyl., 20 bhp, 2F1R (epicyclic), wb 8'4", 11'5"×5'7"×6'11". Model T Fords were used from WWI (Army, RAF) until the 1930s. Shown is a 1923/24 Van of the RASC. Also with platform body (known as 10-cwt Float). Tyres 31×4. Speed 35 mph.

Van, 5/7-cwt, 4×2 (Trojan 10 HP) 4-cyl., 11 bhp, 2F1R (epicyclic), wb 8'0". 1488-cc duplex 2-stroke engine under seat. Chain drive to live rear axle without differential. Solid tyres. Produced by Leyland at Ham, Surrey, from 1923 (following prototypes from 1913). Special body for RAF, *c.* 1925.

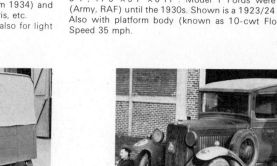

Van, 12-cwt, 4×2 (Bedford BYC) 6-cyl., 64 bhp, 4F1R, wb 8'10½". Basically a civilian type van, produced by Vauxhall Motors during 1936–37 with 19.8 HP 2392-cc (73×95.25 mm) OHV engine. Tyres 5.50-17. Also supplied as Utility Van with rear side windows and collapsible seats. (IWM photo O830).

GB

TRUCKS, 6- to 15-CWT
4×2 and 6×4

Makes and Models: (15-cwt and 4×2 unless stated otherwise) Arrol-Johnston (1914). Austin 20HP (1914–18), BYD (1937/38), etc. Bedford MW (from 1937). Buick D4 (1916/17, US). Commer Beetle (from 1937), etc. Crossley 15/20 and 20/25 HP (1914–18), etc. Daimler 20 HP (1914–18), etc. Ford T (8-cwt, from 1914), B (10/12-cwt, 1932), V8 (from 1936), etc. Guy Ant (from 1935/36). Jowett 7 HP (6-cwt, 1929). Lanchester 19B (c. 1914). Morris-Commercial B (12-cwt, 1929), R (12–15-cwt, 1930), DCTM (1936–39), PU (8-cwt, from 1936), CS8 (from 1934), etc. Napier T55, T70, T70A etc. (from 1914). Rolls-Royce 40/50 HP (1914–18). Talbot/Clement-Talbot (1914–18). Trojan 10 HP (10-cwt, 6×4, 1929). Vulcan VMT (1925/26). Wolseley CA (12-cwt, 1914–18), etc.

Vehicle shown (typical): Truck, 15-cwt, 4×2, Cargo (Austin BYD)

Technical Data:
Engine: Austin 12 HP 4-cylinder, I-L-W-F, 1861 cc, 27 bhp at 2000 rpm.
Transmission: 4F1R.
Brakes: mechanical.
Tyres: 5.50-20.
Wheelbase: 9'4". Track 4'8".
Overall l×w×h: 14'10" × 5'11½" × 7'9".
Weight: 27½ cwt, gross 47½ cwt.
Note: known as the 'Ascot type' these vehicles, which were based on a worm-drive Austin taxi cab chassis, were produced in 1937/38.

General Data: Listed above are light trucks with payload capacity of up to 15 cwt. During World War I these were usually based on passenger car chassis, suitably reinforced to carry a heavier load. In addition to equipping the British armed forces the British industry supplied light trucks to several Commonwealth and other countries. Body types were numerous and included conventional GS (general service) load carriers, vans, machine gun carriers, ambulances (see separate section), etc. During the early 1930s the special 15-cwt WD-type truck was designed and developed. This type was built up, as far as was possible, from current commercial components such as engines, transmissions, axles, steering gear, etc., with a short wheelbase, oversize military-pattern tyres on special wheels, open cab and standard bodywork. In order to have the maximum load space on the short wheelbase it was necessary to mount the cab as far forward as possible. This, together with low-silhouette requirements, meant that the driver's and assistant's legs had to be at the sides of the engine, resulting in the peculiarly shaped front end styling of most of these vehicles. The various models were tested, amongst others, at annual trials in the mountainous districts of North Wales.

Truck, 15-cwt, 4×2, Cargo (Crossley 20/25 HP) 4-cyl., 44 bhp, 4F1R, wb 11'3", 36 cwt. Produced in large numbers during WWI for the RAF (originally RFC) where it was known as Light Tender. Several other body styles, incl. Wireless Tender, etc. Rear suspension used ¾-elliptic leaf springs.

Truck, 8/10-cwt, 4×2, Machine Gun (Ford T) 4-cyl., 20 bhp, 2F1R (epicyclic), wb 8'4". Desert patrol/reconnaissance vehicle with Vickers MG for service in the Middle East (Mesopotamia, etc.). From 1916 Fords had restyled front end with faired scuttle, rounded wings, radiator and bonnet.

Truck, 15-cwt, 4×2 Cargo (Lanchester 19B) This unusual-looking vehicle was based on the 1911–13 Lanchester 38 HP car chassis which had many unorthodox features, including cantilever-type springs. In addition to the truck there were ambulances, armoured cars and other body styles, used chiefly by the Admiralty (RNAS).

Truck, 10-cwt, 4×2, Cargo (Napier 15/20 HP) In 1909 the AA organized the 'Hastings Run', transporting a complete battalion of the Guards from London to Hastings. 249 cars and 28 trucks, including 21 of these Napiers, carried 1000 men and 30 tons of arms and equipment in this successful experiment, which aroused great military interest both at home and abroad.

Truck, 10-cwt, 4×2, Machine Gun (Napier) D. Napier & Son Ltd of Acton, London, built a variety of luxury cars, including some types with 6-cyl. engine. This experimental MG carrier was based on a c. 1905 chain-drive car chassis. In WWI about 2000 Napiers of over 30 types saw service.

Truck, 15-cwt, 4×2, Cargo (Rolls-Royce 40/50 HP) In addition to passenger cars, several other body styles were used on the 1914–18 Silver Ghost chassis, incl. medical cars, armoured cars and tenders (as shown; Middle East). By war's end 321 Rolls-Royces were in military use. (IWM photo Q59378).

Truck, 15-cwt, 4×2, Personnel (Talbot) Clement-Talbot Ltd of North Kensington, London, produced a large variety of cars, ambulances, light trucks, etc. for service in World War I. Many were used by the Royal Navy. 1919 vehicle records showed 1194 units of about 25 models.

Truck, 12-cwt, 4×2, Cargo (Wolseley CA) WD stock figures in 1919 showed 404 Wolseley medium and heavy trucks and 1096 cars, ambulances, light trucks, etc., the latter figure representing some 40 models. Shown is a 15.6 HP 4-cyl. car-based truck used for carrying personnel, supplies, etc.

Truck, 10/12-cwt, 4×2, Machine Gun (Ford B) 4-cyl., 52 bhp, 3F1R, wb 8'10". The 1932 Model B succeeded the 1928–31 Model A. This desert patrol truck with oversize tyres, special body and MG mounting was used by the Sudan Defence Force, who also had the Model AA 1-tonner and, later, V8 trucks.

Truck, 12-cwt, 4×2, Cargo (Morris-Commercial B) 4-cyl., 11.9 HP, 3F1R, wb 9'6", 13'6"×5'4" approx., 18¼ cwt. 1550-cc (69.5×102 mm) SV engine. Tyres 32×4. Conversion of 1929/30 Model B Van, with 7'2"×5'4" load platform on which driver's seat was fitted. Used in Egypt. Similar 'float' on Ford T chassis.

Truck, 12–15-cwt, 4×2, Cargo (Morris-Commercial R) 4-cyl., 35 bhp, 4F1R, wb 10'8", 13'9"×5'10"×7'8", 17¾ cwt. 2513-cc (80×125 mm) SV engine with Smith carb. and Lucas magneto ignition. Worm-drive rear axle. 34×7.50 tyres. Late models had 85-mm bore engine and 7.50-20 tyres. Produced 1930–31.

Truck, 15-cwt, 4×2, Driving Instruction (Morris-Commercial L2/DCTM) 4-cyl., 31 bhp, 4F1R, wb 9'6", 13'5"×5'9½"×7'6". Model CQ 1802-cc (75×102 mm) engine. Equipped with dual controls. Tyres 5.50-20. Mk I (shown) supplied in 1936/37. Mk II (1938/39) had 5.50-18 tyres on well-base disc wheels and hyd. brakes.

Truck, 15-cwt, 4×2, Cargo (Bedford) In 1936 Vauxhall entered a regular truck with oversize tyres in the North Wales WD trials. In 1937 the 'square face' model, which eventually became the mass-produced Bedford MW, appeared. The wide bonnet was necessitated by the large air cleaner specified.

Truck, 15-cwt, 4×2, Cargo (Commer) During the early and mid-1930s Commer Cars Ltd designed and built several experimental 15-cwt trucks, utilizing commercial major mechanical components. This pilot model eventually became the limited-production Beetle infantry truck. Note the half cab top.

Truck, 15-cwt, 4×2, Compressor (Guy Ant) Guy Motors Ltd commenced production of their Ant Infantry truck in 1935/36. Several specialist bodies were developed for this class of vehicle, incl. the air compressor shown here. Guy used a 4-cyl. Meadows engine. Tyres were of size 9.00-16.

Truck, 15-cwt, 4×2, Office (Morris-Commercial CS8) Morris-Commercial Cars Ltd produced most of the WD 15-cwt trucks of the 1930s, starting with the Mk I in 1934. It had a 6-cyl. engine and appeared with many body types. Illustrated is a 1937 Mk III office truck in travelling position.

GB

TRUCKS, 1-ton, 25- and 30-cwt

Makes and Models: (30-cwt and 4 × 2 unless stated otherwise) AEC/LGOC B (1914). Albion A3 (1914). Subs. and India types (from 1923). Autocar UF21 (1915, US). Bedford (from 1938). Belsize (1914). Burford Subs. type (1923). Commer MC and WP3 (1914). B30 Raider (from 1932). N1 (c. 1937; also 4 × 4), etc. Crossley Subs. and India types (from 1923), 25/30 HP (6 × 4, 1927), 20/60 HP (6 × 4, 1929), 26/60 HP (6 × 4, 1930), etc. Daimler CB (1914). Dennis 18 and 20 HP (1914). Fiat 15 ter (1915, I). Ford(son) V8, various models (1935–40). Garford 75 (1-ton) and 66 (1915, US). Garner TW6-O (6 × 4, 1929). Gramm-Bernstein WIL (1-ton, 1915, US). Guy Subs. type (1924/25). Halley Subs. type (c. 1925). Karrier Subs. and India types (1923/24). Kelly-Springfield K30, K31 (1915, US). Lancia Z (1915, I). Leyland, S3, etc. (1914). Maudslay Subs. type (1923). Morris-Commercial C11/30 (1934), CS11/30 (1935–39), CS11/30F (1936–39), D (6 × 4, 1927–32), CD (6 × 4, 1933–39), CDF and CDFW (6 × 4, 1934–40), CDSW (6 × 4, 1935–40), etc. Napier B62, B62A, B72 (1914) Packard 1½D (1915, US). Star UE (25-cwt, 1916). Z (1914). Straker-Squire (1914). Straussler G1 (1-ton, 4 × 4, c. 1938). Talbot Subs. type (1923). Thornycroft Subs. type (1924/25). A3 (6 × 4, 1927), etc. Vulcan S (1914), Subs. type (1923), Vulcan/Holverter (4 × 4, 1926), RSW (6 × 4, 1927), etc. Willys 65XT (1-ton, 1915, US). Wolseley B Subs. CL and CP (1914), etc.

General Data: The above list does not include the many makes and models of which only very small numbers were used in World War I or the not-accepted entries for the Subsidy trials in the early 1920s. Generally speaking there were the following categories of 30-cwt (1½-ton) trucks : (1) the 'heavy' solid-tyred models of World War I (replicas of the Subsidy A type 3-tonner but built on lighter lines) ; (2) the 'light' pneumatic-tyred models of World War I (mainly Fiat and Packard) ; (3) the Subsidy 4 × 2 types of the 1920s (variants of which were built for the India Office) ; (4) the 'Light Six-Wheelers' of the late 1920s and 1930s; (5) the 4 × 2 WD-types of the 1930s (modified conventional vehicles).

In addition there were various commercial types and a number of experimental vehicles including some private ventures.

Vehicle shown (typical): Truck, 30-cwt, 4 × 2, Cargo (Karrier CY)
Technical Data:
Engine: Karrier 25.6 HP 4-cylinder, I-L-W-F, 4114 cc (4″ × 5″), 39.7 bhp at 1800 rpm.
Transmission: 4F1R (RH control).
Brakes: mechanical.
Tyres: front 100 × 720 mm, rear 8″ × 720 mm (Super Cushion).
Wheelbase: 11′0″.
Overall l × w × h: (chassis) 16′10″ × 6′1″ × 8′4″.
Weight: 46½ cwt.
Note: supplied about 1923 to the India Office, also with Workshop bodies and as recovery vehicle. Similar trucks with pneumatic tyres were built for the British WD (Subsidy type).

Truck, 30-cwt, 4×2, Cargo (Albion 24 HP) In 1922/23 the British WD instituted a Subsidy scheme whereby civilian owners of approved vehicles received a £40 grant for three years. The Albion was the first to be accepted as eligible. Shown is its 1925 development as purchased directly for military use.

Truck, 30-cwt, 4×2, Cargo (Crossley 40/50 HP) 4-cyl., 29.7 HP, 4F1R, wb 11'8", track 4'9". Rudge-Whitworth triple-spoked wire wheels (also with disc wheels and solid tyres). Bevel or worm-drive rear axle. 1923. Several other types were in RAF service (Tender, Wireless Van, etc.).

Truck, 30-cwt, 4×2, Water Tank (Autocar UF21) HO-2-cyl., 18 HP, 3F1R, wb 8'1", 14'4"×6'0"×8'10", 32 cwt (chassis). 4¾"×4½" engine below seat. Imported American chassis, bodied in Britain and used mainly in France (265 in 1918). Truck-bodied model was used in East Africa (189 in 1918).

Truck, 30-cwt, 4×2, Cargo (Napier B72) 4-cyl., 20/24 HP, 4F1R, wb 10'6", 40 cwt approx. 3½"×5" engine. Worm-drive rear axle. Essentially a lighter edition of the World War I WD Subsidy-type 3-tonner (*qv*). In 1918 there were over 400 30-cwt Napiers (B62, B72) in British military service.

Truck, 30-cwt, 4×2, Cargo (Guy 30/40 cwt) 4-cyl., 44 bhp, 4F1R, wb 10'10", 16'10"×6'1"×7'10", 50 cwt approx. Tyres, front 36×6, rear 38×7. From 1923 Guy produced several variants for the WD, the RN and for civilian customers (subsidized). Full-floating worm-drive rear axle.

Truck, 30-cwt, 4×2, Office (Halley W) 4-cyl., 25 HP, 4F1R, wb 11'0", chassis length and width 17'5"×6'0". Tyres 34×7. Halley's Industrial Motors Ltd of Yoker, Glasgow, produced this 'Subsidy' chassis for the WD and for civilian operators during the mid-1920s.

Truck, 30-cwt, 4×2, Cargo (Thornycroft A1) 4-cyl., 40 bhp, 4F1R, wb 11'6". Tyres 36×6 (other sizes optional). Model FB4 3¾"×5" SV engine (36 bhp at 1500, up to 40 at higher rpm). Accepted for enrolment under 1923 Subsidy scheme in 1924. Also supplied directly to WD (RASC).

Truck, 30-cwt, 4×2, Cargo (Vulcan VSC) 4-cyl., 20.1 bhp, 4F1R, wb 11'5", chassis length 18'0". Various tyre options, solid and pneumatic. Introduced in 1920. Throughout the 1920s Vulcan of Southport supplied a variety of 30-cwt chassis to the WD, incl. six-wheelers (*qv*).

Truck, 30-cwt, 4×2, Cargo (Bedford WH) One of the first Bedfords supplied to the British Army was this modified 1938 Model WH. Modifications included the fitting of larger tyres, single rear. Cab and front end were new for 1938. The commercial Model WH was a 2-tonner with dual rear tyres.

Truck, 30-cwt, 4×2, Cargo (Commer Raider B30) During the 1930s Commer supplied variants of their Raider, Centaur and LN5 trucks for service in the Middle East, chiefly with the RAF and the Transjordan Frontier Force. They were modified conventional trucks with oversize tyres and special bodywork.

Truck, 25-cwt, 4×2, Office (Commer N1) 4-cyl., 40 bhp, 4F1R, wb 10'0". Commercial type chassis of c. 1937 with oversize tyres on special wheels and house-type van bodywork, built integrally with special cab. A tilted truck was also supplied (with standard cab). Bodywork by Normand Garage.

Truck, 30-cwt, 4×4, Chassis (Commer Spider) Designated Spider this was a four-wheel drive conversion of the standard late-1930s Commer Model N2 30-cwt chassis/cab. Tyres were 9.00-20 India Super Traction. Engine: side-valve 6-cyl., 3181-cc (75×120 mm), 70 bhp at 3000 rpm.

Truck, 30-cwt, 4×2, Cargo (Fordson 79) V-8-cyl., 85 bhp, 4F1R, wb 11'2". Commercial 1937 2-ton chassis with 9.00-20 sand tyres and special bodywork (with standard doors) for RAF in Middle East. Officially known as 'Fordson Armed Tender, Type F'. MG mount on cab roof. Also on 1938 chassis.

Truck, 25-cwt, 4×2, Cargo (Fordson E88W) V-8-cyl., 63 bhp, 3F1R, wb 8'10", 15'1"×5'10"×6'6" (cab), 29 cwt. The E88W Model was a British design, built during 1936–39 (from 1938 also with 52-bhp Four engine). The RN and RAF used commercial Fordson vans and trucks of several types and years.

Truck, 30-cwt, 4×2, Cargo (Morris-Commercial CS11/30) 6-cyl., 60 bhp, 4F1R, wb 11'2". Produced during 1935–39 with open and closed cab. Preceding Model C11/30 had 55-bhp 4-cyl. engine. Model CS11/30F (1936–39) had semi-forward control. All these models had 10.50-16 tyres on WD-pattern divided-disc wheels.

Truck, 1-ton, 4×4, Cargo (Garner/Straussler G1) Experimental truck produced in 1938 by Garner of Sunbury-on-Thames to the design of Mr Nicholas Straussler. It had a Ford V8 engine, transmission, etc. Driven front axle was a modified Ford rear axle with transversal spring. A heavier model (2-ton, G2) was also made.

Truck, 30-cwt, 6×4, MG Carrier (Crossley BGV2) 4-cyl., 15.6 HP, 4F1R×2, wb 10'6" (BC 3'3½"). Chassis designed to carry 30 cwt on roads, 20 cwt across country. Shown as carrier for two MGs and crews, 1928. Many other body and chassis variations. Note belly support wheels and overall tracks.

Truck, 30-cwt, 6×4, Cargo (Garner 17.9 HP) 4-cyl., 17.9 HP, 4F1R×2. Produced by Garner Motors (then at Tyseley, Birmingham), from 1929 until the early 1930s. Dry-sump engine lubrication. Front axle centrally pivoted, with quarter-elliptic transverse springs. WD standard rear bogie.

Truck, 30-cwt, 6×4, Aircraft Refueller (Morris-Commercial D) 4-cyl., 40 bhp, 4F1R×2, wb 10'2" (BC 3'4"). Equipment by Zwicky. One of many applications of this 1927–32 chassis which was sold for civil and military use in many countries. Various tyre and other options. 2837-cc engine (early models 35-bhp 2413- and 2513-cc).

Truck, 30-cwt, 6×4, Cargo (Morris-Commercial D) 4-cyl., 40 bhp, 4F1R×2, wb 10'2" (BC 3'4"), 16'3"×5'11"×8'5". Tyres 32×4½ (32×6 or 34×7½ single all round optional). Well-type body with lath sides, panelled corner frames and tilt. Also with 12' wb. Of the normal-control version an experimental 8-wheeler was made (see page 175).

Truck, 30-cwt, 6×4, Cargo (Morris-Commercial CD) 4-cyl., 55 bhp, 5F1R, wb 10'7½" (BC 3'4"), 17'4"×6'2"×9'0", 52 cwt. Model EB 3519-cc (100×112 mm) SV engine. Produced during 1933–39. Model CDSW (1935–40) had 6-cyl. engine and military pattern front end.

Truck, 30-cwt, 6×4, Wire Laying (Morris-Commercial CDF) 4-cyl., 55 bhp, 5F1R, wb 10'7½" (BC 3'4"), 17'4"×6'2"×8'8", 54 cwt. Produced during 1933–40. Also with winch (Model CDFW, 1934–40). Hydraulic brakes. Tyres 7.50-20. Worm-drive rear axles. (IWM photo H5738).

Truck, 30/50-cwt, 6×4, Cargo (Thornycroft A3) 4-cyl., 35–40 bhp, 4F1R×2, wb 10'4" (BC 4'0"), 18'3"×7'1"×8'10", 75¼ cwt. Built for military and civilian (subsidized) use and for export, from 1927. Tyre size 38×6 or optional oversize (40×8 shown). Also FC version.

Truck, 30-cwt, 6×4, Cargo (Vulcan RSW) 4-cyl., 17.9 HP, 4F1R×2. Weight 60 cwt. One of several types of 6×4s made by Vulcan for the War Office, using their patented articulated rear bogie. Various tyre sizes. Also NC models. Note interconnected front springs. 1930.

GB

TRUCKS, 2-TON
4×2 and 4×4

The 2-ton (40-cwt) truck was not a standard class in the British Army but especially during World War I many hundreds of such vehicles were in military service. These were often impressed vehicles and British makes included Albion, Austin, Burford, Commer, Daimler, Dennis, Guy, Hallford, Halley, Karrier, Lacre, Leyland, Napier and Thornycroft. In addition there were imported American Jeffery 4×4s and in East Africa over 100 Reo 2-tonners were used. During the 1920s the India Office and the WD bought a number of 2-ton Thornycrofts and in the 1930s Morris-Commercial produced heavier commercial types which for Army use were somewhat modified and derated to 40-cwt. Several 2-ton Bedford and Fordson trucks were used by the Army and the RAF respectively. Garner built an experimental Straussler-designed 4×4, the G2.

The unusual SD 2-ton Freighter (with 'tramcar-type controls') also appeared in military livery.

Truck, 2-ton, 4×2-2, Timber Carrier (Lacre 0) Of just under 100 Lacres used in World War I the majority were of this 30 HP (4.33"×5") type. They were used by the WD and RAF and were among the first tractors with semi-trailers ('articulated six-wheeled lorry').

Truck, 2-ton, 4×2, Cargo (Fordson BB) V-8-cyl., 65 bhp (or 4-cyl., 52 bhp), 4F1R, wb 13'1". Standard commercial truck chassis/cab as produced in Britain during 1932–35, with dropside body for RAF. A Unipower six-wheeler conversion of the Model BB was used also (Cargo, Fire Tender).

Truck, 2-ton, 4×2, Cargo (Morris-Commercial WD10/40) 6-cyl., 70 bhp, 4F1R, wb 10'6". Tyres 10.50-16. Military modification of commercial Model CS10/80, produced during 1936–40. Open or closed cab, the latter (as shown) for use in Egypt. Fuel tank under seat, otherwise basically similar to 30-cwt CS11/30 (*qv*).

GB

TRUCKS, 3- to 5-TON

Principal Makes: *Up to 1918:* AEC, Albion, Belsize, Berna (British and Swiss), British Quad (4 × 4), Clydesdale (US), Commer, Daimler, Dennis, Fiat (I), FWD (4 × 4, US), Hallford, Halley, Karrier, Kelly-Springfield (US), Leyland, Locomobile/Riker (US), Mack (5½-ton, US), Maudslay, Napier, Packard (US), Pagefield, Peerless (US), Pierce-Arrow (US), Saurer (CH and F), Seabrook/Standard (US), Spa (I), Stevens, Straker-Squire, Thornycroft, Tilling-Stevens, White (US), Wolseley.
After 1918, 4 × 2: AEC, Albion, Bedford, Ford, Garner, Guy, Leyland, Morris-Commercial, Thornycroft, Tilling-Stevens, etc.
After 1918, 6 × 4: AEC, Albion, Crossley, Ford, Guy (also 6 × 6 and 8 × 8), Karrier, Leyland, Thornycroft, Vulcan.

Vehicle shown (typical): Truck, 3-ton, 4 × 2, Cargo (AEC YC)

Technical Data:
Engine: Tyler JB4 4-cylinder, I-L-W-F, 7720 cc (5" × 6"), 49 bhp at 1300 rpm.
Transmission: 4F1R; worm-drive rear axle.
Brakes: mechanical.
Tyres: 120 × 180 solid.
Wheelbase: 14'2½".
Overall l × w × h: 23'6" × 6'11" × 10'2".
Weight: 84¾ cwt.
Note: standardized Tyler engine was fitted in accordance with WD instructions. Models YA and YB similar in most respects. Total in service in 1918: 5819, as well as 1082 AEC/LGOC buses and trucks. Also used by US forces.

General Data: Traditionally the 'three-ton lorry' has always been the backbone of motor transport in the British armed forces. During World War I the vast majority of vehicles was in this class and when stock was taken at the end of the war there turned out to be 66,352 medium and heavy trucks (not counting steam wagons), the majority of which were 3- to 4-tonners. Most of these were subsequently sold to civilian operators. 3- and 4-ton trucks had also been produced for Allied Governments (e.g. Caledons for the US and Russian Army, etc.) and, in turn, others were imported, as indicated in the above listing (which only gives the principal makes). During the 1920s development of medium (3-ton) six-wheeled trucks began and this continued throughout the 1930s. These had an articulated rear bogie the design of which was patented by the War Department, who offered its use free to manufacturers who wished to produce vehicles of this type. Many such vehicles were subsequently built for the armed forces as well as for commercial operators at home and abroad. The Fordson 6 × 4 Sussex used a conversion design developed by County Commercial Cars Ltd of Fleet, Hants and was used extensively by the RAF.
Meanwhile batches of 4 × 2 trucks were purchased for non-tactical use. Some 4 × 4 models appeared also (Albion, Crossley, Garner-Straussler, Guy, Hardy, Karrier), but no quantity production of this type took place until World War II.

Truck, 3-ton, 4×2, Dental (Albion A10) 4-cyl., 34.5 bhp, 4F1R, wb 13'1". One of a variety of body types mounted on this type of chassis. Albion produced some 6000 A10 chassis for the WD. Other models (A3, A6, A12, etc.) were also in service, mostly impressed.

Truck, 2/3-ton, 4×2, Field Kitchen (Austin) 4-cyl., 20 HP, 4F1R, wb 11'0". Twin diagonal drive shafts to rear wheels. Dual rear springs. Used by WD, RN (Searchlight, Workshop, Trench Pump, etc.) and Imperial Russian Army. A total of about 2000 was made between 1913 and 1917.

Truck, 5-ton, 4×2, Timber Carrier (British Berna G) 4-cyl., 40 bhp, 4F1R, wb 14'1", 21'3" × 7'4" × 10'5". Fitted with winch for loading heavy timbers. Used by Royal Engineers. Spur gear final drive. In addition to over 300 British Bernas the Government in 1918/19 had 591 original Swiss Bernas.

Truck, 3-ton, 4×2, Cargo (Commer RC) 4-cyl., 40.6 bhp, 3F1R, wb 13'3", 20'9" × 7'3" × 10'6", 70 cwt. Lindley preselector gearbox with steering column control. Final drive by enclosed chains. 5322-cc (110×140 mm) engine. 920×110 solid tyres, dual rear. 2303 in service by 1918.

Truck, 3-ton, 4×2, Cargo (Daimler CC) 4-cyl., 40 bhp, 4F1R, wb 13'0", 21'6" × 7'0" × 10'10". 5702-cc Silent Knight sleeve-valve engine. Also with Workshop body. 366 in service in 1918, as well as 1818 Model CB, 2561 Model Y, and several other types, plus 113 Daimler/LGOCs.

Truck, 3-ton, 4×2, Medical (Dennis A Subs.) 4-cyl., 49.6 bhp, 4F1R, wb 13'2", 20'10" × 7'0". Early models had 110 × 150-mm engine; later the bore was 115 mm. Tyres, front 900 × 120, rear 1050 × 120 dual. Nearly 3500 supplied. Shown is a Mobile Chemical Laboratory van as used in the 1920s.

Truck, 3-ton, 4×2, Road Building (Hallford EID180) 4-cyl., 42 HP, 4F1R, wb 13'3", 20'6" × 7'4". This chassis was the most numerous of 1914–18 Hallfords (639 out of 1638 in 1918). Model shown had rolling floor for unloading road surfacing material. Made by J. E. Hall Ltd, Dartford, Kent.

Truck, 5-ton, 4×2, Tipper (Halley G) 4-cyl., 45 HP, 4F1R, wb 14'1", 26'3" × 7'0". Halley's Industrial Motors Ltd of Yoker, Glasgow, supplied many vehicles to the WD. In 1918 nearly 600 were in service, of which about 200 had been impressed. Model shown had 5" × 6½" engine.

Truck, 3-ton, 4×2, Cargo (Karrier WDS) 4-cyl., 39.7 bhp, 4F1R, wb 14'2", 21'5"×7'2"×10'9", 88 cwt. Wheels, front 880× 120, rear 1050×120. Of 1738 Karriers in 1918, 1557 were of this WD Subsidy type. It had a 5"×6" 45 HP engine, leather cone clutch and shaft drive.

Truck, 3½-ton, 4×2, Cargo (Leyland X3) 4-cyl., 30 or 40 HP, 4F1R, wb 14'5", 23'2"×6'9"×11'2". One of a batch of Leylands supplied to the War Department during 1907–09. Many more of similar type were impressed in 1914. Note low position of radiator.

Truck, 3-ton, 4×2, Tanker (Leyland S4X4, S5X4) 4-cyl., 36 bhp, 4F1R, wb 13'11" (Cargo GS:21'5"×7'4"×9'1", 89 cwt). Of the 6411 Leylands in service in 1918, 4721 were of this Subsidy A type (known as RAF-type Leyland). 5932 had been built, with GS, Workshop and other body types. 1914–18.

Truck, 3-ton, 4×2, Cargo (Maudslay, Rover) 4-cyl., 40 HP, 4F1R, wb 13'8", 21'6"×6'10"×10'8". In 1918 there were 1718 Maudslays in service, incl. 1547 of the unified Subsidy A type (shown) which had been co-produced by The Rover Co. Engine bore and stroke were 5"×5".

Truck, 5-ton, 4×2, Cargo (Milnes-Daimler 35 HP) From as early as 1901 the WD had tested and acquired Milnes-Daimler vehicles, incl. this model of 1907/08. In 1918 there were 32, mostly impressed. Basically they were German Daimler chassis, marketed through G. F. Milnes & Co., Ltd (later the Milnes-Daimler Co.).

Truck, 3-ton, 4×2, Cargo (Pagefield N) 4-cyl., 40 HP, 4F1R, wb 12'4", 20'8"×7'3"×10'6". Typical 'Subsidy A-type lorry' made by Walker Bros (Wigan) Ltd, Pagefield Iron Works, Wigan. In 1918, 519 Pagefields were in service, incl. 492 of these. The first truck impressed in 1914 was also a Pagefield.

Truck, 3-ton, 4×2, Anti-Aircraft (Peerless TC) 4-cyl., 40 HP, 4F1R, wb 12'7", 23'0"×7'10"×10'6", 134 cwt. Chain drive. Most widely used American chassis in the British Army in WWI; over 10,000 were supplied. Several body types incl. GS Cargo, Petrol Tank, Water Tank, etc. Note chain drive. (IWM photo STT 1281).

Truck, 3-ton, 4×2, Workshop (Pierce-Arrow) 4-cyl., 30 HP, 3F1R, wb 14'0". One of two built by Wolseley on US chassis. Contained hand tools, vices, benches, 5½" lathe, grinder, drilling machine, forge, anvil, etc. Tools and body cost £565, in 1915. By 1918 1705 Pierce-Arrow trucks were in service (mainly GS types).

Truck, 3-ton, 4×2, Searchlight (Stevens/Dennis) Supplied by Stevens Petrol-Electric Vehicles Ltd of London this vehicle was based on the Dennis Subsidy A 3-tonner. Engine-driven generator gave continuous output of 350 or 230 Amp. at 70 to 110 Volt resp. By 1918 187 were in service, incl. 74 overseas (France).

Truck, 3-ton, 4×2, Cargo (Straker-Squire CO) Vehicles of this make were supplied in limited numbers for military use from well before World War I. By 1918 several hundred were in service, by WD and Admiralty (RN), including 58 impressed. Illustrated are two at a post-war WD surplus auction sale.

Truck, 3-ton, 4×2, Anti-Aircraft (Thornycroft J) 4-cyl., 44.5 bhp, 4F1R, wb 13'7½", (Cargo GS: 22'2"×7'3"×10'5", 89 cwt). T-head engine. Some 5000 Thornycrofts were made for the WD during 1914–18; most were of the Model J (Subsidy A type) with various body types, incl. this 3.7" AA Gun.

Truck, 3-ton, 4×2, Cargo (Wolseley CR6) Vickers 4-cyl., 34.2 HP, 4F1R, wb 13'9", 21'6"×7'2"×11'10". Produced to WD Subsidy A specification. Chassis price £725 (£610 for 30-cwt Model CL 12'-wb Subs. B variant). Several body types, incl. tanker. 385 produced, 1914–18; 353 in service at end of war.

Truck, 3–5-ton, 4×2, Workshop (AEC 506) The AEC 35/50 HP Model 506 was a commercial truck chassis, made during 1925–32. A few were used by the RASC as '3–5-ton Mobile Workshop (1926 provision)' and '3–5-ton GS Lorry', with pneumatic and solid tyres respectively.

Truck, 3-ton, 4×2, Tractor (Bedford WTH) 6-cyl., 72 bhp, 4F1R, wb 9'3". 3.5-litre OHV engine. 1938/39 tractor with Scammell coupling and semi-trailer. Special coil-sprung fixed-side body for torpedo carrying. Other pre-war Bedfords included horsebox (Vincent body) and fire tender.

Truck, 3-ton, 4×2, Recovery (Garner) Basically a commercial low-loading chassis, developed for municipal work. Tyres 26×6. Rating 4½ tons for body and payload. Wheelbase 12'6". Body length 20'. Turning circle 25'. 4-cyl. engine, forward of front axle. Manual winch. Early 1930s. Load: Morris-Commercial D-type 30-cwt 6×4.

Truck, 3-ton, 4×2, Disinfector (Morris-Commercial CVS 11/40) 6-cyl., 65 bhp, 4F1R, wb 11'6", 19'10"×7'1"×9'9", 82 cwt. Model OR35 3485-cc (82×110 mm) side-valve engine with Treasury Rating of 25.01 HP. Lockheed hyd. brakes. Tyres 6.00-20. Spiral-bevel drive fully-floating rear axle. 1939.

Truck, 3½-ton, 4×2, Cargo (Thornycroft Speedy) Modified Thornycrofts taking part in WD trials in the early 1930s included this 3½-ton Speedy and a 2-ton Handy model, both of 1933. They were adapted from commercial trucks, featuring military pattern cabs, etc. No quantity production.

Truck, 4-ton, 4×2, Searchlight (Tilling-Stevens TS19/8) 4-cyl. SV engine, driving dynamo via Cotal gearbox. Electric motor to drive rear wheels. Tyres 9.00-20. Mech. brakes. These petrol-electric vehicles were built for air-defence searchlight units during 1936–38 and used throughout the war.

Truck, 3-ton, 4×4, Cargo (Guy Lizard) The Lizard was a development of Guy Motors' four-wheel drive Quad-Ant tractor but only a few prototypes were produced in 1938/39. The chassis was, however, used as the basis for an armoured command vehicle. A test vehicle is illustrated.

Truck, 3-ton, 4×4, Cargo (Karrier) Karrier Motors Ltd (then part of the Rootes Group) produced this four-wheel drive truck seen climbing a mountain road (Alt-y-Bady) during WD trials in North Wales in October 1938. Similar prototypes were entered by Albion, Crossley, Guy and Hardy (AEC).

Truck, 3-ton, 6×4, Bridging (AEC 644 Marshal) 4-cyl., 70–80 bhp, 4F1R×2, wb 12'8½" (BC 4'0"). During 1931–41 AEC produced 934 Marshal six-wheelers, incl. many for the Army, who fitted several body types, e.g. Cargo and Bridging (Small Box Girder and Trestle or Sliding Bay; the latter is shown).

Truck, 3-ton, 6×4, Cargo (Albion 30/45 HP) 4-cyl., 45 bhp, 4F1R×2, wb 12'6" (BC 3'9"). Tyres 36×8 (shown) or 36×6 (dual rear). Dewandre vacuum servo brakes on rear wheels. Produced during 1927–31, also with other bodies, incl. Bridging Equipment (Trestle and Sliding Bay).

Truck, 3-ton, 6×4, Bridging (Albion BY1) 4-cyl., 63.5 bhp, 4F1R×2, wb 13'0" (BC 4'0"). Tyres 9.00-20. Raft unit, officially known as 'Lorry, Pontoon, No. 5, Mk I'. Also with other bodies, incl. Trestle and Sliding Bay and GS Cargo. Chassis produced during 1938–40.

Truck, 3-ton, 6×4, Wireless (Crossley 30/70 HP) 4-cyl., 70 bhp, 4F1R×2, wb 12'6" (BC 3'8½"). Tyres 36×6. Body by Mann Egerton. Supplied during the early 1930s to the RAF who until and including World War II operated a large number of 6×4 Crossleys, including fire tenders.

Truck, 3-ton, 6×4, Derrick (Crossley IGL8) 4-cyl., 75 bhp, 4F1R×2, wb 12'10" (BC 4'0"). Derrick bodies were supplied on winch-equipped Crossley, Guy and Leyland 6×4 chassis of the late 1930s. Crossley also supplied vehicles to India and Canada (qv), the Transjordan Frontier Force, etc.

Truck, 3-ton, 6×4, Balloon Winch (Fordson Sussex) In addition to Crossleys, the RAF used many vehicles based on British-built Ford 6×4 chassis of the years 1934–39. Illustrated is a 1937 Model 79 V8 with Wild winch. Other body types: Cargo, Refueller, Fire Tender, Photographic Tender, Floodlight, etc.

Truck, 3-ton, 6×4, Cargo (Guy BAX) 4-cyl., 38 bhp, 4F1R×2, wb 12'6" (BC 3'6"). Tyres 36×6. This chassis, produced in 1927/28, cost £995 and was the largest capacity 6×4 under the contemporary WD Subsidy scheme of £40 per annum for three years. On roads it carried 5 tons.

Truck, 3-ton, 6×4, Workshop (Guy FBAX) Basically similar to the vehicle on left but with forward control, crew cab and Standard House Body Mk II. Also appeared with half cab (driver alongside engine but normal bonnet and front wing on left-hand side; three crew seats behind engine).

P

Truck, 3-ton, 6×4, Cargo (Karrier WO6) 4-cyl., 30 bhp, 4F1R × 2, wb 12′6″ (BC 3′8″). Tyres 36 × 6. Karrier Motors were the first British makers of six-wheeled tandem-drive cross-country trucks. Design had started in 1924, production in 1925. Note 'overall chains' on running board.

Truck, 3-ton, 6×4, Cargo (Karrier WO6/A) 4-cyl., 48 bhp, 4F1R × 2, wb 12′0″ (BC 3′8″). Tyres 40 × 9, single rear. This was a variant of the 1928 WO6, with driver alongside engine, built for the Indian Government. Lowest forward speed was 1.14 mph. 12-Volt lighting and starter.

Truck, 3-ton, 6×4, Cargo (Karrier WO6/B) 4-cyl., 57 bhp, 4F1R × 2, wb 14′3¾″ (BC 3′8″). Tyres 40 × 9. 5185-cc (114 × 127 mm) side-valve engine. Derived from Model WO6/A in 1928, this unusual design (for the Indian Government) featured front springing of transverse spring swivelling type.

Truck, 3-ton, 6×4, Cargo (Karrier FM6) 4-cyl., 32.4 HP, 4F1R × 2. On this model of the early 1930s engine accessibility was improved by fold-away wings (swivelling on extensions of the radiator guard). Some of these chassis had a power-driven winch, mounted amidships.

Truck, 3-ton, 6×4, Cargo (Leyland Terrier) 6-cyl., 62–80 bhp, 4F1R×2. The Terrier was Leyland Motors' WD Subsidy type of 1928. It could carry three tons across country, four on roads. Tyre size was 36×6. Vehicle shown was supplied directly to the WD and had military cab and body.

Truck, 3-ton, 6×4, Cargo/Prime Mover (Leyland Terrier) One of a batch of forward-control trucks, equipped with winch and used for hauling cargo and towing anti-aircraft artillery. Tyre size 9.00-20. Produced in 1937 for British Army. Also used, for same purposes, by New Zealand armed forces.

Truck, 3-ton, 6×4, Workshop (Thornycroft A4) 6-cyl., 70 bhp, 4F1R×2, wb 13′0″ (BC 4′0″), 22′0″×7′7″×10′3″. 143 cwt. Crew 2+2. Tyres 40×9. Forward-control medium six-wheeled chassis of 1929. Overhead-worm rear axles. Max. road speed 35 mph. Westinghouse vacuum servo brakes.

Truck, 3-ton, 6×4, Cargo (Thornycroft A5) 6-cyl., 55.5 bhp, 4F1R×2, wb 12′6″ (BC 4′0″). Chassis length 20′5″, width 6′2″. Various tyre size options. Vacuum-assisted mech. brakes. Eligible for the War Office Subsidy of £120 in Great Britain. Payload on good roads 3½ tons. 1930/31.

Truck, 3-ton, 6×4, Fuel Tanker (Thornycroft A5) Forward-control variant of conventional Model A5, used for special bodywork, incl. Mobile Workshop, Breakdown, etc. Model YB6 5701-cc (95×133 mm) SV engine with 7-bearing crankshaft and Simms high-tension magneto. Early 1930s.

Truck, 3-ton, 6×4, Cargo (Thornycroft LE/SC6) 6-cyl., 70 bhp, 4F1R×2. Also known as Tartar. Model SC6 engine, governed to 2930 rpm or 32 mph. CAV-Bosch starter and electrical equipment. Tyres 9.00-22. Hand-controlled radiator shutters. Also NC variants (KF/AC4/1, KF/DC4/1).

Truck, 3–5-ton, 6×4, Artillery Portee (Thornycroft XB FC) 4-cyl., 58–60 bhp, 4F1R×2, wb 18'0" (BC 4'6"), 130 cwt. Carried field gun, Cletrac tractor, crew and ammunition. Towed limber carrying extra ammunition. Road speed 30 mph. One of a wide range of Thornycroft six-wheeled trucks of the 1930s.

Truck, 3-ton, 6×4, Cargo (Vulcan VSW) The Vulcan Medium six-wheeler of 1927 was basically similar to the contemporary designs of Albion, Karrier and Leyland and featured the same type of WD-pattern articulated rear bogie. It had a 17.92 HP (RAC rating) engine.

GB

STEAM WAGONS

Several of the earliest self-propelled vehicles used (or tested) by the British armed forces were powered by steam and these were still very much in evidence during World War I when there were thousands of steam traction engines (see Tractors), steam rollers and steam wagons. Of the latter there were about 1300 at the time of the Armistice, of the following makes: Aveling & Porter, Clayton & Shuttleworth, Foden, Garrett, Leyland, Mann, Ruston Proctor, Sentinel, Tasker, Wallis & Steevens and Wantage. 43 of these vehicles had been impressed and 786 were overseas, practically all in France. The Foden was by far the most numerous. Some continued in service during the 1920s.

Steam Wagen (Thornycroft) Steam-propelled military vehicles made by Thornycroft were first used in the South African campaign of 1899–1902. Vehicle shown took part in War Office trials of 1904/05, together with paraffin-driven models. It had a vertical boiler.

Steam Wagon (Foden) Rubber-tyred 5-tonner. Wheelbase 14′4″. Dim. 22′0″×7′6″×10′0″. Weight 184¾ cwt. Loco-type boiler. By 1918 over 800 were in service. Body types: GS Cargo, Tipper, Disinfector (shown), etc. Disinfectors were used to remove vermin from clothing.

Steam Wagon (Sentinel) Of the 5/6-ton Sentinel steam wagons (35 HP, bore and stroke 6¾″×10″) there were 131 in 1918 (29 with the Admiralty). Body types: GS Cargo and Tipper. This dropside wagon remained in service with the RAF during the 1920s.

GB

TANK TRANSPORTERS

The Great War of 1914–18 saw the first active use of the tank and with this came many associated problems like transport, supply, recovery, etc. Transport of tanks was then carried out by rail, there being no wheeled transporters of sufficient capacity to carry them over roads. From the railheads the tanks were driven to the front. A transporter for crawler tractors was designed but was not ready in time (unlike its French counterpart, which was hauled by the American Knox tractor). When the Scammell Pioneer tractor appeared in the late 1920s a tank transporter was soon designed and development of this vehicle took place during the 1930s. Some specimens are shown here.

Tractor-Trailer, Tractor Transporter (AEC K/Bauly) Modified AEC truck with fifth wheel coupling resting directly on rear axle. Semi-trailer with crew cab, built by H. C. Bauly Ltd of London. Could be used as full-trailer (with dolly). Designed in 1918/19. Carried Holt Caterpillar and similar loads.

Tractor-Trailer, Tank Transporter (Scammell Pioneer) 4-cyl., 85 bhp (or Gardner 6-cyl. 102-bhp diesel), 6F1R, wb (tractor) 15'1" (BC 4'3¼"). Tyres 13.50-20. The Pioneer 6×4, introduced in 1927, was soon adapted for tractor work. This carrier (with Vickers Medium Mk II tank) appeared in 1932.

Tractor-Trailer, Tank Transporter (Scammell Pioneer) Improved and hauled by a more modern version of the Pioneer this tank carrier of about 1937 still featured a detachable rear bogie for loading and unloading. The next design (1939 20-tonner) had a fixed bogie and hinged loading ramps.

GB

TRACTORS, WHEELED
Pre-1920

The earliest recorded use of British military mechanical transport was in the 1850s when Boydell and Bray steam traction engines were tested. In 1868 the Corps of Royal Engineers acquired their first steam traction engine, 'Steam Sapper'. They got their second, an Aveling & Porter, in 1872. Steam traction engines were still in service in World War I, when the following makes were used: Aveling & Porter, Burrell, Clayton & Shuttleworth, Foster, Fowler, Garrett, Mann, Marshall, McLaren, Robey, Ruston Proctor, Tasker, Thornycroft, Wallis & Steevens. In 1904 the first internal-combustion oil-engined tractor appeared, made by Wolseley, followed by a paraffin-driven Hornsby in 1905 and another Wolseley in 1907. In 1909 Broom & Wade and Thornycroft submitted paraffin-engined tractors for trials and in World War I there were the well-known IC tractors of Foster-Daimler, which were also known as Daimler-Foster (made by Foster in Lincoln).

Steam Traction Engine (Fowler Lion) Traction engine No. 7 of the ASC (Army Service Corps) was this 75 HP Fowler Lion, which could haul 18 tons on three wagons. By 1918 some 75 Fowler 'road locos' and 42 steam rollers were in service, as well as many Fowler trailers.

Steam Traction Engine, Crane (Tasker) William Tasker & Sons Ltd of Waterloo Iron Works, Andover, supplied large and small steam traction engines as well as trailers. Shown is ASC No. 76, with front-mounted crane, purchased about 1910. In 1918 there were 22 tractors, 80 trailers.

Steam Traction Engine (Wallis & Steevens) ASC No. 22 was this light Wallis & Steevens, made at Basingstoke, Hants. It towed about 5 tons. In 1918 there were about 20 tractors of this make (mainly the 24 HP type), as well as 10 steam rollers and 11 trailers (as shown).

Tractor, IC, Artillery (Hornsby) Designed and produced by Richard Hornsby & Sons Ltd of Grantham, Lincs, in 1905. Powered by an internal-combustion engine, running on paraffin. Drawbar pull 8 tons. In 1907 it was converted into a full-track tractor. Scrapped in 1914.

Tractor, Steam (Stewart-Crosbie) Experimental tractor with horizontal 2-cyl. 40-bhp (at 600 rpm) compound steam engine and 2-speed planetary gear set. Engine had camshaft-operated poppet valves. Vertical boiler with superheating coils. Took part in 1909 Aldershot trials. Made in Glasgow.

Tractor, IC, Artillery (Thornycroft) From about 1904 Thornycroft built several paraffin-driven tractors for military and commercial use. Model illustrated won a prize of £750 in the 1909 War Office trials at Aldershot. There was also a short-wheelbase version which had a shorter canopy and no rear seats.

Tractor, IC, Artillery (Foster-Daimler) Daimler 6-cyl., 105 bhp, 2F1R, wb 12'0", 21'6" × 8'9" × 11'0", 13 tons. Overall gear ratios, first 137:1, second (direct) 76.8:1. Sleeve-valve engine. 8' rear wheels. Fitted with winch. Hauled 15" navy guns, etc. 74 in service by 1918 (52 by Admiralty).

GB

TRACTORS, WHEELED
From 1920

Makes and Models: AEC/FWD R6T/850 (6 × 6, from 1932). Alvis-Straussler Hefty (4 × 4, 1938), LAC (4 × 4, c. 1939). Armstrong-Siddeley/Pavesi (4 × 4, 1929/30). Fordson (4 × 2, 1923), etc. Glasgow (3 × 2, c. 1923). FWD R6T (6 × 6, 1929). Garner/Straussler G3 (4 × 4, 1939/40). Guy (6 × 6 and 8 × 8, 1930/31), Quad-Ant (4 × 4, 1938). Karrier (4 × 4, c. 1938). Latil TP (4 × 4, c. 1921, F). Morris-Commercial QW (4 × 4, 1938), C8 Mk I (4 × 4, 1938/39). C8 Mk II (4 × 4, 1939/40), CDSW (6 × 4, from 1935). Pavesi P4 (4 × 4, 1923, I). Scammell Pioneer (6 × 6, 1929), Pioneer (6 × 4, from 1936). Straussler Sturdy (4 × 4, 1936), etc. Thornycroft Hathi Mk II (4 × 4, 1924; 6 × 6, 1927). Trojan (6 × 4, 1929).

General Data: Of the above tractors the majority were purpose-built, there being few commercial designs suitable to meet the various military requirements. Most of the vehicles listed were artillery tractors and included are some experimental models and a few foreign-built ones which were tested by the British Army.

The original Hathi tractor (*hathi* is Hindustani for elephant) was assembled in 1923 by the RASC, using components from captured German four-wheel drive tractors, mainly Ehrhardt and Daimler, of the Great War. In October of the same year the War Department issued specifications for an improved model, the Hathi Mark II, which was built in 1924 by Thornycroft and demonstrated in WD trials in October of the same year. Some 24 were subsequently built and used at home and in Australia and India. A 6 × 6 variant was also made; this was probably the first-ever British-built six-wheel drive vehicle. The Scammell Pioneer, first introduced in 1926/27, was soon tested for artillery towing and among the first to be ordered was an all-wheel drive model for India. FWD (England) launched a 6 × 6 tractor in 1929, followed by Guy Motors in 1930/31. Guy also produced an 8 × 8 variant. During the 1930s several lighter 4 × 4 models appeared (Guy, Karrier, Morris-Commercial), as well as a 6 × 4 by Morris-Commercial and some unusual 4 × 4 models designed by Straussler. Some of the latter were used by the RAF.

Vehicle shown (typical): Tractor, 4 × 4, Artillery (Thornycroft Hathi Mk II)

Technical Data:
Engine: Thornycroft GB6 6-cylinder, I-F-W-F, 11,197 cc (120.7 × 165.1 mm), 90 bhp at 1200 rpm.
Transmission: 6F2R.
Brakes: foot: mech. on front prop. shaft; hand: mech. on rear prop. shaft.
Tyres: 40 × 8.
Wheelbase: 11'6".
Overall l × w × h: 16'4" × 6'10½" × 6'8".
Weight: 117 cwt (unladen 100 cwt).
Note: used for towing various types of artillery and with crane jib as breakdown vehicle. Permanent all-wheel drive. 1924/25.

Tractor, Artillery, 4×4 (RASC 'Hathi') This experimental four-wheel drive tractor was built by the Royal Army Service Corps at Aldershot in 1923, using mechanical components from captured German Ehrhardt and Daimler tractors of World War I. *Hathi* is Hindustani for elephant.

Tractor, Artillery, 4×4 (Armstrong-Siddeley) 4-cyl., 45 bhp, 4F1R (Wilson epicyclic) with emergency low. Articulated tractor, built under Italian Pavesi licence, from 1927. Model shown, with 44 × 10 pneumatic tyres, made in 1929. Air-cooled OHV engine, 4″ × 4¾″. Lockable rear differential.

Tractor, Artillery, 4×4 (Guy Quad-Ant) Meadows 4-cyl., 58 bhp, 4F1R × 1, wb 8′5″. Four-wheel drive tractor variant of the Guy Ant truck, introduced in late 1937. Shown is a prototype with open bodywork. Production models had closed steel body, similar to Morris-Commercial C8 (*qv*).

Tractor, Artillery, 4×4 (Karrier F.W.D.) 6-cyl., 80 bhp, 4F1R × 2. Prototype for Karrier KT4 Field Artillery Tractor (supplied to Indian Army during World War II). 4086-cc (85 × 120 mm) SV engine. Tyres 10.50-16. Tractive effort 10,000 lb. GCW about 8 tons. *c.* 1938.

Tractor, 4×4, Artillery (Morris-Commercial C8 Mk II) 4-cyl., 70 bhp, 5F1R×1, wb 8'3", 14'9"×7'6"×7'9", 65 cwt. Tyres 10.50-20 (Mk I and II; Mk III had 10.50-16 tyres). Permanent all-wheel drive. 3519-cc (100×112 mm) side-valve engine. Towed field gun and limber. 4-ton winch. (IWM photo H20971).

Tractor, 4×4, General Purpose (Straussler Sturdy LT1) Singer 4-cyl., 35 bhp, 4F1R×1, wb 6'9", 10'7½"×5'6", 27 cwt. 1459-cc (66.5×105 mm) OHV engine with fluid flywheel. Modified Singer axles. Tyres 10.50-13. Used by RAF as 'Tractor, Straussler, Type B (Light)'. Tubular frame; pivoting rear axle and ballast body. 1935/36.

Tractor, 6×6, Artillery (FWD R6T) Dorman JUL 6-cyl., 78 bhp, 4F1R×2, wb 10'0" (BC 4'0"), 18'10"×7'6"×6'9", 163 cwt. Tyres 42×10.5. 6597-cc (100×140 mm) SV engine. Planetary gear final reduction in wheel hubs. 7-ton winch. Produced by FWD Lorry Co. Ltd (later AEC) in 1929.

Tractor, 6×6, Artillery (AEC/FWD R6T/850) AEC 6-cyl., 95 bhp, 4F1R×2, wb 10'0" (BC 4'0"), 19'3"×7'6"×8'8", 171 cwt. Tyres 9.00-22. 6126-cc (100×130 mm) OHV engine. After take-over in 1929 the FWD R6T was produced by AEC (as Model 850) until 1936. Also as Recovery Tractor, both for RAOC.

Truck, 6×6, Prime Mover (Guy) 6-cyl., 96 bhp, 4F1R×2, wb 13'0" (BC 4'0"), 23'3"×7'5¾". Conventional worm-drive rear axles. Cross shaft and individual propeller shafts outside frame to front wheels. Guy Model C 4¼"×5½" engine with seven main bearings. Designed late 1920s. Supplied to India.

Truck, 8×8, Prime Mover (Guy) Introduced in 1931 this vehicle was substantially similar to that shown on the left but with an additional driving/steering front axle. Laminated spring mounted towing hooks front and rear. Herbert Morris 7-ton winch with 250 feet of cable. Vacuum servo brakes.

Tractor, 6×4, Artillery (Morris-Commercial CDSW) 6-cyl., 60 bhp, 5F1R, wb 9'7½" (BC 3'4"). This winch-equipped tractor was used from 1935 to tow 25-pounder and other guns until superseded by the 4×4 Field Artillery Tractors. The CDSW tractor was operated also by the Irish Army.

Tractor, 6×4, Artillery (Morris-Commercial CDSW) Semi-armoured variant of the CDSW artillery tractor. Similar armour plating appeared on Model CS8 15-cwt 4×2 trucks. Later CDSWs (Bofors AA gun tractor, etc.) with 'soft skin' front end had full-width fixed windscreens. (IWM photo STT 5799).

Tractor, 6×6, Artillery (Scammell Pioneer) 4-cyl., 65 bhp, 5F1R. One of the first all-wheel drive variants of the Scammell Pioneer was this gun tractor built for the Indian Army about 1929. It had a 7-litre petrol engine, cone clutch and 6-ton horizontal drum winch.

Tractor, 6×4, Artillery (Scammell Pioneer) Gardner 6LW 6-cyl. diesel, 102 bhp, 6F1R, wb 12'2" (BC 4'3¼"), chassis: 20'6¼"×8'3½". Tyres 13.50-20. 8.4-litre engine. OD top gear. Built during the late 1930s. Horizontal (later vertical) power winch. Note machine gun mount.

Tractor, 6×6, Artillery (Thornycroft Hathi) In 1927 the Royal Army Service Corps built this exp. 6×6 tractor with dual tyres all round. It consisted of a Hathi Mk II tractor (*qv*), fitted with the WD-pattern articulated tandem rear bogie of a contemporary Thornycroft 6×4.

Tractor, 6×4, Artillery (Trojan) 4-cyl., 11 bhp, 6F3R, wb 8'6", 13'1"×5'3". Tyres 29×5. Track 4'4¾". Epicyclic 3-speed transmission with 2F1R aux. box. Chain drive to front axle of bogie and from there to rearmost axle. No diffs. Tested as 9/12-cwt load carrier and tractor in 1929. 40 supplied to South African Army.

GB

FIRE TRUCKS

The first 'motor steam fire engine' for the WD was completed in January 1907 (by Merryweather) and delivered to Bulford Camp on Salisbury Plain. The second, another Merryweather steamer, went to Aldershot in 1908. Dennis Bros of Guildford also supplied fire engines before World War I. By 1918 the Army had on its strength the following makes: Commer (Model FCB, 1 in France), Dennis (60 HP, 14 in France, 4 in Mesopotamia, 2 in Salonika), Leyland (various types, 4 at home, 1 in France), and Merryweather (various types, 4 at home, 3 in France, 3 in Mesopotamia). During the 1930s the RAF purchased quantities of 6×4 Crossley-based Fire Tenders for use on airfields, supplemented by some on Ford 6×4 chassis. A few typical examples are shown.

Truck, 10-cwt, 4×2, Fire Tender (Ford T) The ubiquitous Model T Ford was used for a multiplicity of purposes. This neat little Fire Tender was based on a pre-1917 chassis and photographed in 1918. 20-bhp engine drove through 2F1R epicyclic (planetary) gearbox.

Truck, 3-ton, 6×4, Fire Tender (Merryweather/Thornycroft A5) The Greenwich firm of Merryweather & Sons supplied the Army and Navy with several models based on commercial chassis (Albion, Thornycroft). Vehicle shown was produced in the early 1930s and had a 5701-cc 55.5-bhp six-cylinder engine.

Truck, 3-ton, 6×4, Fire Tender (Crossley IGL) 4-cyl., 90 bhp, 4F1R×2, wb 12'6" (BC 4'0"). Tyres 10.50-20. PTO-driven foam pump. Heating compartment fitted over pump. Four foam delivery outlets. Standard RAF type of late 1930s. Also with all-enclosed streamlined bodywork.

GB

CARS, TRUCKS and TRACTORS
HALF-TRACK

Makes and Models: AEC/Roadless 501 (c. 1925). Albion A10 (mod., c. 1925). Burford/Kégresse (various models, from 1924). Citroën/Kégresse (various models, from 1923, F). Clayton & Shuttleworth Caterpillar 110 HP (1917). Crossley/Kégresse (various models, from 1924). FWD/Roadless (various models, from 1925). Guy/Roadless 18 HP (from 1925). Holt Caterpillar 75 HP and 120 HP (1914, US). Morris-Commercial/Roadless (various models, from 1925). Morris-Martel/Roadless (1926). Peerless (mod., c. 1924). Ruston Proctor 75 HP (c. 1917). Vulcan/Roadless (1925).

General Data: In World War I the British used many heavy semi-tracked artillery tractors which had one front wheel for steering purposes. Most numerous were the imported American Holt 75 and 120 HP Caterpillars, of which by the war's end there were over 1500. The British-built Ruston 75 HP was virtually identical to the Holt 75 HP and of these there were 157. Lastly there was the Clayton, made by Clayton & Shuttleworth Ltd of Lincoln, with 250 units. After the war many if not most of these machines were sold to civilian operators.

During the early 1920s, following the successes of the light Citroën/Kégresse, several British manufacturers offered commercial half-track modifications of their regular production trucks. Some used the French Kégresse rear bogies which were of the continuous rubber band type; others featured the Roadless half-track system which was supplied by Roadless Traction Ltd of Hounslow, Middlesex. The Citroën/Kégresse was also sold in Britain. The British Army tested most of these vehicles and placed small orders for several of them. Longtime user of the FWD/Roadless tractors was the Royal National Lifeboat Institution. Some steam-propelled half-track tractors were also made (Foden, Sentinel, 1924/25). Illustrated in this section are some examples of these vehicles. For additional coverage the reader is referred to HALF-TRACKS (Olyslager Auto Library/Warne).

Vehicle shown (typical): Tractor, 3-ton, Half-Track, Field Artillery (Burford/Kégresse MA)

Technical Data:
Engine: Burford 4-cylinder, I-L-W-F, 4819 cc ($4\frac{1}{8}" \times 5\frac{1}{2}"$), 29 bhp.
Transmission: 4F1R × 2.
Brakes: mechanical.
Tyres: 34 × 7.
Overall l × w × h: 16'0" × 6'6" approx.
Weight: 93 cwt.
Note: capable of carrying a crew of 6 men and 100 rounds of ammunition. Late model Kégresse positive drive bogies. Speed, road 25 mph, across country 15 mph. Radius of action 100 miles. Gradability 40°.

Truck, 3-ton, Half-Track (Albion A10) Half-track conversion of 1914–18 Albion 32 HP 3-tonner with chain drive, executed during the early 1920s for the Royal Army Service Corps, reputedly by Armstrong-Siddeley Ltd of Coventry. Rear sprockets were driven by secondary chains.

Tractor, 2-ton, Half-Track (Burford/Kégresse) First half-track produced by H. G. Burford & Co. Ltd, using regular 2-ton chassis with hollow-steel-spoked wheels. Modifications included forward control, two-speed auxiliary gearbox (providing 4F1R × 2 trans.) and friction-drive Kégresse bogies. 1924.

Truck, 30-cwt, Half-Track, Wireless (Crossley/Kégresse) Crossley Motors Ltd of Gorton, Manchester, introduced their first half-track in 1924, using Kégresse rear bogies on a 40–45 HP Crossley car chassis. Shown is a Wireless Tender produced for the RAF about 1925.

Car, 15-cwt, Half-Track, Battery Staff (Crossley/Kégresse) In addition to 30-cwt models Crossley, from 1926, produced 15-cwt chassis with 2388-cc (80 × 120 mm) 4-cyl. engine, rated at 15.6 HP (RAC). It had 4F1R × 2 transmission and 32 × 4½ front tyres. A typical WD Reg. No. was MK 8307. 1927.

Truck, 3-ton, Half-Track (Crossley/Kégresse) Heaviest of the Crossley military half-tracks was this forward-control Model PD of 1928, shown here with a test body. This experimental vehicle had the late type positive-drive Kégresse bogies and was sold in 1934. (IWM photo STT 1258).

Tractor, 3-ton, Half-Track, Medium Artillery (FWD/Roadless) 4-cyl., 60 bhp (42 HP), 3F1R × 2, wb 8'6½", 110 cwt. Tyres 40 × 8. Driven front axle. In 1926 the FWD Lorry Co. Ltd supplied two tractors with Roadless Traction rear bogies. The Army designated them B4E1 and B4E2. (IWM photo STT 1233).

Truck, 1-ton, Half-Track, Water Tank (Guy/Roadless) 4-cyl., 18 HP, 4F1R, track 4'8", length 15'4", chassis weight 34 cwt. Tyres 34 × 7. Six purchased by WD in 1926. Roadless Traction patent tracks could flex laterally which improved steering. Two were used in trans-Australia expedition.

Tractor, Half-Track (Morris-Martel/Roadless) The 'Morris Tractor' had Roadless Traction bogies and a solid-tyred rear axle for steering and to prevent 'rearing' and tipping over backwards when climbing. Derived from Major Martel's one-man tank. 4-cyl. 15.9 HP engine with 4F1R × 2 transmission. 1926.

Q

Truck, 1-ton, Half-Track, MG Carrier (Morris-Commercial/Roadless) During the latter half of the 1920s there appeared half-track conversions of Morris truck chassis with truck and touring car bodywork. This machine gun carrier was delivered to the WD in April 1926 (Reg. No. ML 9901), sold in 1933.

Tractor, Half-Track, Artillery (Morris-Commercial/Roadless) After several exp. Morris tractors had been tested, all with Roadless Traction bogies, a batch of this model was made in 1929/30 for military service in India. The tracks were rubber-jointed, for longer life. (Photo IWM STT 1653).

Tractor, Half-Track, Artillery (Clayton 110 HP) Produced in 1917/18 by Clayton & Shuttleworth Ltd, Lincoln. Engine (6-cyl., $5\frac{3}{4}'' \times 6\frac{1}{2}''$) by National Gas Engine Co. Ltd. 3F1R transmission. Steering by spring-loaded front wheel and differential brakes. Max. speed 5 mph. Overall width 7'1". Weight 13 tons approx.

Tractor, Half-Track, Artillery (Ruston 75 HP) With the exception of the name on the radiator this tractor was virtually identical to the American Holt 75 HP Caterpillar. It was made by Ruston, Proctor & Co. Ltd of Lincoln; the engine (4-cyl., $7\frac{1}{2}'' \times 8''$) by Perkins Engineers Ltd of Peterborough.

GB

TRUCKS and TRACTORS
FULL-TRACK

The first tracked vehicle taken into British Army service was a 1905 Hornsby which in 1907 was converted from a wheeled into a full-track tractor. It was followed in 1909/10 by the Hornsby 'Little Caterpillar', which still exists (in the RAC Tank Museum). In the 1920s a large number of full-track military vehicles appeared. Many remained in the experimental stage but some, like the Dragons, were produced in quantity. The Dragon (from 'drag gun'), an open-top gun tractor, was approved in January, 1923, as the initial means of mechanizing the RA. Dragons Mk I and II were issued to Field Brigades, while the Mk III and its developments were intended for use with medium artillery. There was more progress during the 1930s, when track, suspension and steering designs reached a higher standard of quality, reliability and life. Carden-Loyd vehicles of various types were particularly successful, also in AFV roles.

Truck, 30-cwt, Full-Track, Cargo (Vickers-Armstrong) 6-cyl., 50 bhp, 5F1R, 12'3½" × 6'2" × 6'2", 49¼ cwt. Known as (Vickers-) Carden-Loyd Tractor-Truck this vehicle was built to the design of Carden-Loyd and offered commercially by Vickers-Armstrongs Ltd. 1930. (IWM photo STT1712).

Truck, 3-ton, Full-Track, Cargo (Peerless/Vickers 'Caterlorry') Designated B5E1, this was a 1914–18 Peerless, converted by Vickers in 1924. It weighed 112 cwt and with a 3-ton load exerted a ground pressure of 11 lb/sq. in. The vehicle was designed for a life of 300 miles.

Truck, Heavy, Full-Track, Cargo (DMT) This unusual vehicle, known as Roller Track Wagon, was constructed by The Humphrey Pump Co. Ltd and rode on three inside-frame track bogies (one at front, two at rear). All-steel body had top-hinged tailboard.

Tractor, Full-Track, Artillery (Hornsby 60 HP 'Little Caterpillar') Ordered by MT Committee, War Office, in 1909. Delivered May 1910. Designed to draw 60-pdr gun. Weight 169 cwt. Second chain-track tractor of British Army. Paraffin engine later replaced by 120 HP petrol unit.

Tractor, Full-Track, Artillery (OF Dragon Mk I) Leyland 4-cyl., 60 HP, 16'7" × 9'3" × 7'0", 9 tons. The first Dragons appeared in 1922 and were built by the Ordnance Factory at Woolwich. They towed the 18-pdr field gun and limber. Crew 10+1. Max. road speed 12 mph. Dragon Mk II (with V8 engine) followed in 1925

Tractor, Full-Track, Artillery (Armstrong Whitworth Dragon Mk III A) Armstrong Siddeley V-8-cyl., 82 HP, 16'5" × 7'9" × 6'2", 10 tons. One of the further developments in the line of Dragons. Air-cooled 7.8-litre (4" × 4¾") engine, as used in Vickers Medium Tanks, was originally (1915) designed as an aircraft engine.

Tractor, Full-Track, Artillery (ROF Light Dragon Mk IID) Meadows 6-cyl., 59 bhp, Wilson epicyclic gearbox, 12'10" × 6'9" × 6'3", 4.2 tons approx. Produced during 1933–35, mainly by the Royal Ordnance Factory, for towing 18-pdr field gun, 3.7 howitzer, etc. Horstmann suspension with four bogie wheels. Speed 30 mph.

GB

ARMOURED VEHICLES

This section shows a random selection of AFVs (Armoured Fighting Vehicles), except tanks and similar machines. Armoured cars have always been used by the British in considerable quantities. Well-known makes of World War I and long afterwards were the Rolls-Royce and the Lanchester but many other car and truck chassis were fitted with armoured bodywork. Certain types were built for export, e.g. the Austin and Sheffield-Simplex for Russia. During the late 1920s more sophisticated six-wheeled designs appeared, followed by 4×4 types in the late 1930s and World War II. The latter were pioneered by Nicholas Straussler in 1933–35. Also developed during the 1930s were various types of Carden-Loyd/Vickers-Armstrong light tracked vehicles which culminated in the famous Universal Carrier.

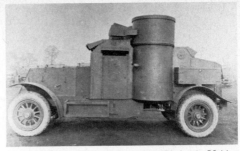

Armoured Car, 4×2 (Austin 30 HP) 4-cyl., 50 bhp, 4F1R, wb 11'6", 16'0" × 6'7" × 9'4", 4½ tons approx. Duplicate steering controls at rear. Austin during 1914–18 built 480 armoured cars on modified 30 HP Colonial chassis. About 400 were supplied to Russia. Shown is 1918 pattern of British Army.

Armoured Car, 4×2 (Lanchester 38 HP) 6-cyl., 65 bhp, 3F1R (epicyclic), wb 11'7", 16'0" × 6'4" × 7'6", 4.8 tons approx. Crew 4. Driver's feet alongside engine. One Vickers MG. Cantilever springs, dual at rear. 4800-cc OHV engine. Oil-cooled multi-disc brake on transmission. Shown in Russia.

Armoured Car, 4×2 (Wolseley CP) 4-cyl., 16/20 HP, 4F1R, wb 12'0". Tyres, solid 1000 × 100, dual rear. Vickers turreted body of 0.196" armour plate. 3 supplied in 1915. Wolseley also produced 16 armoured cars on Peerless and 48 on Pierce-Arrow imported American chassis.

Armoured Car, 4×2 (Rolls-Royce 40/50 HP) 6-cyl., 80 bhp approx., 4F1R, wb 11'11½", 16'7"×6'3"×7'8", 4 tons approx. Crew 3. Based on Silver Ghost chassis with dual rear tyres. Shown is a 1920 pattern model, which was an improved edition of the original 1914 pattern.

Armoured Car, 4×2 (Crossley 1923) Produced for service in India and South Africa these 5-ton Crossleys had a 6-cyl. 50-bhp engine and solid tyres. When the vehicles were worn out in the late 1930s the armoured hulls were transferred to commercial Chevrolet and Ford V8 truck chassis.

Armoured Car, 4×2 (Morris-Commercial CS9/LAC) 6-cyl., 98 bhp, 4F1R, wb 9'8½", 15'7"×6'8"×7'3", 4¼ tons approx. Armoured body by Royal Ordnance Factory on modified 15-cwt truck chassis. Several variants, including Light Reconnaissance Car (prototype shown). 100 built, 1937/38.

Armoured Car, 4×4 (Straussler AC II) Nicholas Straussler, consulting engineer of Hungarian origin, designed his first rear-engined 4×4 IFS/IRS A/C in 1933. It was built by Manfred Weiss in Hungary (*qv*). The second (shown) was tested by the RAF and led to orders from the British, Dutch and Portuguese Govts.

Armoured Car, 6×4, AA (Crossley 20/60 HP) 4-cyl., 26 HP, 4F1R × 2, 15'3" × 6'2" × 8'6" approx., 4½ tons approx. Built in 1928 but turret later replaced by twin Vickers .50" MGs (shown), intended for anti-aircraft defence of mobile formations. One of several Vickers-bodied Crossley A/Cs.

Armoured Car, 6×4 (Lanchester 40 HP) 6-cyl., 88 bhp, 3F1R (epicyclic) × 2, 20'0" × 6'7" × 9'3" approx., 7½ tons approx. Known as 'Armoured Car, Lanchester, Mk I', this was the first British cross-country type. It featured geared turret traverse and telescope sights for the three MGs. 1927/28.

Carrier, Half-Track, AA (Burford/Kégresse) 4-cyl., 28 HP, 4F1R × 2. 3¾ tons approx. 30-cwt chassis, armoured by Vickers, 1928. Designated 'Carrier, MG, Armoured, 30 cwt, Burford/Kégresse'. Twin .303" Vickers MGs. Speed on/off roads 20/10 mph. Gradability 30°.

Tankette, Half-Track (Morris-Martel/Roadless) This vehicle was also known as the Morris-Martel One-Man Tank and was mechanically similar to the Morris Tractor (*qv*), except for the radiator. Like the Kégresse-equipped Crossley-Martel it had tail wheels for steering. Eight built in 1926.

Armoured Car, Wheel-cum-Track (Vickers/Wolseley) Wolseley 6-cyl. engine, 16'8" × 7'3" × 7'0", 7½ tons approx. During 1926–29 Vickers produced several experimental tracked vehicles which for use on roads could be raised on wheels. Max. speed of this specimen: 15 mph on tracks, 45 mph on wheels.

Carrier, Full-Track, Mk VI (Vickers-Armstrong/Carden-Loyd) Ford T, 22.5 bhp, 2F1R (epicyclic) with aux. low gear, 8'1" × 5'7", 1½ tons approx. Conceived by Carden-Loyd Tractors Ltd which was taken over in 1928 by Vickers-Armstrongs. Shown is one of several variants built during 1928–30, with 1 Vickers MG.

Carrier, Full-Track, MG (Vickers-Armstrong) Ford V-8-cyl., 65 bhp, 4F1R, 12'0" × 6'11", 3 tons approx. 1 Vickers 40-mm MG. Developed in 1934 this was the prototype for a long line of Machine Gun and other carriers. Like contemporary Light Tanks and Light Dragons it had Horstmann coil spring suspension.

Carrier, Full-Track, Bren (Thornycroft) The first Bren Carrier was a conversion of the MG Carrier (No. 2, Mk I) by Thornycroft in 1938 and subsequently made in large numbers by this and several other firms. The Bren was a light machine gun of Czech design, adopted by the British Army at that time.

GB

MISCELLANEOUS VEHICLES

In addition to cars, trucks, tractors, armoured vehicles and other types there are military vehicles which are of a specialist nature as well as experimental vehicles which are best described as 'Miscellaneous'. The Royal Air Force, in particular, operated several types of special equipment and special purpose vehicles such as aircraft starters, torpedo carriers, etc. This, of course, still holds true today. Experimental vehicles like the Armstrong Siddeley articulated multi-wheelers, which were developed from the Italian Pavesi design, have been variously described as reconnaissance cars and tractors but did not reach series production stage for either purpose. These, as well as wheel-cum-track vehicles, served mainly as test-beds for the evaluation of various propulsion systems.

Truck, Light, 4×2, Aircraft Starter (Crossley 15/20 HP) The RAF used Airco and Hicks aircraft engine starting equipment, mounted on Crossley and Ford chassis. Shown is an Airco of the mid-1920s. The starter shaft was driven by the vehicle engine after manual engagement with the aircraft's propeller.

Truck, 3-ton, 4×2, Torpedo Carrier (SD Freighter) Delivered to the RAF in 1934 was this special platform version of the SD Freighter municipal chassis with a hydraulically-operated torpedo-handling crane. The complete vehicle was made by Shelvoke & Drewry Ltd of Letchworth, Herts.

Tanker, 3×2, Aircraft Refueller (Thompson) The Thompson three-wheeled tanker for refuelling aircraft was introduced in 1935. It was powered by a Ford engine and operated by the RAF Civil Training College at Desford. Later models for the RAF were of increased capacity.

Vehicle, Multi-Wheeled, 8×8, B10E1 (Armstrong Siddeley)
Genet radial 5-cyl., 75 HP, 4F1R, 15'5"×7'8½"×5'4", 84½ cwt. Air-cooled 4"×4" engine. Tyres (16) 35×5. Steering by articulation. Max. safe road speed 20 mph. Two 6" transmission brakes. Development of Pavesi 4×4 design. 1929.

Vehicle, Multi-Wheeled, 8×8, B10E2 (Armstrong Siddeley)
4-cyl., 45 bhp, 4F1R (epicyclic) with aux. low gear. Air-cooled 4"×4¾" OHV engine. Tyres 40×9, single all round. Frame divided in centre with steering rack and pinion on each unit. Tested at Farnborough in April, 1930.

Car, Full-Track (Overland/Johnson) Before founding Roadless Traction Ltd, Lt-Col. P. Johnson, a member of the Tank Design Department, designed bogie systems for high-speed track-laying vehicles. This test rig of c. 1922, known as Light D Tank, featured cable suspension. (IWM photo Q14668).

Car, Wheel-cum-Track (Wolseley/Vickers) In 1926 Vickers converted a Wolseley touring car to run on wheels or tracks at will. The track assemblies could be raised and the wheels lowered for normal use and vice versa for off-road operation. Although attractive in theory the idea proved impracticable.

HUNGARY

Before 1918 Hungary was part of the Austro-Hungarian Empire and most of the vehicles used then are covered under Austria and Czechoslovakia, these being the territories where the major vehicle manufacturers were situated. Under the Treaty of Trianon, Hungary was only permitted limited military forces but this treaty was gradually renounced and by the late 1930s various types of military vehicles were in production. These were supplemented by imports. Main producers of wheeled vehicles were Magyar Vagonés Gépgyár (Hungarian Railway Carriage and Machine Works) at Györ which, under the marque name of Rába pronounced Rava) licence-produced Praga, Austro-Fiat and Krupp vehicles, and Manfred Weiss RT of Budapest (marque names MW and Csepel) which also supplied a variety of vehicles (incl. licence-built Pavesi tractors). The latter company produced several Straussler-designed prototypes which were later made in quantity in Britain (trucks, trailers, A/Cs). During World War I Büssing and Fiat trucks were licence-produced by Danubia of Budapest. Another early military truck was the Marta, made in Arad.

Ambulance, 4-stretcher, 4×2 (Rába Grand) 4-cyl., 35–46 bhp, 4F1R, wb 3.15 m, track 1.35 m, weight 1600 kg. Tyres 895 × 135. Speed 70 km/h. 3817-cc (90 × 150 mm) SV engine. Chassis produced under Praga licence, c. 1914. Also with command car bodywork. Another car producer was MAG of Budapest.

Truck, 2-ton, 4×2, Cargo (Rába L) 4-cyl., 35 bhp, 4F1R, wb 3.20 m. Shaft-drive Praga Model L truck, produced under licence by the Hungarian Railway Carriage and Machine Works of Györ. Used in light motor transport columns of Austro-Hungarian Army during World War I. Also known as PM2.

Truck, 3-ton, 4×2, Cargo (Rába V) 4-cyl., 35 bhp, 4F1R, wb 3.30 m, track 1.36 m, weight 3600 kg. Front wheels 840 × 120, rear wheels (dual) 1055 × 150 mm. Max. speed 16 km/h. 6840-cc (110 × 180 mm) SV engine. Chain final drive. Produced in World War I under Praga licence. Also known as PS3.

Truck, 3½-ton, 4×2, Cargo (Rába Super) Produced from 1936 until 1944 under Austrian Austro-Fiat licence. 3.6-litre petrol engine, 65 bhp at 2700 rpm. Wheelbase 3.65 m. Company produced Austro-Fiat designs from 1925, starting with 1½-ton truck. From 1936 until 1942 Krupp trucks (and bus chassis) were also made.

Truck, 3-ton, 4×4, Chassis (Straussler/Manfred Weiss) Prototype four-wheel drive truck, designed by Nicholas Straussler and built by Manfred Weiss RT of Budapest. Two Ford V8 engines, side by side, each driving one axle through common transfer case. 1937. Series production by Garner in England (G3, 1938/39).

Truck, 3-ton, 6×4, Cargo (Manfred Weiss Csepel) Typical of various central European cross-country trucks of the mid-1930s. Beam-type front axle, independent rear suspension with swing axles, both with semi-elliptical leaf springs (inverted centrally-pivoted at rear). Spare wheels on idler hubs.

Truck, 1½-ton, 6×4, Cargo (Rába Botond 38M) 4-cyl., 65 bhp, 5F1R, wb 3.03 (BC 1.12) m, 5.75×2.20×2.55 m, 4000 kg. Payload on roads 2000 kg. Tyres 210-20. 3.77-litre OHV engine. Max. speed 60 km/h. IRS with wishbones, balancer beams and coil springs. Track front/rear 1.65/1.76 m. Design Winkler. 1938.

237

Tractor, 4×4 (Tlaskal/Rába) Experimental tractor with four-wheel drive and steering, 4.50-m wb and 40–60-bhp engine, built in 1904 to the design of Capt. Ludwig von Tlaskal-Hochwall. It towed a road train of five self-tracking four-wheeled trailers, all with iron-shod wheels.

Armoured Car, 4×2 (Romfell) Built in Budapest in World War I by Captain Romanic and Ltd.-Col. Fellner (hence Romfell). Engine 75 bhp. Weight 7 tons. Crew 4. Armament one MG in revolving turret. After 1918 another 35 to 40 were produced. A/C on right is an Austin, captured on Russian front.

Armoured Car, 4×4 (Straussler/Manfred Weiss AC I) This advanced rear-engined design with IFS and IRS became the prototype for British A/Cs of World War II. It was designed by the Hungarian-born consulting engineer Straussler in London in 1933 and the pilot chassis was manufactured by Manfred Weiss.

Armoured Car, 4×4 (Straussler 39M Csaba) Following some pilot models for Britain, Hungary in 1938/39 produced Straussler's armoured car for its own use. It weighed about 5900 kg and measured 4.52×2.20×2.27 m. In Britain a somewhat different model was produced by Alvis-Straussler Ltd.

ITALY

Like most countries with a national motor industry, Italy commenced military motorization at an early date. The first Army-owned motor vehicle was a 1902 Fiat 12 HP, bought in 1903. It was followed by several more in subsequent years. During the Italo-Turkish war in Libya in 1911–12 a large quantity of Fiat trucks was used, together with a number of other vehicles. During the First World War, which Italy joined on the side of the Western Allies in 1915, the Italian industry produced large numbers of vehicles for the Italian forces as well as for several other Governments. In fact, Italy was the only allied nation to be almost entirely self-sufficing in the matter of motor vehicles. Unlike England, France, Belgium, Russia, etc., the Italians imported hardly any foreign vehicles for war purposes. This state of affairs was not primarily due to the fact that Italy had an important motor industry, but was brought about by a plan which left the industry relatively undisturbed when war broke out. It has been suggested that France could have been as independent if the Allies had premeditated the war as much as Germany would like the world to believe. Italy was in the fortunate position of coming into the war at her own time and having the experience of her Allies as a guide. Thus the factories were speeded up and left at their ordinary work, while all her neighbours had to go through a revolution involving the stripping of their factories to meet the army call for soldiers and a re-organization of the plant to meet the new condition.

Compared with Britain, which had in her Army about every known make of vehicle, or with France which employed practically all her home makes and a large number of imported vehicles, the situation in Italy was ideal. All the factories were in Turin and Milan, which towns were very conveniently situated for supplying the Armies in the field. From Turin to Gorizia or Montfalcone—extreme points of the front and advanced deep into the enemy's territory—was not more than a 20-hour journey. The Army had the whole of its motor industry right at its back, thus securing prompt supply of vehicles and spares, simplifying the keeping of stocks to a maximum degree.

Fiat headed the list of Italian manufacturers with probably 50 to 60 per cent of the total output of motor vehicles. It was claimed at the time that Fiat, by 1917, had produced more military vehicles than any other firm in the world. With only a few other manufacturers producing trucks it was possible to standardize to a high degree and for the various firms to produce vehicles to a common general layout, as will be seen in the following pages. By October 1918 just over 36,000 vehicles were in military service, some 25,000 of which were trucks of various types. Obviously, few new vehicles were purchased during the 1920s, apart from certain special types. During the 1930s the situation changed drastically. Like Germany, Italy carried out a big re-armament programme and war started again in 1935 when the Italians invaded Ethiopia.

Owing to this situation the coverage in this book overlaps the Italian section in *The Observer's Fighting Vehicles Directory—World War II*, which includes a number of vehicles produced during the 1930s. The two books are therefore supplementary in regard to vehicles of the Italian forces in World War II.

Examples of many of the vehicles shown are exhibited in the Museo Storica della Motorizzazione Militare, Viale dell'Esercito 86, Rome.

Fiat trucks in transport depot in Tripoli, 1912.

Fiat 20B tractor with two heavy artillery pieces, 1917.

Who's Who in the Italian Automotive Industry

The following manufacturers were the main suppliers of motor vehicles during the 1920s and 1930s.

Alfa Romeo	Alfa Romeo SpA, Milano.
Ansaldo	Ansaldo-Fossati, Genova-Sestri.
Bianchi	Edoardo Bianchi-Moto Meccanica SpA, Milano.
Breda	Soc. Ernesto Breda, Milano.
Ceirano*	SA Giovanni Ceirano, Torino.
Fiat	Fiat SpA, Torino (earlier known as F.I.A.T.).
Isotta Fraschini	Fabbrica Automobili Isotta Fraschini, Milano.
Lancia	Lancia & Co., Fabbrica Automobili-Torino-SpA.
OM	Officine Meccaniche SpA, Brescia & Milano.
Pavesi	SA La Motomeccanica, Milano.
Spa*	Soc. Ligure Piemontese Automobili, Torino.

* part of Fiat-Spa-Ceirano consortium.

Between the two World Wars the Italian authorities acquired relatively few foreign vehicles. In 1932 a British Atkinson 4-ton 4 × 2 forward-control chassis/cab (the makers' second diesel truck) was purchased, probably for evaluation, and some American Chevrolet, Dodge and Ford trucks were also obtained. In Abyssinia (Ethiopia) in 1935/36 quantities of American, British and other trucks were used, albeit mainly civilian-owned vehicles, working under military contracts. In the mid-1930s specifications were drawn up for medium and heavy trucks to be standardized for military use. These vehicles were known as *Autocarro Unificato Medio* and *Pesante* (unified trucks, medium and heavy) and were subsequently produced by Alfa Romeo, Bianchi, Fiat, Isotta Fraschini, Lancia and OM. The requirements were as follows:

Autocarro Unificato	Medio (Medium)		Pesante (Heavy)
Gross vehicle weight	6500	kg.	12,000
Payload, not less than	3000	kg.	6000
Engine type	Petrol or Diesel		Diesel
Overall vehicle width, maximum	2340	mm	2350
Body, internal dimensions	4000 × 2000 mm		4750 × 2200
Ground clearance, minimum	200	mm	200
Turning radius, minimum	7000	mm	7000
Max. road speed not less than	60	km/h	45*

* 38 km/h with 12-ton trailer.

Aircraft crash/rescue vehicle on Ceirano 47C chassis, one of various types of special Air Force vehicles used in East Africa in the late 'thirties.

ITALY

MOTORCYCLES and MOTORTRICYCLES

During the first World War the Italian forces used several makes of motorcycles, with and without sidecar. In 1918 the total number of units in service was 6420. Well-known were the *Motociclo* Bianchi *tipo A* of 1914, a belt-driven single-cylinder solo machine, and the Frera which was supplied in single- and V-twin-cylinder form. Motortricycles were used in fair numbers during the 1930s and in World War II. They were made mainly by Benelli and Guzzi. Gilera was another important manufacturer of solo machines and sidecar combinations. Motorcycles and motortricycles were often adapted for special purposes, particularly gun mounts. Some typical examples are shown.

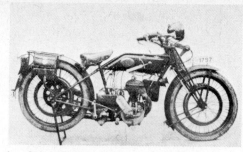

Motorcycle, Solo (Frera) Rigid-frame 250-cc *Motociclo*, superseded by sprung-frame 500-cc model in 1934. During World War I SA Frera of Tradate supplied solo and sidecar machines with 4-bhp single- and 8/10-bhp V-twin-cyl. engines. These were basically commercial machines.

Motorcycle, Solo (Guzzi) 1-cyl., 13.2 bhp, 3F, wb 1.52 m, 2.23 × 0.92 × 1.09 m, 196 kg (with pillion 202 kg). Tyres 3.50-19. Speed 53 km/h. Known as Moto Guzzi *Motociclo Militare 32*. machine gun (Moto Guzzi GT17 *Motociclo porta mitraglia*). 1930s.

Motortricycle, 3 × 2, Cargo (Guzzi) 1-cyl., 13.2 bhp, 3F, wb 1.81 m, track 1.09 m, 338 kg. Payload 300 kg. Tyres 3.50-19. Speed 53 km/h. Known as Moto Guzzi *Motociclo Militare 32*. Model GT17 horizontal engine, developing 13.2 bhp at 4000 rpm. 1930s.

ITALY

CARS and FIELD CARS

Makes and Models: Alfa (1915–18). Alfa Romeo RL (1923), 6C2500 (1939), etc. Bianchi (1915–18), S4, S6M, S9 (1936–39), etc. Diatto (1915–18). Fiat 12 HP (1903), Mod. 4 (1910–12), 3A (1912–15), 70 (1915–20), 501 (1919–26), 509 (1925–27), 518C/Ardita 2000 (1933–37), 518C and L Col. (1933–38), 508 Mil. (1934–36), 508 Mil. Col. (1934–37), 508C Col. (1937–39), 508C Mil./1100 (1939–45), 508C Mil. Col. (1939–43), 2800C Mil. Col. (1939–41), etc. Itala (1915–18). Isotta Fraschini (1915–18). Lancia Col. (1914–18), Aprilia *Torpedo Mil.* (1937–40), etc. Nazzaro (1915–18). Pavesi P4 (*c.* 1925). Rapid (1915–18). Scat (1915–18). Spa (1915–18). Züst (1914), etc.

General Data: Prior to World War II the Italian car industry was almost entirely concentrated in two towns, namely Turin and Milan. Situated in Turin were Fíat (originally FIAT, which stood for Fabbrica Italiana d'Automobili, Torino), Lancia, Itala, Spa (SPA), Scat (SCAT), Diatto, Nazzaro and Rapid; in Milan were Bianchi, Isotta Fraschini, Züst and Alfa (ALFA).

All these firms supplied four-door soft-top touring cars for military use by the Italian Army (as well as several others, including the US forces in France) in World War I and they are listed in approximate order of importance, i.e. volume of production. At the time of the Armistice the Italian forces were operating just under 2500 cars. ALFA became Alfa Romeo after the end of the war and with the exception of Fiat, Alfa Romeo, Lancia and Bianchi all the other makes have now disappeared. During the 1930s Alfa Romeo, Bianchi, Lancia and Fiat developed 'Colonial' cars; these were four-door soft-top models based on components of their regular cars but modified for arduous use on the rough roads of Libya and other African territories. Of some there were special military editions as well, with purpose-built bodywork (see also *The Observer's Fighting Vehicles Directory—World War II*).

Vehicle shown (typical): Car, 5-seater, 4 × 2 (Fiat 3A)

Technical Data:
Engine: Fiat 53A, 4-cylinder, I-L-W-F, 4398 cc (100 × 140 mm), 35 bhp at 1700 rpm.
Transmission: 4F1R. Shaft-drive.
Brakes: mechanical on transmission (hand brake on rear wheels).
Tyres: 820 × 180.
Wheelbase: 3.14 m. Track: 1.40 m.
Overall length: 4.38 m.
Weight: 1600 kg (gross).
Note: produced during 1912–15. Maximum speed 75–85 km/h. 12-Volt electrical equipment. Magneto ignition.

R

Car, 5-seater, 4×2 (Fiat 70) 4-cyl., 18 bhp, 4F1R, wb 2.71 m, track 1.25 m, length 3.82 m. Model 40 2001-cc (70×130 mm) engine with magneto ignition. Multi-disc clutch. Sankey wheels with 765×105 tyres. Speed 70–75 km/h. 1915–20.

Car, 5-seater, 4×2 (Lancia Colonial) Used by the Italian Army in Libya. Made in 1914 by the Fabbrica Automobili Lancia & Co. SA, of Turin, which had been founded in 1908. Built to heavy-duty specification and with complete electrical equipment.

Car, 2-seater, 4×2 (Fiat 508 Mil.) 4-cyl., 36 bhp, 4F1R, wb 2.30 m, 3.73×1.40×1.37 m, 820 kg (gross). Tyres 4.40-19. Derived from 508S civilian Spider which had 30 bhp engine and 3F1R trans. Used by field commanders of small motorized units. 1934–36.

Car, 4-seater, 4×2 (Fiat 508 Mil. Col.) 4-cyl., 20 bhp, 4F1R, wb 2.30 m, 3.14×1.40 m, 1080 kg (gross). Tyres 4.40-19. Speed 72 km/h. Model 108 995-cc (65×75 mm) engine. Track, front 1.18, rear 1.20 m. Hydraulic brakes. Handbrake on transmission. 1934–37.

Car, 5-seater, 4×2 (Fiat 518C/Ardita 2000) 4-cyl., 45 bhp, 4F1R, wb 2.70 m, 4.24×1.67 m, 1650 kg (gross). Tyres 5.00 or 5.50-20. Speed 85 km/h. Model 118A 1944-cc (82×92 mm) engine. Supplied for colonial and military use, 1933–37. Also 3.00-m wb *Berlina* (saloon).

Car, 4-seater, 4×2 (Fiat 508C Col.) 4-cyl., 32 bhp, 4F1R, wb 2.43 m, 4.05×1.45×1.65 m, 1235 kg (gross). Tyres 5.00-18. Speed 100 km/h. Model 108C 1089-cc (68×75 mm) OHV engine. Derived from contemporary Fiat 1100 Balilla. 1937–39. Note RHD and twin spare tyres.

Car, 4-seater, 4×2 (Bianchi S6M) Produced in 1939, exclusively for the armed forces, with open (shown) and closed bodywork. The engine was a Model M6 6-cyl. of 2179 cc (68×100 mm) and 52 bhp at 4000 rpm, with 4F1R gearbox. Tyres 5.50-16. Model S4 (4-cyl.) had similar body.

Car, 5-seater, 4×2 (Lancia Aprilia) V-4-cyl., 48 bhp, 4F1R, wb 2.85 m. Weight 940 kg. Tyre size 13·45. *Torpedo Militare Coloniale* derivation of late-1930s Aprilia car. 1486-cc (74.6×85 mm) OHV engine. Max. speed 113 km/h. IFS. RHD. Also made with truck bodywork and hard-top cab. 1939.

ITALY

AMBULANCES

The majority of Italian Army ambulances in World War I were based on Fiat chassis, notably the Model 15 ter. Of the 1918 total of 945 units, 710 were on the 15 ter chassis. The standard ambulance body of that period was of composite wood and metal construction with double rear doors and four louvre windows on each side. They were built to carry six stretchers in two vertical rows of three each, resulting in rather tall bodywork and making the vehicles top-heavy. The stretchers were carried on spring-mounted hooks attached to the body sides and canvas slings hanging from the roof. Wooden benches were used when sitting patients had to be carried. The Italian Red Cross Society, all members of which were militarized, used a large variety of ambulances, including converted touring cars.

Truck, 1½-ton, 4×2, Ambulance (Fiat 15 bis Libia) 4-cyl., 16 bhp, 4F1R, wb 3.01 m, length 4.43 m, GVW 3200 kg approx. Tyres 880×120. Track 1.40 m. Speed 40–45 km/h. 15–20 HP 3052-cc L-head engine with max. output of 16 bhp at 1500 rpm. Built for use in Libya, 1911–12.

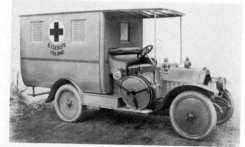

Truck, 1-ton, 4×2, Ambulance (Fiat 2F) 4-cyl., 20 bhp, 4F1R, wb 2.84 m, length 4.08 m, GVW 2500 kg approx. Tyres 820×120. Track 1.40 m. Speed 55 km/h. 2813-cc (80×140 mm) Model 52B engine with magneto ignition. Model 2F chassis was produced during 1911–22.

Truck, 1½-ton, 4×2, Ambulance (Fiat 15 ter) 4-cyl., 40 bhp, 4F1R, wb 3.07 m, length 5.00 m approx. Tyres 880×120. Model 53A L-head engine of 4398 cc (100×140 mm). Standard Italian type of World War I (see General Data). Note fuel tank, protected by wooden slats, at rear.

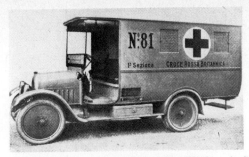

Truck, 1½-ton, 4×2, Ambulance (Fiat 70) One of a fleet of Fiat ambulances operated on the Italian front by the British Red Cross, who also used Buicks (C4), GMCs, a few Fords (T) and a variety of converted touring cars of British, French and German origins.

Truck, 1-ton, 4×2, Ambulance (Spa 25C/10) Italian Army (*Regio Esercito Italiano*) motor ambulance (*Autoambulanza*) of the mid-1920s, based on a Spa light truck chassis. Like the World War I type ambulances and light trucks it had dual tyres at rear and carried two spares.

Truck, 4½-ton, 4×2, Medical (Fiat 18 BLR) 4-cyl., 40 bhp, 4F1R, wb 3.65 m, length 6.00 m approx. Heavy chain-drive truck chassis with van-type body. Cast steel wheels, 750×120 front, 900×140 rear, with solid tyres. Model 64DA 6230-cc side-valve engine. 1914–21.

Truck, 1¼-ton, 4×2, Ambulance (Fiat 618) 4-cyl., 43 bhp, 4F1R, wb 3.05 m, length 4.80 m approx. Regular commercial truck chassis as produced during 1934–37. Model 118A 1944-cc engine. 6.00-18 tyres (6.00-20 on 618 Colonial chassis). Maximum speed about 70 km/h.

ITALY

TRUCKS, LIGHT

Makes and Models: Bianchi (1½-ton, 4×2, 1915–18). Diatto (1-ton, 4×2, 1915–18). Fiat 2F (1-ton, 4×2, 1911–21), 15 bis Libia (1-ton, 1911–12), 15 ter (1½-ton, 4×2, 1911/12 and 1913–22), 508 (0.45-ton, 4×2, 1932–37), 618 and 618 Mil. Col. (1¼-ton, 4×2, 1934–37), 508C Mil. (0.35-ton, 4×2, 1939–45), etc. Itala 17 (1½-ton, 4×2, 1915–18), Lancia 1Z (1½-2-ton, 4×2, 1912–16), Jota, Djota (1½–2-ton, 4×2, 1915–18), Trjota, Tetrajota (4×2, from 1921), etc. Nazzaro (1½-ton, 4×2, 1915). OM Autocarretta 32, 36P, 36M, 36DM, 37 (4×4, from 1932). Spa (2-ton, 4×2, 1912), (1½-ton, 4×2, 1915–18), 25C/10 (1¾-ton, 4×2, 1925), AS37 and TL37 (1-ton, 4×4, 1937–48), CL39 Col. (1-ton, 4×4, 1939–45), etc.

General Data: By far the most widely used Italian light truck of World War I was the Fiat 15 ter, which was also used by other governments, including the French and the British. A lower payload derivative of the 15 ter was the Model 2F which was supplied mainly to the French but also to Britain, where the Royal Navy had a number of them. Of the Fiat 15 ter the British, in 1918, had a total of 386; the majority (294) in Italy itself, the remainder in Salonika (54), Mesopotamia (37) and Russia (1).

Both of these Fiat chassis had an interesting rear axle design; it employed two large T-shaped steel pressings, the rear portion forming the casing for the pinion, crown wheel, differential and drive shafts whilst the forward end contained the propeller shaft and acted as a torque tube. This forward end was hinged to a cross member of the chassis frame.

The Lancias of the same period were also used in considerable numbers, both by the Italians themselves and the Allies. Like the Fiats they had pneumatic tyres (dual rear) and were supplied with a variety of body types, the majority of which were made by Farina (cargo, ambulance, searchlight, etc).

After the war a number of Lancias was armoured and used by the RUC in Ireland, some later being reconverted to load carriers.

During the 1930s some interesting four-wheel drive designs were introduced by OM and Spa; these were used in considerable numbers in East and North Africa in the second World War.

Vehicle shown (typical): Truck, 1½-ton, 4×2, Cargo (Fiat 15 ter)

Technical Data:
Engine: Fiat 53A 4-cylinder, I-L-W-F, 4398 cc (100×140 mm), 40 bhp at 1800 rpm.
Transmission: 4F1R. Shaft-drive.
Brakes: mech. on transmission (hand brake on rear wheels).
Tyres: 880×120 pneumatic, dual rear.
Wheelbase: 3.07 m. Track 1.40 m.
Overall l×w×h: 4.55×1.74×2.65 m.
Weight: 3950 kg (gross).
Note: produced in large numbers during 1913–22. Maximum speed 47 km/h. 12-Volt electrical equipment. Magneto ignition. Various body types. (IWM photo Q 25954). Total number in Italian service in Oct. 1918: 7496.

Truck, 1-ton, 4×2, Cargo (Diatto) Automobili Diatto of 21 Via Frejus, Turin, manufactured various types of trucks for the Italian Government, as well as for export. Shown is a 1-ton shaft-drive truck with pneumatic tyres, dual rear, and 2815-cc (80×140 mm) 4-cyl. engine. c. 1915.

Truck, 1½-ton, 4×2, Cargo (Fiat 15 ter) 4-cyl., 16 bhp, 4F1R, wb 3.01 m, length 4.40 m, GVW 3050 kg approx. Tyres 880×120, pneumatic, dual rear. 3052-cc engine. Maximum speed 40 km/h. Introduced in 1911; superseded in 1913 by an improved version (see preceding page).

Truck, 1½-ton, 4×2, Searchlight (Fiat 15 ter) Long-wheelbase Fiat light truck chassis, fitted with special bodywork to carry 90-cm searchlight, crew and equipment. 80-Volt 100-Amp. Marelli dynamo was driven by the vehicle's engine. 105 of these units were delivered in 1915.

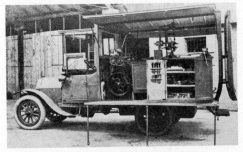

Truck, 1½-ton, 4×2, Workshop (Fiat 15 ter) Another special vehicle based on the Fiat chassis was this *Autofficina Mobile*, a comprehensively equipped workshop for searchlight servicing in the field. Body sides were split horizontally and hinged so as to extend floor and roof area.

Truck, 1½-ton, 4×2, Cargo (Itala 17) Light truck with cranked frame and pneumatic tyres, 4-cyl. monobloc engine of about 30 bhp at 1500 rpm and shaft drive. Max. speed 50 km/h. Kerb weight 3000 kg approx. Produced during World War I by Fabbrica Automobili Itala, Turin. At war's end 427 were in service.

Truck, 1½-ton, 4×2, Cargo (Lancia 1Z) 4-cyl., 35 HP, 4F1R, wb 3.35 m, 4.87×1.67×2.65 m, 2200 kg. Pneumatic tyres, 935×135, dual rear. Acetylene lighting (electric optional). Engine 4942 cc (110×130 mm), max. bhp 70 at 2200 rpm. Maximum payload 2000 kg, speed 60 km/h. 1915.

Truck, 1¾-ton, 4×2, Cargo (Spa 25C/10) 4-cyl., 18 HP, 4F1R, wb 3.50 m. Weight 4300 kg. Tyres 895×135. 2724-cc (85×120 mm) side-valve monobloc engine with Zenith carburettor and high-tension magneto ignition. Full-floating rear axle. Propeller shaft in torque tube. c. 1925.

Truck, 1¼-ton, 4×2, Cargo (Fiat 618 Mil. Col.) 4-cyl., 43 bhp, 4F1R, wb 3.05 m, 4.70×1.94×2.50 m, 3455 kg (gross). Tyres 6.00-20. Model 118A 1944-cc engine with magneto ignition (battery ignition on regular model). Payload 1340 kg. Max. speed 65 km/h. Price 25.250 lire. 1934–37.

Truck, Light, 4×2, Machine Gun (Fiat 508C Mil.) 4-cyl., 28 bhp, 4F1R, wb 2.43 m, 4.00×1.51 m, 1520 kg (gross). Tyres 6.00-16. Model 108C 1089-cc OHV engine. Payload 350 kg. Derived from 1100 Balilla car. Equipped with twin anti-aircraft machine guns. 1939.

Truck, Light, 4×4, Personnel (OM Autocarretta 36P) 4-cyl., 21 bhp, 4F1R, wb 2.00 m, 3.80×1.30 m, 1700 kg approx. Personnel carrier with seats for 10+1. Derived from *Autocarretta 35* truck. 1616-cc air-cooled engine. Lockable differentials. Four-wheel steering. 1936.

Truck, Light, 4×4, Cargo (Spa AS37) 4-cyl., 57 bhp, 5F1R×1, wb 2.50 m, 4.70×2.10 m approx. One of a range of special vehicles produced by the Fiat-controlled Spa concern. Known as *Autocarro Sahariano*. Coil-spring independent suspension. Four-wheel steering. Late model shown.

Truck, Light, 4×4, Cargo (Spa CL39 Col.) 4-cyl., 25 bhp, 5F1R×1, wb 2.30 m, 3.89×1.54×2.30 m, 2790 kg (gross). Payload 1000 kg. Model CLF 1628-cc (72×100 mm) engine. Max. speed 40 km/h. Some had Celerflex semi-pneumatic tyres. Designed in 1939; produced throughout World War II.

ITALY

TRUCKS
MEDIUM and HEAVY

Makes and Models: Bianchi (3½-ton, 4×2, 1915–18). Mediolanum (3- and 4.3-ton, 4×2, from 1935), Miles (3-ton, 4×2, 1939), etc. Breda (heavy six-wheeler, c. 1937). Ceirano 47CM (3-ton, 4×2, c. 1926), 50CM (5-ton, 4×2, c. 1926), etc. Diatto 2½- and 3½-ton, 4×2, 1915–18). Fiat 17 (2-ton, 4×2, 1911–13), 18BC, P (2½-ton, 4×2, 1915–19), 18BL (3½-ton, 4×2, 1914–21), 18 BLR (4½-ton, 4×2, 1914–21), 632N (4-ton, 4×2, 1931–32 and 5-ton, 1933–39), 634N (6-ton, 4×2, 1931–32 and 7½-ton, 1933–39), 633N (5-ton, 4×2, 1933–36), 621N (2½-ton, 4×2, 1934–37), 621PN (3½-ton 6×4, 1934–39), 633NM (5-ton, 4×2, 1935), 633GM (3½-ton, 4×2, 1935–36), 633BM (5-ton, 4×2, 1936–38), 626N and N Col. (3-ton, 4×2, 1939–40), 666N (6-ton, 4×2, 1939–46), etc. Isotta Fraschini (2-ton, 4×2, 1910), (4-ton, 4×2, 1915–18), D70 (3-ton, 4×2, 1937), D80 (5–6½-ton, 4×2, from 1937), etc. Itala (2½-ton, 4×2, 1915–18). Lancia Ro, Ro-Ro, 3Ro, etc. (4×2, up to 6½-ton, from 1932), etc. OM 3B0D (5-ton, 4×2, from c. 1935), etc. Titano (7-ton, 4×2, 1937), etc. Scat (2- to 4-ton, 4×2, 1915–18). Spa (2½- and 3½-ton, 4×2, 1915–18), 36, 38R and 38RA (2½-ton, 4×2, from 1936), Dovunque 32 (2-ton, 6×4, 1933), Dovunque 35 (3-ton, 6×4, from 1936), etc. Züst (2½- to 4-ton, 4×2, 1912–14).

General Data: As in most other classes of Italian vehicles, the Fiat concern was the principal supplier of trucks with payload ratings of over 1500 kg. During the First World War most of the heavy Italian military trucks employed carried 3½ tons. Heavier (5-ton) types had been made before the war but they were found too heavy for work on the Italian front. Another noteworthy point is that the Italians at that time employed hardly any trailers, the roads to their front lines being too narrow and winding.

Practically all 2½- and 3½-ton trucks then had solid-tyred cast-steel wheels, while light trucks were equipped with steel disc wheels and pneumatic tyres.

During the 1930s the Italian Army instituted two classes for military medium and heavy trucks, namely *Autocarro Unificato Medio* and *Pesante* respectively. *Unificato* (unified) indicated that the truck chassis was standardized, albeit not to the extent of two manufacturers building identical models. (See page 239 for more details.)

Vehicle shown (typical): Truck, 2½-ton 4×2, Cargo (Fiat 18BC).

Technical Data:
Engine: Fiat 64BA 4-cylinder, I-L-W-F, 4578 cc, 25 bhp at 1200 rpm.
Transmission: 4F1R. Chain-drive.
Brakes: mech. on transmission (hand brake on rear wheels).
Tyres: solid, front 900×100, rear 1050×140.
Wheelbase: 3.25 m.
Overall length: 4.85 m.
Weight: 5900 kg (gross).
Note: produced during 1915–19 with truck and bus bodywork as *Autocarro Medio* and *Omnibus* or *Autobus* resp. Vehicle was of conventional design and superseded the 1913–15 Model 18A.

Truck, 2-ton, 4×2, Cargo (Fiat 17) 4-cyl., 25 bhp, 4F1R, wb 3.00 m, length 4.52 m, GVW 3950 kg. Built for military use during 1911–13, simultaneously with *Tipo 18* (35-bhp, wb 3.60 m, length 5.52 m, GVW 5750 kg). Engines were Fiat 63 3222-cc and 64 5130-cc respectively.

Truck, 2½-ton, 4×2, Cargo (Fiat 18P) 4-cyl., 40 bhp, 4F1R. Weight 2300 kg. Wheels, front 810×90, rear 900×160. Basically similar to Model 18BL (*qv*) but more suitable for service in mountainous areas. Model 53A 4398-cc engine, developing 40 bhp at 1800 rpm. Nearly 5000 of these trucks were built.

Truck, 3½-ton, 4×2, Van (Fiat 18BL) 4-cyl., 38 bhp, 4F1R, wb 3.65 m, 5.55×1.85 m approx. GVW 7320 kg. Model 64CA 5650-cc engine. Multi-disc clutch. Speed 25 km/h. Van version with hard-top cab shown. Also with open cab and canvas-covered general service/cargo body. 1914–21.

Truck, 4½-ton, 4×2, Cargo (Fiat 18BLR) 4-cyl., 40 bhp, 4F1R, wb 3.65 m, 5.85×1.85 m approx. GVW 8500 kg. Uprated edition of 18BL (18BL*R*, *Reinforzato*) with Model 64DA 6230-cc engine, smaller wheels, wider tyres and revised gear ratios. 1914–21. By Oct. 1918 6315 18BL and BLR trucks were in service.

Truck, 2½-ton, 4×2, Cargo (Diatto) Automobili Diatto of Turin produced two types of medium trucks, the 2½-ton (shown) with shaft drive and a 3½-tonner with enclosed chain drive. The former had an 89×180-mm engine, the latter 94×122 mm, both four-cylinders. Note large spotlight.

Truck, 4-ton, 4×2, Cargo (Isotta Fraschini 16A) This *Autocarro Pesante* weighed 3800 kg and was of conventional contemporary design with enclosed chain final drive, four-cylinder side-valve engine and four-speed gearbox. Radiator guard was fitted on most Italian Army vehicles.

Truck, 3½-ton, 4×2, Workshop (Spa 9000C) Spa produced three types of truck chassis during World War I: 1½-, 2½- and 3½-ton, all with chain drive, cast steel wheels and solid tyres. Shown is comprehensively equipped mobile workshop for self-propelled 102-mm artillery battery (*qv*).

Truck, 2½-ton, 4×2, Cargo (Züst 35) The Fabbrica Automobili Züst of Milan (with offices in Brescia), at one time produced cars, trucks, buses and marine and aero engines. Car production ceased in 1909; truck production was reputedly continued until about 1916. Works were taken over by OM.

Truck, 3-ton, 4×2, Cargo (Bianchi Mediolanum) 4-cyl. diesel, 60 bhp, 4F1R×2, weight 3748 kg. Tyres 34×7. Speed 67 km/h. Model OM65 4961-cc (110×130 mm) engine with 17:1 compression ratio and pre-combustion chambers. Troop carrying capacity 20–25 men. Mid-1930s.

Truck, 5-ton, 4×2, Workshop (Ceirano 50 CM) *Autofficina* or *Carro Officina Mod. 35* on military chassis, which was also used with cargo body and as SP mount for 75-mm AA gun. Engine was 4-cyl. 4712-cc (100×150 mm) petrol, developing 53 bhp at 1750 rpm. Semi-pneumatic tyres. 1935.

Truck, 3½-ton, 4×2, Cargo (Fiat 633GM) 4-cyl., 54 bhp, 4F1R, wb 3.75 m, 6.77×2.20 m, 10,030 kg (gross). Model 350G 6647-cc engine, designed to run on producer gas. Solid tyres, size 195×720.5. Max. speed 34 km/h. Payload capacity 3640 kg. Produced during 1935–36.

Truck, 5-ton, 4×2, Cargo (Fiat 633NM) 4-cyl. diesel, 50 bhp, 4F1R, wb 3.75 m, 6.88×2.20×2.94 m, 10,645 kg (gross). Model 350C 5570-cc engine. Solid tyres, size 195×720.5. Mechanical brakes. Max. speed 29 km/h. Inertia starter unit in front of engine. 1935.

Truck, 7½-ton, 4×2, Cargo (Fiat 634N) 6-cyl. diesel, 80 bhp, 4F1R, wb 4.30 m, 7.43×2.40 m, 14,000 kg (gross). Tyres 42×9. Speed 40 km/h. Model 355C 8355-cc engine. Produced 1933–39. Early models had flat and squarer radiator grille. Cargo or special purpose bodywork.

Truck, 6½-ton, 4×2, Cargo (Lancia Ro NM) 2-cyl. diesel, 64 bhp, 4F1R×2, wb 4.30 m, 7.00×2.34 m, 5500 kg approx. Model 89 3180-cc engine with four opposing pistons, built under Junkers licence; bore and stroke 90×100–150 mm. Model 3Ro had 93-bhp 5-cyl. diesel. Note inertia starter unit. 1933.

Truck, 5-ton, 4×2, Cargo (OM 3B0D) Typical *Autocarro Medio* of the 1930s. Powered by 5702-cc (110×150 mm) 4-cyl. diesel engine, produced under Saurer licence and driving through a 4-speed gearbox with overdrive. Early OM trucks were based on the Swiss Saurer.

Truck, 2½-ton, 4×2, Workshop (Spa 38R) 4-cyl., 55 bhp, 4F1R, wb 3.50 m, 5.78×2.07 m, GVW 5860 kg. *Autofficina mod. 37* on Spa chassis which was in production from 1936 until 1944 and used with a variety of body styles incl. cargo, ambulance, van, etc. 1937.

Truck, 3-ton, 4×2, Cargo (Fiat 626N) 6-cyl. diesel, 70 bhp, 5F1R, wb 3.00 m, 5.79×2.16×2.59 m, 3520 kg. Payload 3140 kg. Tyres 7.50-20. Model 326 5750-cc engine. Air-assisted service brakes. Fiat 626 N Col. (colonial version) was similar but had 8.25-20 tyres. 1939–40.

Truck, 3½-ton, 6×4, Cargo (Fiat 621 PN) 4-cyl. diesel, 53 bhp, 4F1R, wb 3.75 m, 6.14×1.95×2.70 m, 3720 kg. Payload 3640 kg. Tyres 34×7 (dual rear optional). Model 324 4580-cc engine. Vacuum brakes. One of several versions in the 621 Series (621 L, P, R, N, PN, RN, etc.) 1934–39.

Truck, 3-ton, 6×4, Cargo (Spa Dovunque 35) 4-cyl., 60 bhp, 4F1R×2, wb 3.20 (BC 1.00) m, 5.03×2.07×2.90 m, 7400 kg (gross). Tyres 32×6. Hydraulic brakes. Engine Model 18D 4053-cc (*Dovunque 32* of 1933–35 had 46-bhp Model 122B 6-cyl. 2953-cc engine). Produced from 1936.

Truck, 3-ton, 6×4, Cargo (Breda) *Autocarro Campale Pesante* cross-country truck, produced by Soc. Ernesto Breda of Milan. As on the Spa *Dovunque* (='anywhere'—cross-country) the spare wheels were mounted on stub axles, acting as support wheels in rough terrain, as shown.

ITALY

TRACTORS
WHEELED and TRACKED

Makes and Models: Alfa Romeo TM40 (4×4, c. 1939). Breda Tipo 32 (4×4, 1932), 33 (4×4, 1933), 40 (4×4, 1940). Fiat 16 (4×2, 1911–12), 20B (4×2, 1915–20), 30 (4×2, 1914/15), P4 (see Pavesi). OCI 708 CM (full-track 1934/35), OCI A40 (full-track, 1940). Isotta Fraschini (4×2, c. 1915–17). Pavesi P4, P4-100, P4-110, etc. (4×4, various models, from 1914; built by Fiat/Spa from 1926). Pavesi-Tolotti (La Moto-Aratrice) Tipo A (4×2, 1915), Tipo B (4×2, 1916). Soller (4×2, c. 1915–17). Spa TL 37 (4×4, from 1937), TM40 (4×4, 1940), etc.

General Data: In 1909 Fiat designed one of the first, if not the first, Italian tractors intended specifically for artillery towing. It was succeeded in 1915 by two improved models, the *Trattore Tipo 20B* and *Tipo 30*. The latter was of the forward-control type and both carried track devices which could be fitted over the rear wheels. They were used to tow one (or sometimes two) heavy artillery pieces. Other tractors used during World War I, according to a 1917 listing, included various types made by La Moto-Aratrice, Isotta Fraschini and Soller, all produced in Milan. The former were made to the design of Ingg. Pavesi & Tolotti and were also used to tow trains of up to eight drawbar trailers. Ing. Pavesi also designed a revolutionary four-wheel drive articulated tractor, which was initially built by the SA La Motomeccanica. The Fiat-controlled firm of Spa was engaged in its series production, from 1926, and production licences were sold to manufacturers in England (Armstrong-Siddeley), France, Hungary and Sweden. There were light and heavy versions, as well as load-carrying and armoured variants. Production continued until World War II.

During the 1930s several other four-wheel drive tractors were designed and produced by Breda and Spa. Of these, too, there were several variants and like the Pavesi they appeared with solid as well as pneumatic or semi-pneumatic tyres. All were employed extensively during World War II, particularly in North and East Africa. Full-track tractors were relatively rare in the Italian Army. In the invasion of Ethiopia in 1935 an unknown quantity of US Caterpillar tractors was used for artillery towing. Alfa Romeo produced some prototype half-tracks, using French Kégresse track bogies.

The Italian words *trattore* and *trattrice* are synonymous but *trattore* is used mainly for agricultural tractors and *trattrice* for industrial types.

Vehicle shown (typical): Tractor, Heavy, 4×4, Artillery (Breda 32) (*Trattrice Pesante Breda 32*).

Technical Data:
Engine: Breda T5 4-cylinder, I-I-W-F, 8136 cc (120×180 mm), 84 bhp at 1450 rpm.
Transmission: 5F1R.
Brakes: mechanical on transmission.
Tyres: Celerflex semi-pneumatic, 205×980, dual rear.
Wheelbase: 2.65 m.
Overall l×w×h: 5.15×2.08×3.00 m.
Weight: 8450 kg. Payload 3500 kg.
Note: independent front suspension. Separate drive shafts to front wheels. Front-wheel steering. Towing capacity 5000 kg (max. 7200 kg).

257

Tractor, Heavy, 4×4, Engineers (Breda 33) Similar in most respects to Breda 32 artillery tractor; main difference was its longer wheelbase and body. The Breda 33 was used by Army Engineers to carry and tow bridging equipment. Max. speed about 30 km/h. 1933.

Tractor, Heavy, 4×4, Artillery (Breda 40) This tractor superseded the Breda 32 in 1939/40. It had a 110-bhp diesel engine, 50 × 9 pneumatic tyres and other modifications. Note transmission brake just ahead of the rear wheels. Produced also with closed cab and tilt-covered body.

Tractor, Heavy, 4×2, Artillery (Fiat 16) 4-cyl., 45 bhp, 4F1R, wb 3.70 m, length 5.70 m, weight 11,900 kg (gross). Payload 7000 kg. Model 62 6842-cc engine. Wheels, front 900 × 160, rear 1100 × 300. Designed in 1909, produced during 1911–12. Note power winch at rear.

Tractor, Heavy, 4×2, Artillery (Fiat 20B) 4-cyl., 65 bhp, 4F1R, wb 3.52 m, 5.54 × 2.34 × 2.89 m, weight 11,500 kg (gross). Payload 3500 kg. Model 67 10,618-cc engine. Speed 12 km/h. Towing capacity 40 tons. Tracks for rear wheels carried on channels on body sides. Winch at rear. 1915–20.

S

Tractor, Heavy, 4×2, Artillery (Fiat 30) 4-cyl., 60 bhp, 4F1R, wb 3.60 m, length 6.20 m, weight 12,000 kg (gross). Payload 7 tons. Wheels, front 900×160, rear 1100×300. 10.6-litre engine. Fitted with tracks over rear wheels (design Bonagente). Winch at rear 1914/15. By late 1918 the Italians had 1151 artillery tractors.

Tractor, Light, Full-Track, Artillery (Fiat OCI 708 CM) 4-cyl., 30 bhp, 4F1R, 2530 kg. Towing capacity 2200 kg. Model 708C 2520-cc (87×106 mm) petrol engine with high-tension magneto ignition. Mechanical brakes. Metal tracks. Max. speed 16 km/h. Gradability 58%. 1934/35.

Tractor, Medium, 4×2, Artillery (Pavesi-Tolotti A) Built in Milan by La Moto-Aratrice to the design of Ing. Pavesi and Ing. Tolotti. Known as *Trattrice Brevetti Ingg. Pavesi e Tolotti Tipo A*. Front axle was centrally-pivoted. Note track shoes on rear wheels and fuel tank above engine. 1915.

Tractor, Medium, 4×2, Artillery (Pavesi-Tolotti B) Improved edition of *Tipo A*. 1916. 4-cyl. engine, cone clutch, chain final drive. 12 permanently-mounted hinged track shoes on each rear wheel. Speed 10 km/h. 3-ton power winch. One was used by British forces with A-frame crane jib at front.

Tractor, Light, 4×4, Artillery (Pavesi P4) Designed about 1914 by Ing. Pavesi this tractor steered and kept all wheels on the ground by articulation. It had a flat-twin engine and was tested from about 1920 by several armies, incl. the British (shown). Note capstan winch. Also truck and other versions.

Tractor, Heavy, 4×4, Artillery (Pavesi P4-100) Fiat P4/1 4-cyl., 55 bhp, 4F1R, wb 2.42 m, 4.15×2.05×1.65 m, 5800 kg (gross). 4724-cc (100×150 mm) engine. Speed 22 km/h. Dual solid tyres, size 150×1160. Crew 6. Built by Fiat (Spa) during 1936–42; used by Italian and German forces.

Tractor, Light, 4×4, Artillery (Spa TL37) 4-cyl., 52 bhp, 5F1R×1, wb 2.50 m, 4.25×1.80×2.10 m, 3300 kg. Model 18TL 4053-cc (96×140 mm) engine. Pneumatic or semi-pneumatic tyres. Four-wheel steering. One of several body types on this standardized chassis which was introduced in 1936.

Tractor, Medium, 4×4, Artillery (Alfa Romeo TM40) In 1938/39 Alfa Romeo and Fiat/Spa developed prototypes for a new medium tractor (for towed loads of 5000 kg), intended also to supplant the Pavesi (*qv*). The Fiat/Spa, which was accepted and produced until 1948, was similar in general appearance to the Alfa Romeo.

ITALY

COMBAT VEHICLES WHEELED

The first Italian armoured vehicles were based on contemporary commercial chassis, notably Bianchi, Fiat, Isotta Fraschini and Lancia. They were all of similar shape. During the 1920s some experimental armoured cars appeared on the Pavesi P4 chassis and in the following decade the old-established engineering firm of Ansaldo built a quantity of large armoured cars on six-wheeled Fiat chassis. By 1939/40 the standard Italian armoured car was the more sophisticated AB40 (*Autoblinda* 1940), a 7½-ton four-wheeled 4×4 rear-engined model produced by Fiat/Spa. What is claimed to have been the world's first heavy self-propelled gun was a joint effort by Spa and Ansaldo in World War I. It mounted a 102-mm gun. Several more truck-mounted guns (*Autocannone*) appeared in later years on Ceirano, Lancia, Breda and other chassis.

Gun Motor Carriage, 102-mm, 4×2 (Spa/Ansaldo) *Autocannoni da 102 mm* on Spa 9000C chassis with armour-plated front end. Battery was accompanied by ammunition carriers on similar chassis and observation ladder trucks on Lancia 1Z, also with armoured front end. 1915.

Gun Motor Carriage, 75-mm, 4×2 (Ceirano/Viberti/Breda) Ceirano Model 50 CMA/75CK with bodywork by Viberti and gun by Breda. Introduced in the late 1920s and used until and including World War II when they were joined by 90-mm guns on Lancia 4×2 and Breda 6×4 chassis, etc.

Gun Motor Carriage, 3×2 (Guzzi Trialce) The Guzzi (or Moto Guzzi) 'Trialce' was a three-wheeled derivation of the Guzzi 'Alce' motorcycle and was produced with various body types including this machine gun mount with armour shield. It was powered by a horizontal 500-cc 1-cyl. engine.

Armoured Car, 4×2 (Isotta Fraschini 40 HP) This *Automitragliatrice* was built in 1911 and presented to the Italian Army by the Automobile Club of Milan. It carried a crew of six and two MGs. The turret revolved 360°. Front tyres were solid, rear tyres dual pneumatic.

Armoured Car, 4×2 (Lancia 1Z/Ansaldo) Constructed by Ansaldo in Turin on standard Lancia light truck chassis. Used by Italian and other forces until World War II. Weight just under 4 tons. Unusual was the small turret, mounted on top of the main turret. 1915.

Armoured Car, 4×4 (Pavesi P4) About 1925 the SA La Motomeccanica designed some armoured combat vehicles on their famous Pavesi articulated tractor chassis. Shown is a light assault vehicle, offered with various armaments. Also offered was an anti-tank model with 75-mm gun.

Armoured Car, 6×4 (Fiat 611/Ansaldo) From 1934 Ansaldo produced heavy armoured cars on Fiat six-wheeler chassis with 56-bhp 6-cyl. engine. Vehicle illustrated had 47-mm gun in turret. Spare wheels revolved on stub axles to prevent bellying in rough terrain. Crew 5.

JAPAN

Japan's motor industry, now one of the world's most comprehensive, took a long time in getting established. American manufacturers, particularly Ford, General Motors and Chrysler, had seized the opportunity to commence vehicle assembly in the 1920s, much to the dislike of the military authorities who were forced to use much of this American equipment for their wars in Manchuria and China. Gradually this state of affairs changed and by the late 1930s most of the automotive industry was under Japanese military control and engaged in a sizeable armament production programme. Most military transport vehicles retained a distinct American resemblance, the main exceptions being the Type 95 scout car and the standardized Type 94 6×4 cross-country trucks. For a more detailed account of pre-World War II Japanese motor vehicle production see *The Observer's Fighting Vehicles Directory—World War II*.

Motorcycle, with Sidecar, Type 97 (Sankyo) Heavy machine, patterned on the American Harley-Davidson. Introduced in 1937. Engine 1196-cc 45° V-twin of 24 bhp, with 3-speed gearbox. Combination measured 2.70×1.70×1.17 m and weighed about 500 kg. It was also used without sidecar.

Car, 2-seater, 4×4 (Kurogane) Small car, produced in 1937 by Nippon Nainenki Seiko Co. Ltd of Tokyo. It was also made with more streamlined sports type body styling and as a 3-seater Scout Car (Type 95, *qv*). The company, founded in 1935 and called New Era until 1937, specialized in load-carrying three wheelers.

Car, 3-seater, 4×4, Type 95 (Kurogane 'Black Medal') V-2-cyl., 25–33 bhp, 3F1R×1, wb 2.00 m, 3.56×1.50×1.67 (1.57) m, 1000 kg approx. Tyres 6.00-18. Air-cooled 1399-cc engine. Independent front suspension. Mechanical brakes, at rear only. 1935–40. Rikuo and Toyota produced prototypes of similar type.

Car, 5–7-seater, 4×2, Command (Chiyoda HF) Touring car of American pattern, produced in 1935 by TGE (Tokyo Gas & Electric Co.), whose products were named Chiyoda from 1931. In 1937 this company merged with Jidosha Kogyo. TGE also produced 6×4 cars (Chiyoda HS).

Car, 5-seater, 4×2, Command (Toyota ABR) Military version of 1936 Toyota AB touring car (which in turn was derived from the Toyota AA sedan). 6-cyl. 65-bhp engine with 3F1R transmission. Wheelbase 2.85 m. Overall dimensions 4.74×1.73×1.74 m. Weight 1500 kg approx.

Car, 7-seater, 4×4, Command (Mitsubishi Fuso BX33) Heavy all-wheel drive command/staff car, introduced in the early 1930s and manufactured by the Mitsubishi Kobe Dockyard Works at Kobe. A similar model was produced by Jidosha Kogyo, called the Isuzu KIJI, Type 98 (1938).

Car, 7-seater, 6×4, Command, Type 93 (Isuzu K) Six-wheeled command/staff cars were made by Isuzu and other manufacturers (TGE/Chiyoda HS, Sumida K93) and were more numerous than the 4×4 type. In 1932 some special 6×2 Hudsons had been acquired from the US for use in Manchuria.

Truck, 4-ton, 4 × 2, Cargo (OAA C) In 1910 the Japanese Army ordered the Osaka Artillery Arsenal to produce some experimental trucks patterned on an imported French vehicle. Three were made, Types A, B and C. The latter was standardized for military use in 1915.

Truck, 2-ton, 4 × 2, Cargo (DAT 61) Produced in 1928 for the Imperial Army by the DAT Auto Mfg Co. of Osaka. Makes of pre-1920 imported trucks for military use included Benz-Gaggenau (D), Laurin-Klement (CS), Packard (US), Schneider (F), Thornycroft (GB), etc.

Truck, 1½-ton, 4 × 2, Cargo (Rokko) Conventional truck built in 1932 by the Kawasaki Motor Vehicle Company or Sharyo Kaisha of Kawasaki. At this time the majority of vehicles operating in Japan were still of American origin. This changed during the late 1930s.

Truck, 1½-ton, 4 × 2, Cargo (Hiziri) The Ford Motor Co. started vehicle assembly in Yokohama in 1925 and continued until forced to stop in the late 1930s when Japanese companies took over. This truck, patterned on the 1933/34 Ford BB, was produced by the Hiziri Motor Co. in 1937.

Truck, 1½-ton, 4×2, Cargo (Toyota G1) 2630 of these Toyotas were produced from Nov. 1935 to 1937. Its successor, the slightly modified GA, appeared in Sept. 1936. Engine: Model A 6-cyl., 65 bhp. Wheelbase 3.59 m. Overall dimensions 5.95 × 2.19 × 2.22 m. Weight 2470 kg.

Truck, 1½-ton, 4×2, Cargo (Toyota GB) Introduced in 1938 nearly 20,000 of these trucks were made until 1942. They were used mainly as GS load carriers and resembled the American Chevrolet truck in many respects. Engine was an OHV Six of 63 bhp. Picture shows abandoned truck found by US Marines on Guam.

Truck, 1½-ton, 4×2, Cargo (Nissan 80) The Nissan Motor Co. of Yokohama produced American type cars (Graham) and trucks. The Model 80 was made from 1936/37 and had a 3670-cc L-head Six engine of 69 bhp. It was superseded by the normal-control Model 180.

Truck, Light, 4×4, Cargo (Kurogane) derived from the Type 95 Kurogane 'Black Medal' Scout Car (*qv*). Main differences were provision of a closed cab and a pickup box which was similar to that used on the Kurogane Type 1 *Sanrinsha* (motortricycle). Hinged tail gate.

Truck, 2½-ton, 6×2, Cargo (Federal DSW) In the early 1930s Japan bought a number of American Federal six-wheelers. They were unusual in having solid rubber tyres. In the background is a 1935 Chevrolet 4×2 which, together with the 1935 Dodge 4×2, was used in considerable quantity. The latter two were made in Japan.

Truck, 2½-ton, 6×4, Cargo (TGE Chiyoda Q) Following the evaluation of some European 6×4 cross-country trucks (British Scammell and Thornycroft, Czech Tatra) the Japanese industry, in co-operation with the Army, designed 6×4 trucks which were subsequently made by several manufacturers. Shown: Chiyoda, 1932.

Truck, 1½-ton, 6×4, Cargo/Prime Mover, Type 94 (Isuzu) Battered example of the widely used standard 6×4 truck. Type 94A had 70 bhp 6-cyl. petrol engine. Type 94B had 70-bhp 4-cyl. diesel engine. Weight 3400 and 3700 kg resp. Wheelbase 3.35 (BC 1.10) m. Overall dimensions 5.43×1.95×2.25 m. From 1934.

Truck, 1½-ton, 6×4, Gas Supply, Type 94 (Isuzu) The type 94 standard truck appeared in several variations and with many types of bodywork, incl. cargo, tipper, tanker, searchlight, etc. Civilian customers could buy them under an attractive subsidy scheme. There was also a four-wheeled (4×2) variant.

Carriage, Half-Track, Anti-Aircraft Gun (KO-HI) The Japanese Army introduced half-track armoured personnel carriers, artillery tractors and SP gun carriages. One of the latter is shown. This type was powered by a 110-bhp 6-cyl. Ikegai diesel engine and also appeared as a gun tractor.

Tractor, 8-ton, Full-Track, Artillery, Type 92A One of many types of light, medium and heavy crawler tractors used by the Japanese forces. It was powered by a Sumida petrol engine (1932). Later models (Type 92B) had a 105-bhp 6-cyl. diesel engine. Note headlight black-out visors.

Armoured Car, 4×2 (Bedford) British Bedford truck chassis fitted with armoured hulls and engine covers, used by the Japanese forces in Manchukuo, 1935. Other British armoured cars included Vickers/Crossleys (Dowa) of the mid-1920s. The latter had two MGs in a hemispherical turret.

Armoured Car, 6×4, Type 2592 Known as the Naval Type this vehicle was introduced in 1932 and had a 85-bhp 6-cyl. petrol engine. It measured 4.80×1.80×2.30 m and weighed 6200 kg. Other 6×4 A/Cs included Sumidas (from 1930), which could be converted for use on rail roads.

THE NETHERLANDS

Of the few motor vehicles used before 1914 several were locally made, e.g. Eysink motorcycles, Omnia cars and Spyker cars and trucks. Most vehicles used in later years were imported or assembled. Pre-1930 cars included Austin, Daimler and Wolseley from Britain, FN and Minerva from Belgium, Berliet and Renault from France, Fiat from Italy and Ford from the USA.

After 1930 the majority of new cars, as well as trucks, were Fords; from 1932/33 these were made in a new assembly plant near Amsterdam. About 1935 DAF (Van **D**oornes **A**anhangwagen**f**abriek NV, Eindhoven) began to produce various types of military vehicles which were based on low-cost Ford, Chevrolet and other chassis. Most important of these was the Trado four-wheel conversion system of standard rear axles (*qv*), a production licence for which was granted to Delahaye in France. DAF also produced various types of trailers and other military equipment, mainly for domestic use and for the Netherlands East Indies. In World War II most Dutch equipment was destroyed or fell into enemy hands.

Motorcycle, with Sidecar, 3×1 (BSA/Werkspoor) Over 1750 BSAs (600- and 1000-cc) were bought in 1938/39, following earlier batches of BMWs. Many had box-type sidecars, as shown, made by Werkspoor. Other makes used in the 1930s: DKW, FN, Harley-Davidson, etc. 1920s makes: Douglas, Excelsior, Indian, Motosacoche, etc.

Car, 4-seater, 4×2 (Spyker) Dutch staff car of conventional design, made about 1913. In 1902 a Spyker 4×4 car with 6-cyl. engine, 3F1R×2 trans. and central differential had been produced, albeit not for military use. In the 1930s the Army used mainly Ford but also Buick, Hudson, Studebaker.

Car, 4-seater, 4×4, Amphibious (DAF MC139) Citroën 7S 4-cyl., 48 bhp, 3F3R, wb 2.50 m, 3.50×1.70×1.60 m, 1250 kg. Power unit mounted transversally amidships, driving front and rear wheels and propeller. IFS, IRS. Two- or four-wheel steering. Driving controls front and rear. Hyd. brakes. One made, 1939.

268

Truck, Light, 4×2, Field Kitchen (Spyker) Produced about 1916 by the Delft Artillery workshops on a Spyker 12 HP chassis for use by Cyclist Company in the field. A larger model, also on Spyker chassis, had been built in 1911. Both were comprehensively equipped and self-contained.

Truck, Light, 4×2, Machine Gun (various makes) During the First World War many requisitioned cars were stripped of their bodywork and fitted with a general purpose box body. A typical example, mounting an anti-aircraft machine gun, is shown, together with a Douglas motorcycle.

Trucks, Medium, 4×2, Cargo (Packard and Saurer-Spyker) In 1914 the Dutch Army acquired 15 American Packard trucks, mostly with cargo bodies (one had enclosed bodywork, fitted out as a mobile cinema and entertainment unit). In the background is a Saurer truck, supplied through the Dutch firm of Trompenburg (Spyker).

Truck, Heavy, 4×2, Anti-Aircraft Commercial truck, mounting AA gun and Vickers MG. Pre-1930 trucks included: American Ford, GMC, Kelly-Springfield, Packard, Pierce-Arrow; French Renault; German Presto; Swiss Arbenz, Berna and Martini, etc. In the Dutch East Indies: American Fageol, Garford, Wichita, etc.

Truck, 1½-ton, 4×2, Tractor (Ford AA) Short-wheelbase tractor of 1930/31 with Photographic (dark room) semi-trailer. There were also open drop-side semi-trailers for gun carrying. Conventional Ford TT and AA trucks were used as field artillery prime movers (with crew seats), etc.

Truck, 1½-ton, 4×2, Cargo (Ford V8 51) Standard 1936 Ford truck chassis/cab, assembled at the Amsterdam Ford plant. Dropside body, shown with pontoon bridging equipment. Similar GS bodywork existed on earlier and later Fords and on 1939/40 Chevrolets.

Truck, 1½-ton, 4×2, Light Wrecker (Ford V8 018T) Shortly before the outbreak of war the Netherlands Army acquired many new trucks, incl. this Werkspoor-equipped twin swinging-boom wrecker. In 1940 Werkspoor also supplied cargo bodies for new Fords and Chevrolets.

Truck, 6×4, Medium Wrecker (Ford/DAF) 1940 Ford chassis with DAF Trado bogie which in this application had additional springs on the centre pivots. Similar wrecker appeared with four in-line driven rear wheels, on two oscillating half axles (suspension as on certain DAF semi-trailers).

Tractor, 4×4, Light Artillery (Chevrolet/DAF) In addition to Trado bogies DAF offered front drive conversions for Chevrolet and Ford which consisted of a transfer case behind the gearbox and two diagonal drive shafts to the front wheels. In the transfer case was a (lockable) differential. Shown: 1940 Chevrolet with RHD.

Tractor, 6×4, Light Artillery (Ford/DAF) For light trucks DAF offered a special rear bogie known as Trado IV and seen here on a 1937 Ford V8 Model 77. This conversion was more complicated than the heavier Trado III. DAF conversions were commercially available.

Tractor, 6×4, Field Artillery (Chevrolet/DAF) Regular 1935 Chevrolet 1½-ton chassis with DAF bodywork and DAF Trado III rear bogie. Tracks for the bogie were carried on reels in compartments ahead of the rear wheels; they could be simply and quickly unrolled and fitted. Also as searchlight carrier.

Tractor, 6×4, Field Artillery (Ford/DAF) One of a series of tractors based on Ford V8 Model 51 trucks of 1936. Trado bogies (see following page) were fitted mainly to Chevrolet and Ford but also to GMC and other chassis. Some Trado-equipped 1937 Chevrolets were supplied to Spain. (See page 372.)

DAF Trado rear bogie This illustration clearly shows how a 4×2 truck's rear axle was converted. The system was developed in 1934/35 by Messrs van der Trappen en van Doorne (hence: Trado), following an earlier design by NV SYS (later NETAM). Note brakes on leading wheels only.

Truck, Ice Cutter Conversion of 1939 Chevrolet/DAF Trado art. tractor. Rearmost hubs fitted with circular saws, which could be jacked up for travelling. After cutting the ice (of the Dutch inundation defence lines) a DAF MC139 amphibian would follow to break it. Winch, automatic tow hook and spade were also DAF products.

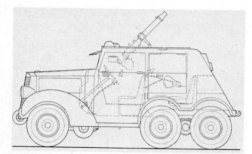

Armoured Car, 6×4 (Ford/DAF) Using the Trado bogies, DAF designed several armoured and other cross-country vehicles. This armoured command car with AA gun, designed in 1938 for a Ford V8 light truck chassis, was one which did not reach the production stage. Note engine armour.

Armoured Car, 6×4 (DAF M39) Ford V-8-cyl. (at rear), 95 bhp, 4F4R, wb 2.50 m (BC 1.12 m), 4.75×2.08×2.16 m, 5800 kg. Speed, forward 75 km/h, rearward 50 km/h. Chassisless construction. 12 built, 1939. Also used were Swedish Landsverk 181 and 182 (M36 and M38) A/Cs (see Sweden).

POLAND

After the First World War the new Polish Army acquired many war-surplus transport vehicles from the West; these were mainly of American and French origin. During the 1920s efforts were made to produce Polish vehicles but it turned out to be a period of uncoordinated activities by people striving to motorize the Army in the face of an indifferent attitude of the authorities. This gradually changed but it was found that it was better to assemble and licence-produce foreign vehicles than to put indigenous designs into series production. The state-run enterprise Panstwowy Zaklad Inzynierii (PZInz.) in particular designed several interesting cross-country vehicles but most remained in the prototype stage. By 1938 Poland possessed 12,182 trucks and tractors of which the German *Blitzkrieg* in the following year caused almost total destruction.

Motorcycle, with Sidecar, 3×1, MG Carrier (CWS-M111) V-2-cyl., 21 bhp (at 3500 rpm). Made in 1932 by Centralne Warsztaty Samochodowe of Warsaw. Polish Army also had open and closed staff cars and ambulances of this make (CWS-T1 61-bhp 3-litre 4-cyl., 1926 and CWS-T8 80-bhp 3-litre 8-cyl., 1929).

Field Car, 4-seater, 4×2 (Polski Fiat 508/IIIW 'Lazik') PZInz. 108 4-cyl., 22–24 bhp, 4F1R, wb 2.25 m, track 1.30 m, 800 kg. Tyres 5.50-16. Speed 90 km/h. 995-cc (65×75 mm) engine. Lockable diff. Used as command car, radio car, MG carrier, etc. Also ambulance (508/III) and light truck (508/IIIW). 1935–36.

Field Car/Tractor, 4×2 (Polski Fiat 508/518T) Basically as 508/III but mostly with 45-bhp 1944-cc engine. Two seats, rear locker, towing hook (for 37-mm AT gun, signals trailers, etc.). Weight 1450 kg. Speed 65 km/h. Variants: light truck (508/518 'Mazur') and exp. 4×4 car. 1936–39.

273

T

Truck, 2½-ton, 4×2, Cargo (Polski Fiat 621L) 6-cyl., 45–50 bhp, 4F1R. Produced by PZInz. under Italian Fiat licence from about 1931 until World War II. Maximum speed 60 km/h. Maximum payload 3000 kg. Model 621R had a lowered chassis frame and was used for bus bodies.

Truck, 3-ton, 4×2, Tank Transporter (Polski Fiat 621) One of several variants of the Polski Fiat 621, carrying TK3 tankette (Polish development of British Carden-Loyd). Another variant on the 621 was an SP Anti-Aircraft gun, the WZ18/24 of 1939 (also on De Dion-Bouton and Ursus chassis).

Truck, 6×4, Ambulance (Polski Fiat 618) After 1918 the Army received war-surplus Fiat, Ford, and GMC ambulances from W. Europe, including 20 GMCs with X-ray equipment. During the 1920s American Model TT Fords were imported and bodied in Poland. This Fiat was licence-produced in the mid-1930s.

Truck, Half-Track, Ambulance (Polski Fiat 621L) Produced by PZInz. in 1935 on a Polski Fiat 621L truck chassis which had been equipped with rear track bogies patterned on the French Citroën-Kégresse (many of which were used by the Polish Army). Known as *samochód sanitarny*.

Tractor, 4×4, Artillery (PZInz. 703) Designed by M.Debicky, with 90-bhp OHV engine designed by J. Werner, 1937. Similar: PZInz. 312, 1938. Prototypes only. Earlier wheeled tractors included WWI-surplus Austro-Daimler, Latil, and Nash Quad and 1924 Pavesi, but none were used in quantity.

Tractor, Half-Track, Artillery (PZInz. 202) In 1937/38 a study group at PZInz. designed this half-track tractor (5 tons, 65 bhp), a 3½-ton 100-bhp full-track tractor (PZInz. 152), an 8-seater 4×4 command car (PZInz. 303), a 4×4 truck (PZInz. 613), etc. None of these were progressed beyond the prototype stage.

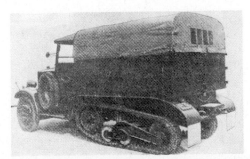

Tractor, Half-Track, Artillery (Polski Fiat C4P) 4-cyl., 47 bhp, 4F1R×2, 4.70×1.90×2.35 m, 3200 kg. Speed 30 km/h. Derived from Polski Fiat 621L. Some 200 were built in 1934/35. Used for towing 75-, 105- and 120-mm field guns, 75-mm AA gun and searchlight equipment.

Tractor, Half-Track, Artillery (Polski Fiat C4P) This half-track was similar to the one shown on the left except that the body was shorter and a clamp was fitted at the rear for the towing of trailing guns. The bogies had positive drive with the driving sprockets at the front. 1934.

Tractor, Full-Track, Artillery (C2P) Polski Fiat 122B 4-cyl. 46-bhp engine. Weight 2750 kg. Based on modified TKS tankette chassis. Towed 75-mm field gun, 100-mm howitzer, 37-mm AT gun and 40-mm AA gun. Crew 4. Speed (with gun) 45 km/h. Towing capacity 2000 kg. 1936–39.

Tractor, Full-Track, Artillery (C7P) Saurer CBLD6 6-cyl. 115-bhp diesel engine, produced under licence. 4.60 × 2.40 × 2.40 m, 8500 kg. Towing capacity 5400 kg. Crew 4–5. Speed (with trailed load) 25 km/h. Used also for towing tank recovery trailers. 71 built, 1934.

Tractor, Full-Track, Artillery (C7P) Tractors with Skoda 220-mm mortars, which were transported in three pieces (base, mount, barrel) on three trailers. Range of the 128-kg projectiles was 15,000 metres. The C7P was used also as basis for a bridge layer (load 9 tons, span 8 metres).

Armoured Car, 4 × 2 (Ursus WZ.29) 4-cyl., 35 bhp, 4F1R, 5.49 × 1.85 × 2.47 m, 4800 kg. Crew 4. Armament one 37-mm gun and three 7.92-mm MGs. Speed 35 km/h. Armour max. 10 mm. Based on Polski Fiat 621L truck chassis. 14 built in 1929. Austin, Ford and Peugeot armoured cars were also used.

RUSSIA

Before and during World War I all motor transport vehicles were imported. Vehicle types were selected during extensive trials like the 5-day long-distance exercise in 1912 when 54 trucks of 19 makes from six countries participated. During the 1920s the Russian AMO (later renamed ZIS) and NAMI vehicles were introduced and in 1932 a large factory was built at Gorkiy with American assistance for the production of the Ford-pattern GAZ vehicles. GAZ and ZIS became the largest producers of cars, trucks, buses, ambulances, armoured cars, etc. At Yaroslavl a new plant had started operating in 1925; it produced heavier type trucks, using German and US engines in certain models. Also during the 1920s several tractor factories were founded. Imports were restricted to certain special types. The Soviet industry was able to cope with domestic demands until World War II when following the German invasion the USSR acquired vast numbers of American, British and Canadian vehicles under the Lend-Lease scheme.

Car, 5-seater, 4×2 (Mercedes) Most cars used by the Russian Army in World War I were imported from Western Europe. This Mercedes sported a metal device which would cause obstructions such as wire ropes set across roads to be lifted up and over the car. Note spotlight and triple horns.

Car, 5-seater, 4×2 (Delaunay-Belleville) Produced by SA des Automobiles Delaunay-Belleville of St. Denis, Seine, France. Tsar Nicholas II owned several cars of this, his favourite make. He is seen here talking to Count Brobinsky, Governor of Galicia, the Grand Duke Nicholas and the chief-of-staff.

Car, 5-seater, Half-Track (Packard Twin Six) Adolphe Kégresse, manager of the Tsar's garages, devised half-track bogies which were fitted to several cars, incl. Delaunay-Belleville, Packard and (later) Rolls-Royce. After the 1917 Revolution Kégresse returned to his native France, where the system was developed further (*qv*).

277

Car, 5-seater, 4×2, Phaeton (GAZ-A) 4-cyl., 40–42 bhp, 3F1R, wb 2.63 m, 3.87×1.71×1.78 m, 1080 kg. Tyres 5.50-19. Engine 3280-cc (98.43×107.95 mm) L-head. Compression ratio 4.2:1. Speed 90 km/h. Produced in Gorkiy during 1932–36. Used by Soviet Army as command car. Patterned on US Ford Model A.

Car, 5-seater, 4×2, Sedan (GAZ-M-1) 4-cyl., 50 bhp, 3F1R, wb 2.84 m, 4.62×1.77×1.76 m, 1370 kg. Tyres 7.00-16. Engine GAZ-M-1 3280-cc with 4.6:1 compression ratio. Mechanical brakes. Speed 100 km/h. Produced 1936–41, also with Pickup bodywork (GAZ-M-415, 1939–40).

Car, 6-seater, 4×2, Sedan (ZIS-101) 8-cyl., 90 bhp, 3F1R, wb 3.60 m, 5.75×1.89×1.87 m, 2550 kg. Tyres 7.50-17. Engine ZIS-101 5750-cc (85×127 mm) L-head with 4.8:1 CR. Speed 115–120 km/h. Mech. brakes. Beam axles with leaf-spring suspension front and rear. Produced 1936–41.

Car, 6-seater, 4×2, Phaeton (ZIS-102) Basically similar to ZIS-101 sedan shown on left. Alternative engine for both cars had aluminium alloy cylinder head (instead of cast iron) and 5.5:1 compression ratio, producing 110 bhp at 3200 rpm. Both were utilized as staff cars in the Soviet Army.

Truck, 2-ton, 4×2, Cargo (AEC) During the First World War the Russians for their transport needs relied entirely on imported vehicles from the US and Western Europe. This British AEC 2-tonner was known as the 'Russian B' and was one of the few AEC types ever built with such a low payload capacity.

Truck, 2/3-ton, 4×2, Cargo (Austin) One of the most numerous Russian Army trucks in World War I was the British Austin 2/3-tonner. It appeared with various body types, incl. workshop and had individual propeller shafts to each rear wheel. Lighter Austin chassis were also used, mainly as ambulances.

Truck, 2-ton, 4×2, Cargo (Burford) British trucks used by the Imperial Russian Government included AEC, Austin, Burford (shown), Caledon, Napier (light, medium and heavy types), Sheffield-Simplex, etc. US imports: FWD (4×4), Garford, Hurlburt, Locomobile, Packard (2-, 3- and 5-tonners), White, etc.

Truck, 1½-ton, 4×2, Cargo (AMO-F-15) First Soviet truck. Some 6000 made, 1924–32. Army had Cargo, Ambulance, Phaeton (staff car) versions. Engine: 4-cyl. 4400-cc (100×140 mm), 35 bhp at 1400 rpm. CR 4:1. Trans. 4F1R. Tyres: 280×135. Weight: 1920 kg. Speed: 50 km/h. Also used at this time: Lancia Pentaiota (1924, I).

Truck, 3-ton, 4×2, Cargo (JA-3 or Ya-3) First truck produced by the Yaroslavl Plant. Prototype 1925, series production 1926–28. Engine: AMO-F-15 (qv). Gearbox: 4F1R. Weight: 4180 kg. GVW: 7180 kg. Tyres: 38×7. Speed: 30/km/h. Fuel consumption: 40 litres/100 km.

Truck, 3½-ton, 4×2, Cargo (JA-4 or Ya-4) This somewhat heavier truck superseded the JA-3 in production and was made during 1928–29. Like its predecessor it was used in the Soviet Army. The engine was a German Daimler-Benz 70-bhp petrol six-cylinder of 7007 cc (100×150 mm). 3-seat cab.

Truck, 2½-ton, 4×2, Cargo (AMO-3/ZIS-3) Produced during 1932–34 this truck had an AMO-3 6-cyl. engine of 4880 cc (95.25×114.3 mm) with 4.6:1 CR, developing 60 bhp at 2200 rpm. Gearbox: 4F1R. Weight: 2840 kg. Tyres: 34×7. Speed: 60 km/h. Fuel consumption: 33 litres/100 km.

Truck, 5-ton, 4×2, Cargo (JAG-4 or YaG-4) Similar in appearance to JA-4 but powered by a 6-cyl. ZIS engine of 5550 cc (101.6×114.3 mm), 73 bhp at 2300 rpm and 4.6:1 CR. Gearbox: 4F1R. Weight: 4750 kg. Payload on/off roads: 5000/3500 kg. Tyres: 40×8. Speed: 40 km/h. Produced during 1934–36.

Truck, 1½-ton, 4×2, Cargo (GAZ-AA) GAZ-A 4-cyl., 40 bhp, 4F1R, wb 3.34 m, 5.33×2.04×1.97 (cab) m, 1810 kg. Track, front/rear 1.40/1.60 m. Tyres 6.50-20. Speed 70 km/h. Cantilever rear springs. Built during 1932–38 it was similar to the US Ford Model AA. Superseded by 50-bhp GAZ-MM.

Truck, 1½-ton, 6×4, Cargo (GAZ-AAA) Based on GAZ-AA but with tandem-drive rear bogie and payload capacity (on roads) of up to 2500 kg. Usually fitted with the more advanced GAZ-M1 engine which had 4.6 vs 4.22:1 CR and developed 50 bhp at 2800 rpm vs 40 at 2200. Various body types.

Truck, 2½-ton, 6×4, Cargo (ZIS-6) 6-cyl., 73 bhp, 4F1R×2, wb 3.90 (BC 1.08) m, 6.06×2.23×2.16 m, 4230 kg. Tyres 34×7. Payload on roads up to 4000 kg. Six-wheeler version of ZIS-5 3-ton 4×2 (same overall dimensions but weight 3100 kg). Both introduced in 1934.

Truck, 8-ton, 6×4, Cargo (JAG-10 or YaG-10) Hercules YXC 6-cyl., 93.5 bhp, 4F1R×2, track front/rear 1.45/1.83 m, 6.98×2.34×2.54 m, 5430 kg. Imported American engine, 7020 cc (111×120.6 mm), CR 4.6:1. Tyres 40×8. Produced at Yaroslavl, from 1932 to 1941. Also as SP mount for AA gun.

Truck, 12-ton, 8×8, Cargo (JAG-12 or YaG-12) Continental 6-cyl. 93-bhp engine of 7020 cc (111.1 × 120.6 mm) with 4.6:1 CR, imported from the US. Speed 40 km/h. Fuel consumption 52 litres/100 km. 3-seat cab. Produced at Yaroslavl, reputedly from 1932 to 1941. Picture taken in 1933.

Truck, 1½-ton, Half-Track, Cargo (GAZ-60) Modification of GAZ-AA 4 × 2 truck (*qv*) with GAZ-M1 50-bhp engine and 2-speed aux. gearbox (4F1R × 2 trans.). Wheelbase 3.34 m (to bogie centres). Overall 5.33 × 2.36 × 2.10 m, 3520 kg. 1938. There were also half-track versions of GAZ cars and ZIS trucks.

Tractor, Full-Track, Artillery (Kommunar TG-G-50) One of many Soviet full-track artillery tractors, this model was introduced in 1923. It weighed 8250 kg and had a 4-cyl. 48-bhp paraffin engine. 75- and 90-bhp engines were also used. In World War I Russia used American Holt and Lombard (Vaubleok) tractors.

Tractors, Full-Track, Artillery (STZ-NATI 1-TA) 4-cyl. engine with 4F1R transmission. Track width 310 mm. Engine developed 52 bhp at 1250 rpm. Piston displacement 7460 cc (125 × 152 mm). CR 4.0:1. Magneto ignition. 1937. Also produced with forward control, truck cab and dropside body (STZ-5-2TB).

Armoured Car, 4×2 During and after World War I the Russians employed a great variety of armoured cars, most of which were based on car and truck chassis imported from Britain and the US. Many were improvised. Vehicle shown was captured by the Germans and exhibited at the Berlin Zoo in 1916.

Armoured Car, 4×2 (Austin) The Russians imported complete British Austin armoured cars as well as chassis, which were armoured in Russia (Putilov). One of the latter is shown here on the Petrograd front during the Civil War of 1919/20. There were also half-track variants. Note the diagonally-placed turrets.

Armoured Car, 4×2 (GAZ BA-20M) Light armoured car based on the 1936 GAZ-M1 passenger car chassis with 4-cyl. 50-bhp engine. It weighed about 2500 kg, had 10-mm armour, one 7.62-mm MG and a speed of 85 km/h. It superseded the earlier GAZ-A-based 'Bronieford'.

Armoured Car, 6×4 (GAZ BA-10M) Heavy armoured car based on the GAZ-AAA truck chassis with 4-cyl. 50-bhp engine. Several variants were produced (BA-3, BA-6, etc.), starting in the early 1930s. The improved BA-10M (or BA-32-3) appeared in 1937 and had a 45-mm gun and two MGs.

SWEDEN

Swedish Army transport vehicles were usually products of the two main national producers Scania-Vabis and Volvo and the smaller Tidaholms Bruks AB of Tidaholm. In addition imported American Chevrolet and Ford chassis were used. Artillery tractors included some imported types e.g. Czech Skoda 6V (6×6) and Praga T VI Sv. (full-track) and Italian Pavesi (also produced under licence). About 1939/40 some Demag semi-tracks were bought and other German imports included Tempo G1200 cross-country cars and Klöckner-Humboldt-Deutz A3000 4×4 trucks (*tgbil* m/42M). Passenger cars in the 1930s included Volvo and US Chevrolet, Ford and Plymouth. Sweden remained neutral during World War II and some vehicles introduced during the early part of that period are included (see also *The Observer's Military Vehicles Directory— from 1945*).

The two Swedish armaments firms Bofors and Landsverk designed and produced a variety of vehicles, many of which were available commercially; some of these are also shown.

Car, 4-seater, 4×2 (Scania-Vabis) AB Scania-Vabis was formed in 1911 by the merger of the firms of Scania of Malmö and Vabis of Södertälje. In 1916 the company supplied this 4-cyl. 20-hp staff car, as well as trucks. From 1924 the company concentrated on truck production.

Ambulance, 4-stretcher, 4×2 (Horch/Scania-Vabis) Tilt-covered field ambulance on an imported German Horch 4-cylinder shaft-drive car chassis. The bodywork was produced by Scania-Vabis and the vehicle was delivered to the Swedish Army authorities in October, 1914.

Ambulance, 4-stretcher, 4×2 with Trailers (Scania-Vabis) Scania-Vabis Model 2134 field ambulance train consisted of a car-based towing vehicle and two single-axle trailers, all with similar bodies for accommodation of up to four stretcher cases each. They were produced in 1916.

285

Truck, 4-ton, 4×2, Cargo (Scania-Vabis Ela) 4-cyl., 40–45 bhp, 4F1R. Enclosed chain final drive. Solid tyres, dual rear. Dropside body. Open cab with removable folding top. Note radiator guard and hinged vent doors in bonnet. Model 1545 engine. Built in 1921.

Truck, 2-ton, 4×2, Field Bakery (Scania-Vabis) Basically similar to Model Ela but with shaft drive and pneumatic tyres. A 4-wheel trailer with similar bakery outfit was towed. One of a fleet of at least 7 delivered in 1924. Same chassis also with cargo body and cab with folding top and windscreen.

Truck, 3–4-ton, 4×2, Cargo (Scania-Vabis) A number of these trucks were delivered to the Swedish Army in 1938. They had a 110-bhp engine and a wooden dropside body with very short rear overhang. Similar GS trucks were supplied on Volvo, Chevrolet and Ford commercial chassis.

Truck, 3-ton, 4×2, Signals Van (Volvo/Hägglund) *Radiobil TMR8*, produced in 1941 by AB Hägglund & Söner of Örnsköldsvik on Volvo LV81DS chassis. A similar vehicle (*TMR9*) was based on Scania-Vabis 8116 chassis with single rear tyres. Both carried radio transmitting/receiving equipment.

Truck, 3–4-ton, 4×4, Cargo (Scania-Vabis F11) 4-cyl., 115 bhp, 4F1R×2, wb 3.80 m, 5.95×2.10×2.40 m, 4400 kg. GVW 8500 kg. Introduced during World War II. Officially known as *tgb* m/42 SLT (*Scania Last Terrängbil*). Model 402 5.65-litre (115× 136 mm) engine. Prototype (105-bhp) shown.

Truck, 3–4-ton, 4×4, Cargo (Volvo TLV131D) 6-cyl., 90 bhp, 4F1R×2, wb 3.80 m, 6.35×2.10×2.45 m, 4350 kg. GVW 8080 kg. Designated *tgb* m/42 VL. Model FCT 4.4-litre engine. 4-ton winch. *tgb* m/42 VLT (Volvo TLV141D) had FET 105-bhp engine, *tgb* m/42 (Volvo TLV142M) had FET engine and 4.10-m wheelbase.

Truck, 8-ton, 4×4, Tractor (Scania-Vabis) This 70-bhp-engined tractor truck featured four-wheel drive and four-wheel steering with duplicate driving controls at rear. The prototype was designed and built, reputedly in 7–8 weeks, in 1916, for the Swedish Army Materièl Administration.

Tractor, 4×4, Artillery m/28 (licence Pavesi) Sweden was one of several countries to use and later (1928) licence-produce the Italian Pavesi articulated tractor. The Swedish version differed from the original in many details (wheels, front end, windscreen, etc.). It was used mainly for towing AA guns.

287

Tractor, 6×6, Artillery (Skoda 6V) Just before the outbreak of World War II Sweden bought a number of heavy Skoda tractors (*tgdb—terrängdragbil*) for the coastal artillery. Some of these had an extra crew cab, as shown, which was added in Sweden. Note the bumper wheels at front.

Tractor, 6×4, Artillery, m/40 (Volvo TVB) 6-cyl., 130–140 bhp, 4F1R×2, wb 3.44 (BC 1.28) m, 6.60×2.38×2.30 m, 6825 kg. GVW 11,500 kg. Pneumatic-tyred (9.00-6) support wheels fore and aft of front wheels. Also with closed bodywork and as SP mount for anti-aircraft gun (with single cab and frontal armour).

Tractor, Full-Track, Artillery, fm/32 (Bofors) In 1932 the well-known armaments firm of Bofors produced this full-track tractor which was derived from a conventional four-wheeled agricultural type. The engine was a 46-bhp diesel and the maximum speed was 6 km/h.

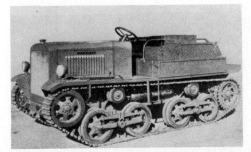

Tractor, Full-Track, Artillery (Landsverk 132) This tractor had a 90-bhp engine, weighed 5.5 tons and was commercially available. A smaller version (Landsverk 131, 60 bhp, 4.5 tons) was also made. Speed was 22–24 km/h. Customers included Thailand (Model 131 with 75-mm Bofors AA guns).

Armoured Motorcycle, 3×1 (Landsverk 210) Aktiebolaget Landsverk at Landskrona designed and produced various types of armoured vehicles, which were usually offered commercially. About 1933 they built this prototype armoured motorcycle combination with tripod-mounted MG in the sidecar.

Armoured Truck, 4×2 (various chassis) Starting with the *Pansarbil* m/31 in 1931 the Swedish Army employed similar armoured trucks throughout the 1930s. Typical specification: 4.2 tons, crew 5, 37-mm gun, 2 MGs, 60 bhp, 60 km/h. About 1925 a number of Tidaholm TSL armoured cars and trucks had been acquired.

Armoured Car, 4×2 (Landsverk 185) Ford V-8-cyl., 65 bhp, 4F4R, wb 3.70 m; complete vehicle: 4.94 × 2.02 × 2.30 m, 4200 kg. Produced in 1933 on modified Ford V8 truck chassis with special bullet-proof tyres and duplicate driving controls at rear. Speed, both directions, 60 km/h. Crew 4–5.

Armoured Car, 4×4, fm/29 (Landsverk) Scania-Vabis 6-cyl., 85 bhp; complete vehicle: 5.43 × 2.33 × 2.46 m, 8000 kg approx. Introduced in 1929 this heavy A/C had four-wheel drive and steering, rear engine and duplicate driving controls at rear. Armament: 37-mm gun (in turret), 2 MGs. Armour 6 mm.

Armoured Car, 4×4, m/39–40 (Landsverk/Volvo Lynx)
Volvo 6-cyl., 135 bhp, 5.10×2.30×2.20 m, 7800 kg. Crew 6. Armament one 20-mm automatic gun and three 8-mm machine guns. Max. speed 70 km/h. Armour 18 mm. Bullet-proof tyres. Swedish designation: *Pansarbil m/39–40* (or *m/40*).

Vehicle, Wheel-cum-Track (Landsverk 5) 6-cyl., 80 bhp, 4F1R×2, 5.00×2.45 m, 5200 kg. Experimental vehicle, built in 1929. Speed on wheels 75 km/h, on tracks 25 km/h. One of the first Swedish wheel-cum-track vehicles, succeeded by fully-armoured Maybach-engined Landsverk 30 in 1931.

Armoured Car, 6×4 (Daimler-Benz/Landsverk 181) 6-cyl., 65 bhp, 4F1R×2, wb 3.88 m, 5.56×2.00×2.45 m, 6200 kg. Commercially available from 1933. Vehicles shown were supplied to the Netherlands (*paw M36*) and later captured by German *Wehrmacht*. Picture was taken in Amsterdam, Feb. 1941.

Armoured Car, 6×4 (Landsverk 182) Basically similar to the Landsverk 181 this was a heavier type, powered by a 150-bhp Büssing-NAG V-8-cyl. engine. It was bought by, amongst others, the Dutch and Irish Armies. Vehicle shown (Irish Army Cavalry Corps) was later re-engined (Ford 332) and fitted with pneumatic tyres.

U

SWITZERLAND

The Swiss Army started experimenting with motor vehicles in 1898 and in 1907 an arrangement was made with the Swiss Automobile Club (ACS) for the formation of a Volunteer Automobile Corps. Few vehicles were purchased until 1914 when not enough trucks were available for requisition; some 340 were then bought, mainly from Berna and Saurer. These two manufacturers, together with a few smaller ones, have over the years produced many thousands of trucks for the Swiss Army as well as for export. Saurers, for example, were used even in America. On the other hand, Switzerland imported certain types of vehicles, e.g. Italian Pavesi artillery tractors. By the end of the 1930s AG Adolph Saurer in Arbon introduced a family of 4×4, 6×6 and 8×8 cross-country vehicles which were in use until long after World War II. From 1929 Saurer and the firm of Berna in Olten produced several vehicles to common designs.

Motorcycle, with Sidecar, 3×1 (Motosacoche) Built in 1921 this 500-cc twin-cyl. combination took part in a military parade in Zug exactly 50 years later. Until 1939 Motosacoche built 850- to 1000-cc machines for the Army. Sidecar combinations were also used as mobile repair units, with two-wheeled trailer.

Car, 4-seater, 4×2 One of the first cars employed by the Swiss armed forces. It participated in the 1901 Autumn manoeuvres, together with some other cars and two trucks. In the late 1940s the Army was still using civilian cars, e.g. 1932 Buick sedans, for machine gun squads.

Truck, 1½-ton, 4×2, Cargo (Martini JL) This 18/24 HP truck was offered in 1917 by the firm of Martini of St. Blaise. It had a 4-cyl. 36-bhp 3817-cc (90×150 mm) engine with four-speed gearbox and shaft-drive. The 3.50-m wb chassis weighed about 1000 kg and could also be supplied with solid tyres.

Truck, 3½-ton, 4×2, Cargo (Berna C2/GC) Introduced by Motorwagenfabrik Berna AG during World War I the Model C2 had a 4-cyl. engine of 35 bhp at 1200 rpm. Cubic capacity was 5530 cc (105×160 mm). Supplied to the Swiss, British and French armies and to civilian customers until 1928.

Truck, 5-ton, 4×2, Cargo (Saurer A) First established by Franz Saurer in 1853 as a foundry, the Saurer company produced their first truck in 1903. Shown is one of the models built in 1917. It had a 4F1R gearbox and shaft drive. Lighter models were also produced.

Truck, 1½-ton, 4×2, Cargo (Berna 18 PS) One of the lightest models in the Berna production programme in 1937 this 1½-tonner, known as 'Camionettli', was powered by a 4-cyl. petrol engine of 90 bhp. It later appeared also with house-type van bodywork.

Truck, 5-ton, 4×2, Cargo (Saurer 5BL) Typical heavy truck of the Swiss Army, produced in the mid-1930s. It had a 95-bhp 6-cyl. engine and could also be supplied with solid tyres (Model 5 BLD). Note the hardtop cab with half-doors and side screens.

Truck, 2½-ton, 4×2, Cargo (Saurer 1CR1D) Conventional truck of the late 1930s. Basically a civilian model it was slightly modified to meet Swiss military requirements (radiator brush guard, front towing attachments, etc.). It was powered by a 4-cyl. petrol engine, developing 65 bhp.

Truck, 2½-ton, 4×2, Cargo (GMC AC) In 1939/40 the Army bought a number of American GMC trucks (which unlike other Swiss military vehicles had LHD). This survivor still has its wartime Imbert producer gas equipment and belongs to the Military Vehicle Drivers' Association.

Truck, 5-ton, 6×4, Balloon Winch (FBW) Dating from 1925 this six-wheeler with balloon winch equipment was produced by the firm of Franz Brozincevic & Co. Motorwagenfabrik of Wetzikon, Zürich. It is powered by a 4-cyl. OHV petrol engine of 45 bhp and weighs 10,700 kg.

Truck, 8-ton, 6×4, Signals Van (FBW Gallilea) Heavy six-wheeler, produced in 1938. The two rear axles were driven by a 6-cyl. petrol engine of 38.4 HP (treasury rating). The house-type van body was designed for the installation of radio transmitting and receiving equipment. Weight 11,260 kg.

Tractor, 4×2, Artillery (Berna T5g) Both Berna and Saurer introduced heavy gun tractors in 1918. This Berna appeared in 1932 and had a 4-cyl. 75-bhp engine. It measured 4.60 × 1.82 × 2.49 m and weighed 5500 kg. Wheelbase was 2.90 m, track 1.52 m. Speed 40 km/h. 3-ton winch.

Truck, 2-ton, 6×6, Chassis (Saurer 2M) Of this vehicle only two pilot models were built in 1936. The 90-bhp 6-cyl. diesel engine drove through an 8F2R transmission. The suspension was independent with torsion bars. Compressed air was used for brake application and engine starting.

Truck, 2-ton, 4×4, Cargo/Prime Mover (Saurer 4M) Introduced in 1938 this four-wheel steering vehicle had IFS and IRS with interconnected coil springs. Engine: 4-cyl. 50-bhp diesel. Transmission: 5F1R×2. Chassis weight: 2400 kg, dimensions 5.00 × 1.85 m. Wheelbase was 2.50 m.

Truck, 2½-ton, 6×6, Cargo/Prime Mover (Saurer 6M) 6-cyl. 85-bhp Saurer CTDM diesel engine with 5F1R×2 transmission. Independent suspension. 5-ton winch. Overall dimensions 5.45 × 2.00 × 3.05 m. Wheelbase 2.80 (BC 1.10) m. Produced 1940–48. Prototype shown. Also 8×8 variant (8M, 1943–45).

293

USA

The history of US military motorization goes back to just before the turn of the century when Colonel (then Major) R. P. Davidson designed a gun carrier on an 1898 Duryea three-cylinder car chassis. The gun was a Colt automatic, with an armour shield. In 1900 two more were built, now on steam-powered carriages and further Davidson designs, based on Cadillac car chassis, appeared later. By 1905 several motor vehicles were in US Army service, including a 'battery repair wagon and forge' by the Artillery Division at Fort Myer, Virginia (built by the US Long Distance Automobile Co. in 1902), a 1904 Winton 'automobile telegraph car' and a 1905 Cadillac 'automobile repair car', both by the Signal Corps at Fort Omaha, Nebraska, and Fort Leavenworth, Kansas, respectively. The Signal Corps also had two Winton 1904 touring cars, at other locations. The Medical Bureau of the War Department in Washington, DC, at this time had a steam-powered White ambulance. About 1906 the first efforts were made at replacing the Army's mule-drawn load-carrying wagons by motor vehicles.

Not much progress was made and only very general specifications were drawn up by the War Department covering motor vehicles for Army use. At that time the motor truck had not been developed to a great extent commercially and those that were on the market and in production were deemed not to be of sufficiently rugged construction for military use. No definite policy as regards motor transportation was adopted by the War Department at this time but it was realized that the complexity and expense of the mechanism, as well as the time and skill required to develop it, would make it necessary for a nation to depend for its military motor vehicles on civilian facilities organized in times of peace. This necessity was realized in Europe and attempts were made there to force the trend of design in the direction of the ideal military vehicle by means of subsidies. The best known subsidy scheme was that of the British War Office, developed in 1912, for medium and heavy trucks. The specifications were worded so as to require certain standardization of controls and standardization of interchangeability of pricipal assemblies, thus facilitating operation and the solution of the spare parts problem.

Those manufacturers whose vehicles were accepted after stringent tests were subsidized and their facilities made subject to call on the outbreak of war. The French subsidy plan, started in 1909, was similar to that of the British but acceptance was based on performance rather than on combined performance and design. Germany and Austria had similar subsidy plans.

In 1912 a long-distance road test was undertaken by the US Army under the direction of Capt. A. E. Williams of the QMC. Three Government-owned trucks took part (White, Sampson, FWD) as well as a works-entered Autocar and White. The 1500-mile test route was from Washington, DC, to Fort Benjamin Harrison, Ind., by way of Atlanta and only three trucks completed it.

It can be safely said that the United States had no coherent motor transport policy prior to 1910 and 1911. From 1910 to 1916 attempts were made to develop truck specifications and some progress was made. Although these specifications were in no case

FWD truck, one of the finalists in the 1912 1500-mile test.

USA

drawn up with a standard government vehicle in view but simply as a guide in the purchase of commercial vehicles for military use, they contained nevertheless three principal requirements which have long been considered essential in military vehicles. These requirements were as follows:

(a) They should be capable of construction by all principal manufacturers.
(b) Parts should as far as possible be interchangeable.
(c) They should stand up under a test approximating service conditions.

In other words what was required, was ease of production, ease of maintenance, and ability to operate under the conditions of field service.

On 30 June 1914 the US Army Quartermaster Dept. had in use 35 trucks, as follows:

El Paso: 1 Mais, 1910, 2 Whites, 1913 and 1914.
Fort Huachuca: 1 White, 1912.
Fort Leavenworth: 1 Mais, 1910.
Fort D. A. Russell: 1 Mais, 1910.
Fort Sam Houston: 1 Sampson, 1911.
Fort Sill: 1 White, 1913.
Galveston: 1 Atterbury, 1913, 1 Jeffery, 1913, 2 Whites, 1913, 1 White, 1914.
Honolulu, Hawaii: 1 White, 1913.
Marfa: 1 White 1914.
Philadelphia: 1 Automatic, 1912, 1 Electric, 1907.
San Diego (Signal Corps): 1 FWD, 1912, 1 White, 1912.
San Francisco: 1 Gramm, 1 Mais, 1912, 2 Packards, 1910, 2 Whites, 1910 and 1914.
Vera Cruz, Mexico: 1 White, 1913.
Washington: 1 Mack, 1911, 3 Whites, 1911, 1912 and 1913 (the latter used by the War College).
Washington Barracks: 1 White, 1911.
West Point: 2 Alcos, 1912, 1 White, 1911.

During the fiscal year 1914 the following additional trucks were purchased, bringing the total to 61:
1 Driggs-Seabury (New York, NY), 1 Federal (El Paso, Tex.), 2 Jefferys (Eagle Pass, Tex. and Sam Houston, Tex.), 1 Kelly-Springfield (St. Louis, Mo.), 3 Lippard-Stewarts (Rio Grand City, Tex., Brownsville, Tex. and Fabens, Tex.), 1 Lord-Baltimore (Washington Depot), 2 Macks (El Paso, Tex.), 2 Studebakers (Fort Rosecrans, Cal.), 5 Velies (1 each for Hachita, NM, Douglas, Ariz. and Nogales, Ariz. and 2 for Marfa, Tex.), and 8 Whites (4 for Galveston, Tex. and 2 each for Honolulu, Hawaii and Sam Houston, Tex.). These were all $1\frac{1}{2}$-ton 4×2 trucks, except for the Driggs-Seabury and Kelly-Springfield, which were 2-tonners.

There were also 27 cars (10 bought in 1914) and three ambulances. In addition the QMC had 15 trucks, eight cars and one ambulance in the Philippines. During 1915 some 30 more trucks were bought by the QMC. The Navy Dept. at this time had about 65 motor vehicles. The USMC had bought 72 vehicles during 1909–17. In 1916 came the first opportunity for the motor truck to prove itself in active service. This was during the Punitive Expedition in which General John J. Pershing was sent into Mexico in pursuit of Pancho Villa. Pershing demanded trucks, mule-drawn wagons being too slow for his 'flying columns'. There resulted a scramble to purchase any kind of truck, regardless of suitability. 13 types of trucks, made by eight different manufacturers, were used. In addition there were tank trucks, some machine shop trucks and trailers and a few tracked vehicles. Meanwhile the war in Europe had broken out and the great demand for trucks upon US factories by foreign governments accelerated the development of commercial trucks from a military point of view. It was during this period that the four-wheel drive trucks as produced commercially by FWD and Jeffery proved their suitability for military service. Most of these were shipped to Europe; a smaller number went to the Mexican border.

Concurrently with the operation of the truck trains for the Punitive Expedition, the Quartermaster Corps had proceeded with its development work of which the earlier mentioned road test of 1912 was a part.

Its policy as outlined in 1916 covered interchangeability, a guarantee of spare parts supply from the manufacturer, and the development of specifications for commercial $1\frac{1}{2}$-ton trucks, both two- and four-wheel drive. This was later modified to include

USA

3-ton trucks. In this year the Quartermaster Corps in conjunction with the Truck Standards Division of the Society of Automotive Engineers began work on standard specifications for the design and production of 1½- and 3-ton (Class A and B) trucks, specifications for which were issued in 1917. Before they were issued a state of war had been declared between the United States and Germany.

When war was declared every service in the Army needed a great amount of motor transportation and each service at once, simultaneously, started to develop and obtain it. The principal agents were the Quartermaster Corps, the Ordnance Department, the Corps of Engineers, the Signal Corps, and the Medical Department. Two of the most important items of the truck specifications were standardization and interchangeability, and even before orders for commercial vehicles were placed manufacturers began to report difficulties in meeting these requirements. In July 1917 a committee representing the Quartermaster Corps and the SAE was formed to consider this matter. The Committee soon reported that by January 1918 it would be possible to design, test and place in quantity production completely standardized government models. It was recommended that this be done and that in the meantime no more commercial vehicles be purchased than were necessary for immediate needs.

This standardization policy was approved. Specification, design, and production work of the Quartermaster Corps centred around this programme. New designs and experiments were carried out on three principal trucks. These were the Standard AA, Standard A, and Standard B trucks. The latter, popularly known as the Liberty Truck, reached quantity production, unlike the others which were not placed in production as wartime exigencies required that the full available capacity of facilities on commercial makes of vehicles similar to these types, was taken.

The Ordnance Department, traditionally the principal purveyor for the Artillery, had as far back as 1915 interested itself in the four-wheel-drive type truck as the vehicle best suited for motorized artillery. In 1916 a recommendation of a board of officers, approved by the Secretary of War, called for trucks for ammunition and special artillery vehicles to be furnished by the Ordnance Department.

War development of motor transporation by the Ordnance Department was along three lines—crawler tractors, special body vehicles, and ammunition trucks. Crawler tractors of several types were designed and produced and many sent overseas. The special vehicles were repair and supply trucks for the Artillery and Ordnance Repair Units. Many different trucks and trailers were developed using selected chassis and a few individual bodies differing in equipment. For ammunition trucks the Ordnance Department purchased the Nash and FWD chassis under Quartermaster specifications (May 1917) and designed the well-known ammunition body. This body was later abandoned. Late in 1917 the Ordnance Department took up the standardization idea and developed the standard four-wheel-drive chassis known as the Militor. This however was not put into quantity production and only a few vehicles of this type ever came into the service. Development of this type, however, was carried on.

The Signal Corps, realizing that existing trucks would not meet its needs, developed two standard trucks for the Air Service—the Heavy and Light Aviation trucks. These trucks were produced in quantity.

The Corps of Engineers selected Mack trucks for heavy work and designed a number of special bodies such as dump and lithographic; also a lighter Mack and Cadillac searchlight truck.

The US Army in France, 1917; new motor vehicles ready for issue

USA

The Medical Department developed ambulances. Before the war several types were used but during the war they specialized on the Ford and the GMC. They also developed ambulance trailers.

Early in 1918 motor transportation design was centralized in the Motor Vehicle Board. This Board decided on standardization of vehicles as follows: three types of passenger cars, three motorcycles, a few types of truck chassis and eight trailer chassis. This resulted in a modest total of 103 types of vehicles, including special bodies and equipment demanded by the various services. By 1920 the responsibility for motor transport was returned to the QMC. The Ordnance Department had responsibility for tractors and armoured combat vehicles. After the war, during which reputedly about 275,000 vehicles, of over 200 makes, had been procured, many thousands of vehicles were left behind in Europe, others being handed over to State Highway and US Government departments. Vehicles that would be needed by the Army had been selected before the surplus stock was disposed of. As a result of this ample fleet of vehicles, by 30 June 1929 the QMC had purchased only 763 new vehicles, of which 709 were cars. Of the 17,305 vehicles on hand, 16,542 were of World War I vintage. The Quartermaster Corps pushed hard for standardization: standardization would mean efficient spare parts supply. In 1931 Congress appropriated $406,800 for Army trucks, specifying how much each was to cost. When the vehicles could not be bought for the stipulated price, the Quartermaster General directed the purchase of units and assemblies of the desired truck types. When these were assembled, the result was 129 vehicles of 6 types, ranging in payload capacity from $1\frac{1}{2}$-ton upwards. Most of these types are shown in this book, as USA/QMC trucks. It was not until Congress passed the War Department Appropriation Act for fiscal year 1935, which forbade the use after 1 January 1935 of funds to repair pre-1920 vehicles, that the Corps could begin to eliminate such vehicles from further service.

From this point in time quantities of new vehicles of various types were purchased; most of these were conventional or modified conventional commercial types but there were some special vehicles as well, as will be seen in the following pages. During the 1930s the Ordnance Department designed many types of armoured vehicles, a few of which are also shown.

In 1939, War Department policy on truck procurement was modified to recognize the need for standardization. General purpose tactical motor vehicles would comprise five chassis types, namely $\frac{1}{2}$-, $1\frac{1}{2}$-, $2\frac{1}{2}$-, 4-, and $7\frac{1}{2}$-ton. Body types would be reduced to a minimum and be adaptable to one of the five standard chassis types. All-wheel drive was declared standard for all military vehicles, others being substitute standard. Vehicles were to be procured once a year in the largest possible numbers, for maximum standardization. Maximum interchangeability of major parts and components was needed and the required military characteristics for the $\frac{1}{2}$-, $1\frac{1}{2}$- and $2\frac{1}{2}$-ton tactical chassis were approved for

Typical US Army truck of the mid 1930's: commercial chassis/cab with special cargo and personnel body. Note hinged troop seats.

USA

vehicles procured in 1940. By this time the Second World War had broken out and subsequent developments of US military vehicles are dealt with in the Observer's Directories of World War II and post-1945 vehicles.

Who's Who in the US Automotive industry

Note: owing to the large number of makes and the complexity of some of their histories, only the more important ones which were operative during the late 1930s are listed.

(American) Bantam	American Bantam Car Co., Butler, Pa.
Allis-Chalmers	Allis-Chalmers Mfg Co., Milwaukee, Wisc.
Autocar	Autocar Co., Ardmore, Pa.
Biederman	Biederman Motors Corp., Cincinnati, Ohio.
Brockway	Brockway Motor Co., Cortland, NY.
Buick	Buick Motor Div. of General Motors Corp., Flint, Mich.
Cadillac	Cadillac Motor Car Div. of General Motors Corp., Detroit, Mich.
Caterpillar	Caterpillar Tractor Co., Peoria, III.
Chevrolet	Chevrolet Motor Div. of General Motors Corp., Detroit, Mich.
Cletrac	Cletrac Tractor Co. (The Cleveland Tractor), Cleveland, Ohio.
Clydesdale	Clydesdale Motor Truck Co., Clyde, Ohio.
Coleman	Coleman Motors Corp., Littleton, Colo.
Corbitt	The Corbitt Co., Henderson, NC.
Crosley	The Crosley Corp., Cincinnati, Ohio.
Cunningham	James Cunningham, Son & Co., Rochester, NY.
Diamond T	Diamond T Motor Car Co., Chicago, Ill.
Dodge, Fargo	Dodge Brothers Corp., Div. of Chrysler Corp., Detroit, Mich. (military sales through Fargo Motor Corp., Div. of Chrysler Corp.).
Duplex	Duplex Truck Co., Lansing, Mich.
Federal	Federal Motor Truck Co., Detroit, Mich.
Ford	Ford Motor Co., Dearborn, Mich.
FWD	Four Wheel Drive Auto Co., Clintonville, Wisc.
GMC	General Motors Truck & Coach Div., Yellow Truck & Coach Mfg. Co., Pontiac, Mich.
Harley-Davidson	Harley-Davidson Motor Co., Milwaukee, Wisc.
Hendrickson	Hendrickson Motor Truck Co., Chicago, III.
Hudson	Hudson Motor Car Co., Detroit, Mich.
Hug	The Hug Co., Highland, III.
Indian	Indian Motorcycle Co., Springfield, Mass.
Indiana	Indiana Motors Corp., Cleveland, Ohio (Div. of White from 1932–39).
International	International Harvester Co., Chicago, III.
John Deere	John Deere & Co., Moline, III.
Linn	Linn Mfg Corp., Morris, NY.
Mack	Mack Mfg Co., Allentown, Pa.
Marmon-Herrington	Marmon-Herrington Co. Inc., Indianapolis, Ind.
Minneapolis Moline	Minneapolis Moline Power Implement Co., Minneapolis, Minn.
Moreland	Moreland Motor Truck Co., Burbank, Ca.
Nash	Nash Motors Co., Kenosha, Wisc.
Oshkosh	Oshkosh Motor Truck Inc., Oshkosh, Wisc.
Packard	Packard Motor Car Co., Detroit, Mich.
Plymouth	Plymouth Motor Corp., Div. of Chrysler Corp., Detroit, Mich.
Pontiac	Pontiac Motor Div. of General Motors Corp., Pontiac, Mich.
Reo	Reo Motor Car Co., Lansing, Mich.
Sterling	Sterling Motors Corp., Milwaukee, Wisc.
Studebaker	The Studebaker Corp., South Bend, Ind.
Walter	Walter Motor Truck Co., Ridgewood, Long Island, NY.
Ward LaFrance	Ward LaFrance Truck Corp., Elmira, NY.
White	White Motor Co., Cleveland, Ohio.
Willys	Willys-Overland Motors Inc., Toledo, Ohio.

USA
MOTORCYCLES and POWER CARTS

Although motorcycles were widely used, the number of makes was restricted to only a few.

Harley-Davidson and Hendee (Indian) were the main suppliers, with 14,666 and 18,018 resp. (plus 14,332 and 16,804 sidecars resp.) during World War I alone. Cleveland supplied 1476 solo machines.

Power Carts were small cross-country cargo carriers, a few of which were built by RIA in the early 1920s with motorcycle engines. The operator walked or ran at the rear. Some tracked amphibious models were made which the operator could control from inside or at the rear. Caterpillar produced a V-twin Hendee (Indian)-powered model in 1920, known as the Peoria Power Cart. The Syracuse Power Cart, also of 1920, had a 5-bhp single-cylinder engine.

Motorcycle, Solo, Heavy (Harley-Davidson) V-2-cyl., 8.8 SAE-hp, 3F, wb $59\frac{1}{2}$ in, $92 \times 30 \times 37$ in, 400 lb. With sidecar: $92 \times 64 \times 36$ in, 550 lb. Wheeler & Schebler carb., Bosch & Berling high-tension magneto. Bore and stroke 3 15/16 × $3\frac{1}{2}$ in. 1917–18.

Motorcycle, with Sidecar, Heavy (Hendee Indian) V-2-cyl., 61 CID, 3F, wb 60 in, $92 \times 62 \times 40$ in, 586 lb. Without sidecar: $92 \times 30 \times 42$ in, 370 lb. Wheeler & Schebler carb., Dixie high-tension magneto. 28 × 3 clincher tyres; pressure 40 lb. Sidecar payload 350 lb. 1917.

Motorcycle, with Sidecar, Heavy (Hendee Indian) V-2-cyl. Indian machine of *c.* 1915. Standard sidecar chassis with special superstructure for Lewis machine gun on detachable tripod. Some were fitted with an armour shield. Similar outfit on Harley-Davidson.

Motorcycle, with Sidecar, Heavy (Indian) Typical sidecar combination of about 1939. These, as well as solo machines and motor tricycles (with spoke or disc wheels), were supplied by Harley-Davidson of Milwaukee, Wisc., and Indian of Springfield, Mass, during the late 1930s.

Motor Tricycle, Heavy (Delco) In Nov. 1938 the Delco Appliance Division of General Motors was asked to design a new type military tricycle with HO-2-cyl. engine, shaft drive and car-type rear axle. Delivered in April 1939, it was followed by a solo machine. Both remained experimental.

Power Cart, Large (Ordnance) Harley-Davidson V-2-cyl., 7½-bhp, 2F1R, 84½ × 42 × 40 in, 735 lb. Payload 300 lb. Produced in 1920/21. Followed by larger type in 1922 (99½ × 42 × 38 in, 895 lb, payload 450 lb). Normal speed on roads 5–6 mph (max. 7.9 mph; low gear 1½ mph).

Power Cart, Small, M1924E (Ordnance) Harley-Davidson V-2-cyl., 18 bhp, 2F1R, 80 × 42 × 40 in, 840 lb. Payload 450 lb. Steering by articulation. Cargo box at front, engine and controls at rear. Designed and produced by Rock Island Arsenal in 1924/25.

USA

CARS, 4×2

Following the use of a variety of civilian type cars, in 1917 three makes were standardized for Army service: Ford (light open type), Dodge (medium, open and closed) and Cadillac (heavy, open and closed). Up to 1 Nov. 1918 these manufacturers had received orders for a total of 35,974 cars, of which 14,526 had been completed; about half of these had been 'floated' overseas. Ford had orders for 20,201 Model T tourers, 5,379 of which were delivered.

2344 Fords, 3390 Dodges and 1503 Cadillacs were shipped to France. US forces in France also used European cars, including Fiats.

During the 1930s civilian type sedans were purchased from Buick, Cadillac, Chevrolet, Fargo (Plymouth), Ford, etc. Station wagons were supplied on ½-ton and heavier truck chassis.

Car, Heavy, Balloon Destroyer (Cadillac) One of the US Army's first SP guns was this Davidson-Cadillac of 1910. It was reported that 'two Colt automatic guns were used to take pot shots at balloons with little accuracy'. Two were made, following a slightly smaller one in 1909.

Car, Passenger, Medium, Open (Dodge) 4-cyl., 37 bhp, 3F1R, wb 114 in, 153×65×81 in, 2400 lb. Introduced in 1914. Standardized for Army use in 1917. Also supplied with detachable winter top and with all-enclosed bodywork. Total delivered: 7191.

Car, Passenger, Heavy, Closed (Cadillac) V-8-cyl., 70 bhp, 3F1R, wb 132 in, 190×68×84 in, 4670 lb. Also open type (touring; wb 125 in, 4280 lb). Both were standardized for use by US Army (1956 supplied). This particular car was used by General John J. Pershing (shown).

Car, Passenger, Heavy, Closed (Hudson) This staff car was one of several types built by the Hudson Motor Car Co. for the US forces in World War I. Other body styles on the Super Six chassis included open touring and ambulance. Front bumper was special fitment.

Car, Passenger, Heavy, Closed (Locomobile) Built in 1917 by the Locomobile Company of America, this 6-cyl. car (with dual ignition) was used by General Pershing in France and subsequently preserved. It measures 204 × 87 × 71½ in. Equipped with dual rear tyres, four spares.

Car, Passenger, Light, Open (Studebaker, Ford) The British employed many Studebakers and Fords (Model T) and in 1918 possessed 1447 and 18,984 respectively (incl. vans, etc.). The USMC's first motor vehicle was also a Studebaker (Model 30, bought in 1909). (IWM photo Q5745).

Car, Passenger, Heavy, Open (White) 4-cyl., 28.9 SAE-hp, 4F1R, wb 140 in, 208 × 69 × 88 in, 5250 lb. Officially known as Staff Observation Truck, this 9-passenger car was built on a 1-ton White 'TEBO' chassis, with the makers' own L-head engine. Pneumatic tyres, size 36 × 9. 1917.

Car, Passenger, Light, Open (Ford) 4-cyl., 40 bhp, 3F1R, wb 103½ in, 151 × 66 × 71 in, 2215 lb. Standard Phaeton (Body Type 35A) on 1929 Ford Model A chassis. Other military Model As were 1930/31 sedans and Commercial Station Wagons (Body Type 150B).

Station Wagon, Medium (Chevrolet) 6-cyl., 57–60 bhp, 3F1R, wb 112 in, 173 × 76 × 78 in, 3585 lb. Based on 1935 Chevrolet ½-ton 4 × 2 Model EB commercial chassis with some modifications (car-type pressed-steel wheels with oversize tyres, etc.). Used by Coast Artillery.

Car, Passenger, Light, Sedan (Fargo) 6-cyl., 66 bhp, 3F1R, wb 107¾ in, length 181 in, 2925 lb. 201.3 CID L-head engine. Basically this was a 1934 Plymouth PG sedan with some modifications (pressed-steel wheels, light truck type hood (bonnet), brush guard, etc.). Tyres 5.25-17.

Car, Passenger, Medium, Sedan (Cadillac) Not a standard US Army sedan but one of many which were impressed into military service at the beginning of World War II. This is a 1939 Cadillac V8 Series 60, which served as an officers' car in England.

USA

FIELD CARS
4×2 and 4×4

Makes and Models: American Austin (4×2, c. 1933). Cadillac (4×2, 1925). Chevrolet (4×2, 1927, 1929, 1932, etc.). Crosley (4×2, 1939/40). Dodge/Fargo (4×4, 1939/40). Ford T (4×2, from mid-1920s). Ford/Marmon-Herrington (4×4, late 1930s). International (4×2, 1934). Oldsmobile (4×2, 1932). Pontiac (4×2, 1928), etc.

General Data: Grouped under the general heading of Field Cars are passenger car and light commercial chassis which were modified for off-road operation. These modifications varied from the fitting of oversize tyres on open touring cars to the conversion from rear- to all-wheel drive. The earliest examples of American field cars were commercial Chevrolet and Ford chassis, equipped with larger tyres and a simple platform with four bucket seats. These first appeared in the mid-1920s and were the result of a recommendation to the US War Department in 1924 to introduce light vehicles (weighing not over 900 lb and with a capacity of 450 lb) into the army for 'cross-country work of a general military nature'. The vehicles were to be designed to carry machine guns of 27-mm calibre, light mortars, ammunition and signals equipment. Although the Model T Ford of the period could be converted to all-wheel drive (a conversion kit was offered by J. F. Livingood of New Virginia, Iowa; it was claimed that it could be fitted to any Ford chassis in 3 to 4 hours), this was not done on the Army version. Somewhat more sophisticated models were built on Oldsmobile and Pontiac chassis; these featured armour protection for the radiator and an armour shield replacing the windscreen.

The real breakthrough came when during the mid-1930s the firm of Marmon-Herrington introduced a four-wheel drive conversion of the standard ½-ton Ford commercial chassis. At first these were supplied as pick-up trucks (with open and closed cab) but by 1937 a command car version appeared which became the progenitor of the well-known Dodge Command and Reconnaissance cars of World War II. A scaled-down version of these vehicles became known as the 'Jeep' (of which the Marmon-Herrington design was named the grandfather).

Vehicle shown (typical): Field Car, 4-seater, 4×2 (Chevrolet AA)

Technical Data:
Engine: Chevrolet 4-cylinder, I-I-W-F, 170.9 cu. in (3 11/16 × 4 in; 2801 cc), 21.7 SAE-hp.
Transmission: 3F1R.
Brakes: mechanical.
Tyres: oversize balloon.
Wheelbase: 103 in.
Overall l × w × h: N.A.
Weight: N.A.
Note: based on 1927 Series AA Light Delivery chassis, which was a variant of the Chevrolet Series AA Capitol car. Running boards and rear wings were standard equipment, supplied with the chassis.

Field Car, 4-seater, 4×2 (Ford) Generally similar to the Chevrolet on the previous page was this Model T Ford cross-country car, photographed at Aberdeen Proving Ground in 1927. The Ford, however, had a pedal-operated planetary type transmission.

Field Car, AA Gun, 4×2 (American Austin) Based on a modified midget truck chassis, with super-balloon tyres, was this triple anti-aircraft machine gun. Chassis was by American Austin (1930–35, later Bantam). Steering wheel was detachable. c. 1933.

Field Car, 4-seater, 4×2 (Chevrolet) One of several 1929/30 Chevrolets of the US Marine Corps (where they were known as: Automobile, Chevrolet, Cross Country), shown with Browning machine gun mounted in the back. Note alternative gun bracket on cowl.

Field Car, 4-seater, 4×2 (Pontiac) Several 1928 Pontiac car chassis were fitted with special wheels and oversize tyres, bucket seats, armour plating, etc. This version carried two-way radio equipment. Note aerial mount. German *Reichswehr* used similar cars (*qv*).

V

Field Car, 5-seater, 4×2 (Ford) Squad Car, T1, based on 1935 Ford V8 Model 48 112-in wb Phaeton (Body Type 750). Gross wt 4500 lb. Length 184 in. Armour thickness $\frac{1}{4}$- and $\frac{1}{2}$-in. Max. speed 80 mph. Bracket for .30 calibre machine gun. Oversize tyres.

Field Car, 5-seater, 4×4 (Ford/Marmon-Herrington) Command car on 1937 Ford V8 Model 77 Commercial chassis, converted to four-wheel drive by Marmon-Herrington (Model LD1-4, wb $113\frac{1}{2}$ in, tyres 7.50-15). Series production from 1939 on Dodge T202 4×4 chassis (VC1).

Field Car, 4-seater, 4×2 (Crosley) Based on small car with twin-cylinder air-cooled engine, introduced in 1939 by Powell Crosley, Jr. Claimed to be able to carry four fully-equipped soldiers at up to 50 mph and 50 mpg. Experimental only.

Carrier, Machine Gun, 4×2 (Howie-Wiley) Experimental low-silhouette MG carrier, designed by Col. Robert G. Howie and built by Master Sgt Melvin C. Wiley, using 4-cyl. Am. Austin engine. Poor ground clearance and rough ride (unsprung wheels) earned it the nickname 'Belly Flopper'. 1937.

USA

AMBULANCES

Principal makes of ambulance chassis during World War I were Ford and GMC. These were the standard US Army chassis for this purpose. Many other chassis were used, however, both by the Americans and for export, the British using thousands of Buicks, Fords, GMCs, etc. The US Marine Corps used mainly Ford and King chassis. Cunningham was one of the suppliers of ambulance bodies. During the 1930s ambulances were supplied by Chevrolet, Ford, Henney (Packard), International, etc. Most common were Chevrolet and Ford commercial panel vans, converted for ambulance use.

Ambulance, Heavy, 4×2 (White) One of the first ambulances of the US Army Medical Corps at Washington, DC, was this steam-propelled White, delivered in 1906. The White Motor Co. of Cleveland, Ohio, became large-scale suppliers of military vehicles of many types.

Ambulance, Light, 4×2 (Dodge) Based on the widely-used Dodge $\frac{1}{2}$-ton truck chassis of World War I was this extended-wheelbase ambulance for eight sitting patients. It was issued only for domestic use. Manufacturers were Dodge Brothers of Hamtramck, Detroit.

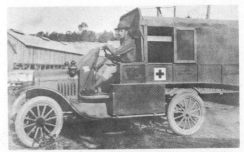

Ambulance, Light, 4×2 (Ford) Special long chassis of the Model T, built in considerable numbers throughout the war for the US Army, USMC (shown), British Army, etc. The US Army ordered over 10,000, 5,340 of which were delivered before the Armistice.

Ambulance, Heavy, 4×2 (GMC) 4-cyl., 38 bhp, 3F1R, wb 132 in, 192×74×94 in, 4540 lb. Wooden body for four stretchers or eight men sitting. Standard ¾-ton chassis with 35×5 tyres. 5553 produced until Nov. 1918; 4001 shipped overseas.

Ambulance, Heavy, 4×2 (King) Fast 8-cyl. car chassis, made by King Motor Car Co. in Detroit, used by USMC for ambulances and armoured cars. Body of standard type, produced elsewhere, arranged to carry eight casualties. 1917/18.

Ambulance, Heavy, 4×2 (Chevrolet) 6-cyl., 57 bhp, 4F1R, wb 131 in, 219×78×118 in, 5900 lb. Based on regular 1935 1½-ton panel van. In 1939 and 1940 similar ambulances were supplied by Chevrolet and Ford, also with panel van body and dual rear tyres, but equipped with radiator brush guard.

Ambulance, Heavy, 4×2 (International) Photographed in January, 1934, this US Marine Corps vehicle was known as: Truck, Ambulance, International AL-3, 10-passenger. It was a long-wheelbase chassis of International Harvester's contemporary A-line.

USA

TRUCKS, LIGHT
Pre-1920

Light trucks were those with payload ratings of up to one ton. Among the first makes used were: Autocar (1910), Cadillac (1905), Federal (1911), Waverly (1908; electric), Winton (1904), etc. Many others were used during 1914–16. For use in World War I (1917–18) the US Army standardized on Dodge and Ford ($\frac{1}{2}$-ton), GMC ($\frac{3}{4}$-ton) and White (1-ton). These chassis were used extensively and with various body types. The USMC used Fords, Studebakers, etc. Many of these trucks remained in service throughout the 1920s, when very few new vehicles were procured.

American light trucks and vans used by the British Army during World War I included Buick, Ford, Garford, Studebaker, etc.

Truck, $\frac{1}{2}$-ton, 4×2, Cargo (Dodge) Chassis same as Dodge car (qv) but stronger springs and 33×4 tyres (plain tread front, non-skid rear). Dim. 153×65×85$\frac{1}{2}$ (w/tilt) in, 3800 lb. 2644 supplied, as well as 1012 Light Repair trucks. 1802 went overseas.

Truck, $\frac{1}{2}$-ton, 4×2, Cargo (Ford) Model T Canopy Delivery truck as used by the USMC (slightly larger version on Studebaker). Army ordered 12,002 Light Delivery trucks, 5492 of which had been delivered by Nov. 1918. 7206 were shipped overseas.

Truck, $\frac{3}{4}$-ton, 4×2, Cargo (GMC) 4-cyl., 38 bhp, 3F1R, wb 132 in, 196×64×106 in, 4450 lb. Model 16 standard chassis, also without hard-top at rear and with ambulance (qv) and 180-gal. water tank bodywork. Tyres 35×5. 1917.

Truck, ½-ton, 4×2, Personnel (Hudson) Based on Hudson Super Six car chassis which was also used with passenger car and ambulance bodywork. Vehicle shown provided accommodation for eight passengers, including the driver. Note lockers under seats. 1917.

Truck, ½-ton, 4×2, Van (Waverly) Electric delivery wagon, purchased in 1908 for use by the War College. Carried mail and records between War Dept. at Washington, DC, and War College at Arsenal Point, covering about 35 miles a day. Capacity 1200 lb.

Truck, 1-ton, 4×2, Reconnaissance (White) 4-cyl., 28.9 SAE-hp, 4F1R, wb 140 in, 214×69×101 in, 5500 lb. Two front and rear seats, all back-to-back, containing equipment lockers, gun chest, ammunition trays. Similar: Machine Gun truck. 2196 chassis supplied, 1917–18.

Truck, ½-ton, 4×2, Telegraph (Winton) Bought by the US Army Signal Corps in 1904 this 2-cyl. 'telegraph car' was used at Fort Omaha, Neb. A similarly-bodied but shorter Cadillac 1-cyl. 'repair car' followed in 1905 for use at Fort Leavenworth, Kan.

USA

TRUCKS
½- to 1-TON
4×2, 4×4, 6×4
1920–1940

Makes and Models: (Note: 4×2 unless stated otherwise) Chevrolet ½-ton (Pickup, Recce, Panel, Carryall, Canopy Express, Telephone Maintenance, Light Repair, Pigeon Loft, Sedan Delivery, etc., 1929–40), ¾-ton (Pickup, Panel, 1938–40), ¾-ton, 6×4 (Cargo, etc., 1927–29), 1-ton (Canopy Express, Dump, Stake, 1938–40), etc. Diamond T ½-ton (Recce, 1935; Cargo, 1940). Dodge ½-ton (Recce, 1924), Dodge/Fargo ½-ton (Pickup, Panel, Recce, Telephone Maintenance, Sedan Delivery, 1934–40), ½-ton, 4×4 (Cargo, Recce, Radio, 1939/40), etc. Fargo (see Dodge). Ford ½-ton (Pickup, 1932, 1935), etc. Ford/Marmon-Herrington ½-ton, 4×4 (Pickup, Cargo, etc., 1937–40). GMC ½-ton (Panel, 1936; Carryall, 1939/40); ½-ton, 4×4 (Pickup, 1940), etc. International ½-ton (Recce, 1934–35); ½-ton, 4×4 (Pickup, 1938), etc. Reo ½-ton (Pickup, Panel, 1935), etc.

General Data: By far the majority of trucks in the ½- to 1-ton payload class purchased by the US government for military use were standard commercial 'off-the-shelf' vehicles with standard pickup, panel delivery, and other body styles. The term Recce (short for Reconnaissance) in the above listing usually denoted a station wagon with certain modifications. Many similar vehicles, particularly Fords, were used by foreign armies. Marmon-Herrington were the main suppliers of four-wheel drive models, usually conversions of standard Ford V8 light trucks. The first of these was built in July 1936, and tested on the famous King Ranch in Texas. The Belgian Army was the first to show an interest in such a vehicle and the first order from the US QMC came in 1937. They were also ordered by the USMC and Coast Guard. The latter ordered a series of similar conversions of the International D-line pickup in 1938. Marmon-Herrington also converted Ford passenger cars, sedan delivery vans and heavier types to four-wheel drive.

During the 1920s the US military authorities carried out some interesting modifications to improve cross-country performance of light 4×2 trucks, e.g. in 1922 and 1923 there appeared tracked 3- and 4-axle conversions of the 1917 Dodge Light Repair truck, followed in 1927–29 by 3-axle Chevrolets.

Vehicle shown (typical): Truck, ½-ton, 4×4, Pickup (Ford/Marmon-Herrington LD3-4).

Technical Data:
Engine: Ford V8 8-cylinder, V-L-W-F, 239 cu. in (3917 cc), 95 bhp at 3600 rpm.
Transmission: 4F1R×2.
Brakes: hydraulic.
Tyres: 7.50-15.
Wheelbase: 113½ in.
Overall l×w×h: 196×73×86 in.
Weight: 3995 lb. GVW: 4995 lb.
Note: 64 supplied in 1939. Conversion of standard Ford Commercial Model 99C pickup. Max. speed 45 mph. Also produced with closed cab.

Truck, ½-ton, 4×2, Light Repair (Chevrolet) Used by Motor Sergeant of Field Artillery Battery this was basically a 1933 Chevrolet Series CB pickup with military open cab, radiator guard, etc. Wheelbase 109 in. Engine 206-cu. in OHV Six. 18-in tyres.

Truck, ½-ton, 4×2, Tel. Maint. and Repair (Chevrolet) 6-cyl., 75 bhp, 3F1R, wb 112 in, 189×71×74 in, 3450 lb. Tyres 6.50-16. Commercial Series HC chassis/cab with 216-cu. in OHV engine and military type telephone maintenance and repair bodywork. Vehicle cost $795.

Truck, ½-ton, 4×2, Bomb Service, T1 (Ford) Prototype of what became the standardized Truck, ½-ton, 4×2, Bomb service, M1 (Diamond T, Ford, GMC, 1940/41). This unit was based on the standard 1939 Ford V8 112-in wheelbase Commercial truck in chassis/cowl form.

Truck, ½-ton, 4×4, Chassis (Ford/Marmon-Herrington) Modified 1937 Ford V8 Commercial Model 77 chassis. Four-wheel drive conversion by Marmon-Herrington for US Army (Ambulance Unit). Most unusual was the forward-mounted cab with fixed windshield and soft top.

Truck, 1-ton, 4×4, Cargo (Ford/Marmon-Herrington) The first Marmon-Herrington 1-ton Ford conversion for the US Coast Guard (in 1938) was basically Ford's newly-introduced 1-ton Model 81Y chassis/cab with Express pickup body (Type 830). Engine was 85-bhp V8; also listed as 90 HP.

Truck, ½-ton, 4×4, Cargo (International/Marmon-Herrington) The Marmon-Herrington Co. concentrated on Ford for their numerous low-cost all-wheel drive trucks, but occasionally converted trucks of other makes, like a series of D-line Internationals for the US Coast Guard in 1938.

Truck, ½-ton, Full-Track (Dodge/Chase) In 1923 this modification of a 1917 Dodge Light Repair truck appeared, featuring two additional axles and Chase endless tracks. In 1922 a tracked tandem-axle conversion had been made (see section on Half-Tracks).

Truck, ¾-ton, 6×4, Cargo (Chevrolet) The Ordnance Department tested a 4-cyl. 6-wheel Chevrolet cross-country truck at APG in late 1927. In 1929 a small number of the new 6-cyl. model were taken into service. Specimen shown (6-cyl.) carried Browning .30-cal. mg and crew.

USA

TRUCKS, MEDIUM, 4×2
Pre-1920

Shown here are some of the medium 4×2 trucks (1½- to 2-ton) of 1917–18 and before. Among the earliest makes used were Atterbury, Autocar, Long Distance, Mack, Sampson, Studebaker, White, etc. The Studebaker was an electric truck, built in 1908 for the US Naval Gun Factory. For the war the US Army standardized the White Model TBC but also used large numbers of Garfords, Packards, Pierce-Arrows, etc. A military Class A 'Liberty' truck was designed but not produced in quantity. The Signals Corps and its Air Service branch did develop and use a standard truck, the 'Light Aviation', of which 3210 were built (out of 3900 ordered) by Denby, GMC, Paige and Republic during 1917–18. US forces overseas also used British, French and other foreign trucks.

Truck, 4×2, Artillery Repair (US Long Distance) 4-cyl., 24 bhp, 4F1R, wb 110 in, 180×72×91½ in, 12,000 lb (gross). Winch-equipped 'battery repair wagon and forge' with lathe, grindstone, and other tools and stores for machinists, carpenters, saddlers and farriers. 1902.

Truck, 1½-ton, 4×2, Cargo (Atterbury) This 4-cyl. shaft-drive medium truck was supplied to the US Quartermaster Corps in 1913. Manufacturers were the Atterbury Motor Car Company of Buffalo, New York, who operated from 1910 till the mid-1930s.

Truck, 2-ton, 4×2, Cargo (Mack) In 1913 the International Motor Company (later renamed Mack Manufacturing Corp.) delivered this 4-cyl. 2-ton chain-drive hard-top 'delivery wagon' to the QM Dept. of the USMC Navy Yard in New York, NY.

Truck, 1½–2-ton, 4×2, Cargo (Garford) 4-cyl., 29 SAE-hp, 3F1R, wb 142 in, 222 × 71 × 61 in, 5225 lb. Solid tyres, front 36 × 4, rear 36 × 7. Wisconsin 312-cu. in (4¼ × 5½ in) L-head engine. HT magneto ignition. 1010 delivered, 1917–18, of 5010 ordered.

Truck, 2-ton, 4×2, Chassis (GMC) Built by the General Motors Truck Co. in Pontiac, Mich. since 1916, the GMC Model 41 was a worm-drive chassis, many of which were used by the Air Service. They were also made with pneumatic tyres.

Truck, 1½–2-ton, 4×2, Cargo ('Light Aviation') These trucks were assembled mainly for the Signal Corps, from known and tried units (engines, transmissions, axles) by Denby (488), GM Truck (1888), Paige (480) and Republic (354). 1829 were shipped overseas. There was also a 'Heavy Aviation' (*qv*).

Truck, 1½–2-ton, 4×2, Cargo (Packard) 4-cyl., 25.6 SAE-hp, 4F1R, wb 144 in, 220 × 69 × 114 in, 5890 lb. Full-floating worm-drive rear axle. Tyres 34 × 3½, dual rear. 276-cu. in (4 × 5½ in) L-head engine. 1917–18. Packard also supplied many trucks to the British Army.

Truck, 2-ton, 4×2, Cargo (Pierce-Arrow) 4-cyl., 25.6 SAE-hp, 4F1R, wb 150 in, 228×72×126 in, 7950 lb. GVW 11,950 lb. 276-cu. in T-head engine. Worm final drive. 34×4 solid tyres, dual rear. Speed 15 mph. 1916/17. Heavier types supplied to British and French.

Truck, 1½–2-ton, 4×2, Cargo (USA/QMC 'Liberty') 4-cyl., 44 bhp, 4F1R, wb 144 in, 220½×73×120 in, 7765 lb. 312-cu. in (4¼×5½ in) L-head engine. Prototype for military Class A medium truck which, unlike the Class B 3-ton (qv), was not produced in quantity. 1918.

Truck, 1½-ton, 4×2, Ordnance Repair (White) 4-cyl., 22.5 SAE-hp, 4F1R, wb 145½ in, 220×70×62 in, 6000 lb approx. Basically similar to White Model TBC 1½–2-ton cargo truck. Own 3¾×5 1/8-in L-head engine. Tyres front 38×7, rear 40×8, on wooden artillery wheels. 1917.

Truck, 1½-ton, 4×2, AA Gun (White) The Model TBC chassis was used for several body types, incl. Cargo, Repair, Radio Operating, Wire Reel, and SP Gun mount (shown). Some Whites were still in use by the late 1930s, albeit converted to have pneumatic tyres.

USA

TRUCKS, 1½- to 2-TON, 4×2
1920–1940

By far the majority of new vehicles in this class were standard commercial type chassis purchased from the 'big three' (Chrysler, Ford, General Motors) during the 1930s. They were usually modified to meet military requirements by such additions as radiator brush guards, tow hooks, brackets for pioneer tools, and fitted with special bodywork varying from open cargo bodies with canvas tilt to shop vans. Large numbers of 1½-ton chassis, mainly Chevrolet, Dodge and Ford, were supplied to foreign nations (for British Commonwealth countries these came from the manufacturers' Canadian subsidiaries); they were usually furnished without cabs and bodies, and offered excellent value for money. Several examples of these are illustrated elsewhere in this book.

Truck, 1½-ton, 4×2, Cargo (Chevrolet) In 1935 the US Army purchased a relatively large number of Model QC 131-in wb Chevrolet chassis which were fitted with Cargo (shown), Canopy Express, Dump, Panel Delivery and Light Repair bodies. Truck shown cost $750.

Truck, 1½-ton, 4×2, Reconnaissance (Diamond T) Hercules JXB 6-cyl., 68 bhp, 4F1R, wb 158 in, 229×76×96 in, 5700 lb. Model 226 Special chassis with 7.00-20 tyres and 12-passenger bodywork, supplied in 1934/35 at $1080. 263-cu. in L-head engine. Four cab doors and side curtains.

Truck, 1½-ton, 4×2, Tractor (Dodge/Fargo) Short-wheelbase tractor truck with Highway van-type semi-trailer. The Fargo Division of Chrysler Corp. also supplied 1½-ton Dodge chassis with or for Cargo, Dump, Stake, Express and Station Wagon bodies. 1935.

Truck, 1½–2-ton, 4×2, Radio (Federal) The Federal Motor Company were among the leading assemblers of trucks, using their own sheet metal but engines, transmissions, axles, steering gear, brakes, etc. from specialist suppliers. This radio van dates from *c.* 1932.

Truck, 1½-ton, 4×2, Cargo (Ford) 4-cyl., 40 bhp, 4F1R, wb 131½ in. Standard swb Model AA chassis/cab of 1930/31 with special bodywork and military pattern radiator brush guard, tow hooks, pintle, pioneer tools (shovels, axe, mattox). Used by Field Artillery.

Truck, 1½-ton, 4×2, Cargo (Ford) Typical medium cargo truck of the mid-1930s (1935 Ford V8 Model 51, 4F1R, wb 131½ in). The steel body was a standard design with canvas tilt (paulin) and hinged wooden bench seats, produced by Hercules of Galion, Ohio.

Truck, 1½-ton, 4×2, Radio (GMC) 6-cyl., 73 bhp, 4F1R, wb 108 in, 201×91×108 in, 6327 lb. Built in 1939 this GMC Model AF311 featured civilian type front end and cab doors with integral radio van bodywork for the US Army Signals Corps. It had Timken-Detroit axles.

319

Truck, 1½–2-ton, 4×2, Cargo (Ward LaFrance) Trucks operated by the US forces during the 1920s were almost exclusively of 1917/18 vintage (often 're-worked'). A rare exception was this Waukesha-engined Ward LaFrance purchased by the QMC in 1928.

Truck, 1½-ton, 4×2, Anti-Aircraft (White) During the early 1930s several types of truck-mounted artillery appeared. This was one of two .30-cal. dual AA machine gun mounts, T3, based on commercial White chassis of about 1931. There was a removable canvas cab top.

Truck, 1½-ton, 4×2, Anti-Aircraft (White) 4-cyl., 61 bhp, 5F1R, wb 143½ in, GVW 14,000 lb. Commercial chassis/cab with hydraulically-operated T5 Multiple Machine Gun Mount. It used four .50-cal. M2 Machine Guns and T2 Power Control. Produced in 1932.

Truck, 1½-ton, 4×2, Desert Patrol (Chevrolet) Many foreign governments used American truck chassis, particularly during the late 1930s. This 1938 Chevrolet TA of the Transjordan Frontier Force (IWM photo E2971E) was a typical example. Ford V8s were also used extensively.

USA

TRUCKS, MEDIUM, 4×4
Pre-1920

Very few four-wheel drive trucks were used until 1912 when the famous FWD Model B was introduced. Of this truck many thousands found their way to Europe when war broke out in 1914. The US forces also used the FWD, first at the Mexican border, along with the Jeffery Quad. The latter, introduced in 1913, was renamed Nash Quad in mid-1917, and was also exported (Britain, France, etc.).

Both were used with several types of body. An attempt to mass-produce a standard military 4×4, the Militor, came too late. The FWD and the Jeffery/Nash Quad remained in service in the USA and several other countries (Argentina, Britain, France, Spain, etc.) up to well into the 1930s. The Quad also appeared with various types of armoured hulls (see Combat Vehicles, Wheeled).

Truck, Medium, 4×4, Searchlight (Avery) Centrally-mounted four-cylinder dual-ignition engine driving dynamo to generate power for propulsion (3 HP motor in each dual-tyred wheel) and searchlight. Speed $5\frac{1}{2}$–$9\frac{1}{2}$ mph. Made by Avery Co. of Peoria, Ill., in 1910.

Truck, Medium, 4×4, Cargo and Prime Mover (Duplex) The Duplex Power Car Co. (later Duplex Truck Co.), founded in 1909, were among the first American all-wheel drive vehicle makers. This unit was supplied to the USMC about 1913. Final drive was by pinions and internal gears at each wheel.

Truck, Medium, 4×4, Chassis (GMC) In 1915 the General Motors Truck Co. built one experimental four-wheel drive chassis. It was a conversion of a conventional solid-tyred truck and the front-drive propeller shaft is just visible under the frame, below the driver's seat.

Truck, 3-ton, 4×4, Cargo (FWD) In 1912 the Four Wheel Drive Auto Co. of Clintonville, Wisc., launched their well-known Model B truck, thousands of which were subsequently built for the US and other governments. During 1917–18 Kissel, Mitchell and Premier were co-producers. Peerless built 500 for Britain (1916).

Truck, 3-ton, 4×4, Ordnance Repair (FWD) Wisconsin 4-cyl., 36 bhp, 3F1R×1, wb 124 in, 222×76×126 in, 12,220 lb (gross). Tyres 36×6. Artillery or steel disc wheels. Silent-chain sub-transmission with lockable centre differential. Chassis standardized by QMC for various body types in 3–5-ton class.

Truck, 3-ton, 4×4, Balloon Winch (FWD) Self-contained balloon winch sets (by Cunningham and others) were mounted on the FWD Model B chassis. Other special bodies included air compressor, artillery supply, heavy repair shop, office, power saw, small arms repair, searchlight, etc. 1917–18.

Truck, 2-ton, 4×4, Cargo (Jeffery Quad) In 1913 the Thomas B. Jeffery Co. of Kenosha, Wisc., introduced their Quad trucks which drove, steered and braked on all wheels. The first Quad, shown, was supplied to the US QMC. During the war it was co-produced by Hudson, National and Paige.

Truck, 2-ton, 4×4, 240-mm Trench Mortar (Jeffery Quad)
Buda 4-cyl., 32 bhp, 4F1R, wb 124 in, 200×74×77 in, 7900 lb. Typical application of the Jeffery Quad, which became Nash Quad in 1917. Engine bore and stroke were $4\frac{1}{4} \times 5\frac{1}{2}$ in (early models $3\frac{3}{4} \times 5\frac{1}{4}$ in).

Truck, 2-ton, 4×4, Ammunition (Nash Quad) Standard ammunition body on 1917 Model 4017A chassis. In 1918 11,490 Quads were produced, the majority for the US Army and US Marine Corps. Production then gradually decreased until 1928 when the last of a total of 41,674 was delivered.

Truck, 3-ton, 4×4, Tractor (Militor) Wisconsin 4-cyl., 36 bhp, 4F1R, wb 128 in, 216×80×124 (105) in, 9000 lb. Ordnance design (Model TTH) to replace commercial FWD and Nash Quad. 150 supplied by Sinclair Motor Corp. of New York City and Militor Corp. of Jersey City, N.J.

Truck, 3-ton, 4×4, Ammunition (Militor) The Militor M1918 was basically similar to the French Latil and the American Walter, with radiator behind engine and internal spur gear drive on all wheels. Engine B×S $4\frac{3}{4} \times 5\frac{1}{2}$ in; governed speed 1200 rpm (13 mph). Hele-Shaw multiple disc clutch.

USA

TRUCKS, 1½- to 3-TON, 4×4
1920–1940

Makes and Models: Chevrolet (1½-ton, 1936). Clydesdale 80D (2½-ton, 1937). Corbitt 10FB6 (1½-ton, 1934), 12FD6 (2½-ton, 1934/5), F18 (2½-ton, 1937), CF18 and 80D (2½-ton, 1938), F12 and F14 Spec. (1½-ton, 1939), etc. Dodge T9/K39X4 (1½-ton, 1934), T9/K45 (1½-ton, 1935), T200/RF40X (1½-ton, 1938), T201/TF40X (1½-ton, 1939), etc. Dodge/Marmon-Herrington (1½-ton, 1936). Fargo (see Dodge). Federal Q9 (2½-ton, 1935). Ford/Marmon-Herrington B5-4 (1½-ton, 1935/36), E5-4 (1½-ton, 1938), etc. Franklin/QMC (1½-ton, c. 1930), etc. FWD HH6 (2-ton, 1930), etc. GMC 4272 (1½-ton, 1936), 4409 (1½-ton, 1937), ACKX353 (1½-ton, 1939), etc. Indiana (see White). Marmon-Herrington 33T1 (2-ton, 1931), A10 (2-ton, 1933), etc. Reo 4WD (1½–2-ton, 1934/5). USA/QMC TT-L (1½-ton, 1923), TT-H3 (3-ton, 1928), (3-ton, 1931), etc. White Indiana (1½-ton, 1933, 1935), 14X4 (2–2½-ton, 1939), etc.

General Data: In the late 1920s the US QMC (Quartermaster Corps) began experimenting with a variety of medium-weight all-wheel drive trucks. These were mostly built by the QMC at Fort Eustis, Virginia, using engines and other components purchased from several contemporary automobile manufacturers, including Franklin, Marmon, etc. These vehicles followed a number of earlier experiments. In March 1931 Walter C. Marmon and Colonel Arthur W. Herrington (who had been deeply involved in military motor transport from World War I) founded the Marmon-Herrington Company in Indianapolis, Ind.
  This company specialized in the design and manufacture of 4×4 and 6×6 vehicles for military and other heavy-duty purposes (oilfields, state highway departments, etc.) and produced many vehicles for the US QMC, Air Corps, Ordnance Dept., Coast Guard, USMC, etc., as well as for foreign armies (Belgium, Persia, Russia, etc.). From 1935 this company also converted large numbers of standard commercial trucks (mainly Ford) to all-wheel drive.

Vehicle shown (typical): Truck, 1½-ton, 4×4, Cargo (General Motors Truck, GMC 4272).

Technical Data:
Engine: Chevrolet 6-cylinder, I-I-W-F, 206.8 cu. in (3 5/16×4 in), 76 bhp at 3200 rpm.
Transmission: 4F1R×2.
Brakes: hydraulic.
Tyres: 6.50-20.
Wheelbase: 133 in.
Overall l×w×h: 204×82×110 in.
Weight: 5875 lb.
Note: first 4×4 GM truck supplied to US Army QMC, derived from commercial 1936 Chevrolet Model RB 131-in wb chassis (with early style cab). Standard military body and fitments.

Truck, 2½-ton, 4×4, Cargo/Light Repair (Corbitt) Continental E602 6-cyl., 98 bhp, 5F1R×2, wb 165 in, chassis wt 5730 lb. Tyres 7.50-20. 360-cu. in (4 1/8 × 4½ in) L-head engine. Commercial chassis (Model 12FD6, 1934/5). USA registration (W-0076) indicates Light Repair truck.

Truck, 1½-ton, 4×4, Cargo (Dodge/Fargo) 6-cyl., 80 bhp, 4F1R×2, wb 143 in, 209×76×108 in, 4665 lb. Tyres 7.00-20. 241.5-cu. in L-head engine. Dodge T9/K45 modified commercial truck, supplied through Fargo Div. in 1935 ($1260; with closed cab $1250). Timken axles and transfer case.

Truck, 1½-ton, 4×4, Chassis (Dodge/Marmon-Herrington) In 1936 Marmon-Herrington supplied their first 4×4 Dodge conversion, for the Argentine Army. It was a Model VF with 160-in wb. Illustrated is a 1937/38 Model RF, with right-hand drive, also for Chrysler Export.

Truck, 2½-ton, 4×4, Cargo (Federal) Waukesha 6MK 6-cyl., 85 bhp, 4F1R×2, wb 165 in, 242×92×118 in, 9500 lb. Tyres 9.00-20. Built in 1934/5 to QMC specification No. 9, hence model designation Q9. Vacuum-assisted hyd. brakes. Timken double-reduction axles.

Truck, 1½-ton, 4×4, Cargo (Ford/Marmon-Herrington)
The first Marmon-Herrington all-wheel drive conversion of a standard Ford V8 truck was the Model B5-4 of 1935 (based on the 131½-in wb Ford Model 51). It also appeared as an open Squad Car with single rear tyres for the Ordnance Dept. (B5-4 SQC).

Truck, 1½-ton, 4×4, Earth Borer (Ford/Marmon-Herrington)
Model E5-4 of 1938 (converted Ford V8 85 HP Model 81T) fitted with post hole digger unit for US QMC. Note dual front tyres and cut-away front wings/fenders. One of many Ford/M.-H. trucks of the late 1930s.

Truck, 1½-ton, 4×4, Cargo (Franklin/QMC) Franklin 6-cyl., 91 bhp, 4F1R×2, wb 112 in, 176×82×88 in, 5350 lb. 272-cu. in OHV air-cooled engine. 7.50-20 tyres. Several trucks were built by US QMC using Franklin engines and other components. Warner trans., Wisconsin transfer. 1930/31.

Truck, 2-ton, 4×4, Cargo (FWD) Waukesha 6MS 6-cyl., 73 bhp, 4F1R×2, wb 130 in, 210 (approx.)×94×114 in, 9600 lb. GVW 13,600 lb. 315-cu. in L-head (side-valve) engine. Brown-Lipe transmission, own axles. Lockheed hydraulic brakes. Maximum speed 35 mph. Several supplied in 1930.

Truck, 2-ton, 4×4, Aircraft Refueller (Marmon-Herrington) Hercules WXC 6-cyl., 94 bhp, 4F1R×2, wb 155 in, 240 in (approx.) ×93×106 in, GVW 13,500 lb. The first vehicles produced by Marmon-Herrington were 33 of these Model 33T-1 refuelling trucks, supplied in 1931 to the QMC at Holabird for use by the Air Corps.

Truck, 2-ton, 4×4, Chassis (Marmon-Herrington) In 1932 this TL30 pilot truck was demonstrated (at Bofors in Sweden) to officials of the Persian Army. It resulted in the first order from that country: a series of Model A30S-4 3–3½ ton chassis for Bofors AA gun tractors, recce cars, etc.

Truck, 2-ton, 4×4, Cargo (Marmon-Herrington) Group of Model A10 trucks, built in 1933 and claimed to be 'the first drop frame trucks, establishing new standards for the Army'. During this year the firm also produced several types of six-wheel drive trucks.

Truck, 1½–2-ton, 4×4, Cargo (Reo) 6-cyl., 82 bhp, 4F1R×2, wb 142 in, 209×76×108 in, 6819 lb. Tyres 7.00-20. Own 268-cu. in L-head engine. Wisconsin transmission. Ross steering. Hydraulic brakes. Chassis weight 4910 lb. price $1695. Delivered 1934/35.

Truck, 1½-ton, 4×4, Cargo (White Indiana) Waukesha 6BK 6-cyl., 82 bhp, 4F1R, wb 148 in, 219×85×108 in, 7500 lb. Tyres 8.25-20. 282-cu. in L-head engine. Max. speed 45 mph. Supplied in 1934/35 with Cargo (shown) and Radio body at $1745 and $1825 resp.

Truck, 1½-ton, 4×4, Cargo (USA/QMC) Hinkley 4-cyl., 40 bhp, 4F1R×2, wb 147 in, 228×76×122 in, 6600 lb. Tyres 38×7. Internal spur gear final drive. 312-cu. in L-head engine. Designed and built by MT Div. of QMC in 1923 and designated TT-L. Experimental only.

Truck, 3-ton, 4×4, Cargo (USA/QMC) Continental 21R 6-cyl., 105 bhp, 12F1R, wb 172 in, 268×98×124 in, 16,000 lb. Tyres 44×10. Double-reduction spur gear final drive. 428-cu. in OHV engine. Air brakes. Speed 50 mph. QMC design, produced in 1928, designated TT-H3.

Truck, 3-ton, 4×4, Cargo (USA/QMC) Hercules YXC 6-cyl., 93 bhp, 4F1R×2, wb 160 in, 242×96×118 in, 12,685 lb. Tyres 9.00-20. 428-cu. in L-head engine. Air brakes. Brown-Lipe transmission. Wisconsin transfer. Produced by QMC in 1931 (also Continental-engined 4×2 version).

USA

TRUCKS, 1½ to 4-TON 6×4 and 6×6

Makes and Models: Autocar C7066 (4-ton, 6×6, 1939). Biederman 110 (4-ton, 6×6, 1936/37). Coleman (3-ton, 6×6, 1929). Corbitt 168FD8 (2½-ton, 6×6, 1933), (3-ton, 6×4, 1935), etc. Diamond T (2½-ton, 6×4, 1930/31). Federal 75K-131 (2½-ton, 6×4, 1939). Ford/Couse (6×4, 1938). Ford/Marmon-Herrington (2½-ton, 6×4 and 6×6, from 1937). Garford (2½-ton, 6×4, 1929–31). GMC 4929 (3-ton, 6×6, 1937/38). TK (3-ton, 6×6, 1938). AFWX-354 (2½-ton, 6×4, 1939). ACKWX-353 (2½–3-ton, 6×6, 1939). Mack NB (2½-ton, 6×4, 1939). Marmon-Herrington TL29-6 (2½-ton, 6×6, 1933), A30-6 (3-ton, 6×6, 1934), etc. USA/QMC TT-SW (1½-ton, 6×6, 1926), various exp. models (2½- to 4-ton, 6×4 and 6×6, 1930–32). White Indiana 16X6 (4-ton, 6×6, 1934/35), 950X6 (4-ton, 6×6, 1939).

General Data: One of the first six-wheel drive trucks of the US Army was an experimental model designed and built in the early 1920s by the Engineering Section, Motor Transport Division of the Quartermaster Corps, under supervision of Chief Engineer A. W. Herrington (co-founder of the Marmon-Herrington Co. in 1931). This truck, as well as a 4×4 variant, had separate axles and drive shafts and drive to the front wheels was via bevel gears which were concentric with the steering knuckle pivots. These two vehicles had several other unusual features, including rear wheel brakes running in oil with the lining attached to the drums. A winch was mounted under the frame and driven from the transfer case.

In later years several commercial 6×6 trucks were acquired and about 1930 the QMC assembled some more special military vehicles, largely from standard components. During the 1930s more commercial designs became available and many of these were adopted for military service.

The 2½-ton 6×6 ('deuce-and-a-half') type truck became the backbone of the US Armed Forces' motor transport in World War II, when hundreds of thousands were produced, notably by General Motors (GMC), International Harvester and Studebaker.

Vehicle shown (typical): Truck, 4-ton, 6×6, Cargo and Prime Mover (White Indiana 16X6).

Technical Data:
Engine: Hercules YXC3 6-cylinder, I-L-W-F, 478.8 cu. in, 105 bhp at 2200 rpm. CR 4.4:1.
Transmission: 4F1R × 2.
Brakes: air (Bendix-Westinghouse).
Tyres: 8.25-22.
Wheelbase: 160 in.
Overall l × w × h: 236½ × 92 × 108 in.
Weight: 12,395 lb.
Note: specially produced for US Army in 1934/35 by Indiana Motors Corp. (a division of White since 1932). Brown-Lipe transmission, Wisconsin transfer, Timken axles.

Truck, 3-ton, 6×6, Cargo (Coleman) In 1925 the QMC first tested a commercial 4×4 Coleman tractor with 51-bhp 4-cyl. Buda engine. This is a 6-wheeled truck variant with non-articulating rear bogie. Steering pivots were inside front wheel hubs. USA registration No. W46. 1929/30.

Truck, 2½-ton, 6×6, Cargo (Corbitt) Lycoming AEF 8-cyl., 113 bhp. Model 168-FD8 6×6 chassis with 7.50×20 tyres and 126×70-in cargo body. Gross weight 15,000 lb. Registration prefixed '3', indicating 'light' truck. Built in 1933 the vehicle cost $4334.36.

Truck, 2½-ton, 6×6, Anti-Aircraft (Diamond T) Hercules YXC 6-cyl., 93 bhp, 4F1R×2. Undersize auxiliary tyres at front, stub-axle mounted spare wheels. Multiple AA Gun Mount shown; same chassis/cab also appeared with overhead gantry wrecking body (Rescue Truck). 1930/31.

Truck, 6×4, Aircraft Maintenance and Repair (Ford/Couse) Comprehensively equipped 'Mobile Airport' with over 3000 pieces of equipment incl. generator, transmitter, lathe, crane, etc. Weight 12 tons. Ford V8 85-bhp engine. Private venture by Couse Laboratories, Inc. of East Orange, N.J., 1938.

Truck, 2½-ton, 6×6, Searchlight (Ford/Marmon-Herrington)
Ford V-8-cyl., 85 bhp, 4F1R×2. Among the first Ford/M.-H. 2½-ton 6×6 trucks, introduced in 1937, were Artillery Tractors for the National Guard and these Searchlight units for the US Marine Corps. They were based on the Ford Model 79 truck.

Truck, 2½-ton, 6×4, Cargo (Ford/Marmon-Herrington)
Ford V-8-cyl., 85 bhp, 4F1R×2, wb 157 in, 236×90×108 in, 8400 lb. Tyres 7.00-20. Based on Ford Model 817T 4×2 truck, fitted with Marmon-Herrington transfer case and rear bogie, military bodywork, etc. 1938.

Truck, 2½-ton, 6×4, 75-mm Gun Mount, T3 (Garford) Buda BA6 6-cyl., 83 bhp, 4F1R×2, wb 145 (BC 45) in, 228×88×121 in. Tyres, front 8.25-20, rear 7.00-20. Built in 1929/30. Similar but longer chassis built in 1931, with 3-in AA Mount, T1, and stub-axle-mounted spare wheels.

Truck, 3-ton, 6×6, Cargo and Prime Mover (GMC) Oldsmobile 6-cyl., 72.5 bhp, 4F1R×2, wb 155½ in, 232×87×86 in, 8950 lb. Tyres 7.00-20. 229.7-cu. in engine. Timken axles. Used for pulling 155-mm howitzer, etc. Occasionally fitted with dual front tyres. 1938 Model 4929.

Truck, 2½-ton, 6×6, Cargo and Prime Mover (Marmon-Herrington) In 1933 Marmon-Herrington produced some Model TL29-6 artillery prime movers for the Ordnance Dept. Vehicle shown (under test) was fitted with dual tyres and tyre chains all round. Large engine, light chassis.

Truck, 1½-ton, 6×6, Cargo, TTSW (USA/QMC) Hinkley HA500 4-cyl., 40 bhp, 4F1R×2, wb 160 in, width 76 in, height 120 in. Tyres 38×7. Unlike the QMC 6×6 of 1923 which had centrally-pivoted inverted rear springs, this 1926 model had a sub-frame. TTSW = Truck Tractor Six-Wheeled.

Truck, 2–2½-ton, 6×6, Staff Car (USA/QMC) Lycoming AEC 8-cyl., 130 bhp, 4F1R×2, wb 162 in, 239×90×94 in, 10,750 lb. 420-cu. in ($3\frac{3}{4} \times 4\frac{3}{4}$ in) in-line L-head engine. Brown-Lipe transmission, Wisconsin transfer and axles. 7.50-20 tyres. Maximum speed over 60 mph. 1931.

Truck, 4-ton, 6×6, Cargo (USA/QMC) Hercules RXB 6-cyl., 4F1R×2, wb 176 in, 260×96×118 in, 15,370 lb. 501-cu. in L-head engine. Wisconsin transmission, transfer and axles. Air brakes. 9.00-22 tyres. Speed 45 mph. Also 6×4 version (weight 14,200 lb). One of several exp. trucks of 1931.

USA

TRUCKS, HEAVY, 4×2
Pre-1920

Before 1917 the US forces had few heavy trucks. By 1917 3–5-tonners were known as 'heavy' types, over 5-ton was 'extra heavy'; all were classified as Class B. During the war the Army ordered 4043 Class B trucks for domestic distribution; 2211 were delivered before the Armistice, the majority by Mack, Peerless and White. For combat service 70,221 were ordered, 24,330 delivered; 17,620 of these were shipped overseas, including 7,655 Standardized B or 'Liberty' trucks. The latter had been designed to QMC requirements by the Society of Automotive Engineers (with committees for each major component) and 43,005 were ordered from 29 sources. 9452 had been completed, by 15 manufacturers, when the war ended.

Truck, 3-ton, 4×2, Photographic ('Heavy Aviation') Assembled from selected components during 1917–18 by Kelly-Springfield (1725), Federal (1000), Velie (700), United (188) and Standard (186) to Signal Corps/Air Service design (A.S. on radiator). They were also used by the Marine Corps.

Truck, 5½-ton, 4×2, Workshop (Mack) 4-cyl., 40 SAE-hp, 3F1R, wb 156 in, 269×78×126 in, 17,100 lb (loaded). 471-cu. in (5×6 in) L-head engine. Tyres, front 36×5, rear 40×6. 'Extra Heavy' chassis, similar to 'Heavy' 3½-ton model but stronger frame, springs, etc. Blacksmith Shop shown.

Truck, 5½-ton, 4×2, Sprinkler, Water (Mack) 2563 5½-ton (and 368 3½-ton) Mack chassis, popularly known as 'Bulldogs', were made for the Engineer Corps during 1917–18. Other bodies: Cargo, Dump, Wrecker, Degassing, Printing Press, etc. Produced commercially until the late 1930s.

Truck, 4-ton, 4×2, Cargo (Moreland) The Moreland Motor Truck Co. of Los Angeles, Cal., produced 1½-, 2½-, 4- and 5-ton trucks. The US QMC ordered 60 of these 4-tonners, 40 of which had been delivered when the war ended and orders were cancelled.

Truck, 3-ton, 4×2, Gun Mount (Packard) One of the first heavy trucks used by the US Army was this Packard regular model of 1909 on which was mounted a 3-pdr automatic gun which could sustain a rate of fire of 100 shots per minute. It was an experimental private venture.

Truck, 3-ton, 4×2, Cargo (Packard) During 1917–18, Packard supplied 4856 3-ton and 60 5-ton trucks to the US QMC (compared with one in 1914). During 1914–16 many hundreds of various types (1½- to 5-ton) went to Britain, France, etc. Shown is one of the French Army.

Truck, 3½-ton, 4×2, Cargo (Riker) 4-cyl., 28.9 SAE-hp, 4F1R, wb 150 in, 259×85×119 in, 9550 lb. Tyres 36×5. 340-cu. in T-head engine. Worm-drive rear axle. Built by Locomobile for US QMC (1690 during 1917–18) and British Army. The British had 1192 in service by 1918.

Truck, 3–5-ton, 4×2, Cargo (USA 'Liberty') 4-cyl., 52 bhp, 4F1R, wb 160½ in, 261×84×75 in, 10,700 lb. 425-cu. in (4¾× 6 in) L-head engine (Continental, Hinkley, Waukesha, Wisconsin). Full-floating worm-drive rear axle. Wood or steel wheels. Tyres, front 36×5, rear 40×6 dual. Note the hefty radiator guard and the coil springs between bumper and chassis.

Truck, 3–5-ton, 4×2, Tanker (USA 'Liberty') Three 250-gallon tanks. The Standardized Class B chassis was 'nationally designed' and produced (1917–18) by Gramm-Bernstein, Selden, Garford, Pierce-Arrow, Republic (about 1000 each), Bethlehem (675), Diamond T (638), Brockway (587), Sterling (479), and six others.

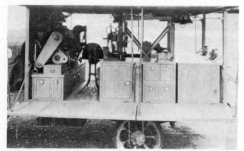

Truck, 3–5-ton, 4×2, Machine Shop (USA 'Liberty') The 'Liberty' was used with many body types, incl. Cargo, Tanker, Machine Shop, Photographic, Radio Repair, Field Lighting, Wrecker, etc. 7655 went overseas. Some were modified in the 1920s with modern wheels and pneumatic tyres.

Truck, 3-ton, 4×2, Cargo (White) The White Co. of Cleveland were among the largest suppliers of trucks in World War I. Customers included the US, Britain, France, Canada, Russia, etc. Note that the radiator was mounted on cylindrical spring barrels, a typical White feature.

USA

TRUCKS, 2½- to 7½-TON, 4×2
1920–1940

The majority of trucks in this class were commercial models of many different makes and models, a complete coverage of which would be of limited use. A random selection of typical vehicles is presented here, by make in alphabetical order. Bodywork in this class included cargo, dump, searchlight, tanker, wrecker, etc. The only purely military types, i.e. those which were produced by or under supervision of the US QMC, were reworked USA 'Liberty' trucks during the 1920s and a 3- and 6- tonner assembled by the QMC from selected components in 1931. Most truck makers of this period assembled their vehicles largely from components bought from specialist manufacturers (engines, transmissions, axles, brakes, steering gear, wheels, etc.).

Truck, 3-ton, 4×2, Tractor (Biederman) Continental E600 6-cyl., 76 bhp, 5F1R, wb 140 in, 198×86×74 in, 6100 lb. Tyres 8.25-20. 288-cu. in L-head engine. Air brakes. Produced in 1934 for towing low-loader semi-trailer, as shown. Basically a commercial truck.

Truck, 2½-ton, 4×2, Tanker (Corbitt) Hercules JXD 6-cyl., 84 bhp, 4F1R, wb 159 in, 233×97×88½ in, 8860 lb. Tyres 7.50-20. 320-cu. in L-head engine. Lockheed hyd. brakes. 500-gallon tank on Model 14B6 chassis, supplied in 1935/36 at $2235.

Truck, 2½-ton, 4×2, Searchlight (GMC) 6-cyl., 69 bhp, 4F1R, wb 160 in, 238×93×95 in, 7443 lb. Tyres 7.50-20. GMC 239 OHV engine. Hydraulic brakes. Commercial COE Model F18 chassis with special body for carrying searchlight and equipment. 1937.

Truck, 2½-ton, 4×2, Sound Equipment (GMC) Signal Corps van with high-power public address system, claimed to 'cover crowds of 100,000 persons with 100% efficiency'. Carried two collapsible tripods, each with three 100-watt loudspeakers, etc. 1938.

Truck, 2½-ton, 4×2, Cargo (GMC) 6-cyl., 80 bhp, 4F1R×2, wb 164 in, 244×85×97 in, 7378 lb. Tyres 7.50-20. 'Militarized' commercial truck, supplied in 1939 for $1405. During 1935–40 GM Truck supplied many types of 4×2 trucks to the armed forces.

Truck, 2½-ton, 4×2, Wrecker (International) Delivered to the USMC in 1933. Single swinging boom crane, suspended from crane tower behind cab. Hinged support legs at either side. A-line chassis, also with cargo, tanker, repair and other body types.

Truck, 3-ton, 4×2, Cargo (International) The US Marine Corps purchased considerable numbers of trucks from the International Harvester Co., both before and during World War II. This is a D-line (Model D40) commercial truck with special steel body. 1939.

337

Truck, 5-ton, 4×2, Cargo (USA 'Liberty') During the 1920s (and in some instances well into the 1930s) the US forces used many ex-World War I 'Liberty' trucks. These had been modernized in various respects. Shown is a 3rd Series reworked Class B of the US Army.

Truck, 3-ton, 4×2, Cargo (USA 'Liberty') The US Marines also employed reworked USA 'Liberty' trucks. This specimen, known as Class D, was photographed in 1933 and featured pneumatic tyres (single rear), closed cab, electric lighting, wheelarch body, etc.

Truck, 5-ton, 4×2, Wrecker (USA 'Liberty') Several reworked 'Liberty' trucks were fitted with lifting and recovery gear, in this case a block and tackle suspended from an overhead gantry. Note the large pneumatic tyres and contemporary fenders. c. 1930.

Truck, 3-ton, 4×2, Cargo (USA/QMC) Continental E603 6-cyl., 96 bhp, 4F1R×2, wb 160 in, 242×96×118 in, 10,900 lb. Tyres 9.00-22. 383-cu. in L-head engine. Brown-Lipe transmission. Wisconsin transfer. Air brakes. Also 4×4 (*qv*). Built by QMC in 1931.

X

USA

TRUCKS, 4- to 7½-TON, 4×4
1920–1940

Manufacturers of heavy four-wheel drive trucks included Coleman (Prime Mover, 1925), FWD (5-ton Cargo, 1930, Model Super-Seven), GMC (4-ton Cargo, 1939, Model ACKX853), Marmon-Herrington (5-ton Dump, 1935; Barrage and Observation Balloon Winch, 1938–40, Model C90-BWS-4), Oshkosh (5-ton Dump, 1934, Model L12), etc. In addition the US QMC built a 5-, a 6- and a 7½-tonner themselves in 1931/32, using commercially-available engines, power trains, axles, brakes, steering gear, etc. Most cross-country trucks in these and higher payload classes, however, were based on 6×4 and 6×6 chassis (see following section).

Truck, 5-ton, 4×4, Prime Mover (Coleman) Buda 4-cyl., 4F1R×2, wb 144 in, chassis wt 7789 lb. Tyres 42×9. Tread 56 in. Commercial truck for work in oilfields, mining, etc. Sample shown was obtained for Army tests in 1925. Note position of spare tyre.

Truck, 5-ton, 4×4, Cargo (FWD) Waukesha RB 6-cyl., 127 bhp, 4F1R×2, wb 165 in, width 99 in, height 126 in, 16,300 lb. 677-cu. in L-head engine. Lockable centre diff. 120 bought by US Army in 1930. Made available commercially as 7½-ton Super-Seven in same year.

Truck, 5-ton, 4×4, Balloon Winch (Marmon-Herrington) Hercules RXC 6-cyl., 131 bhp, 5F1R×2, wb 150 in, 265×100×115 in approx., 20,000 lb approx. Tyres 11.00-20. 529-cu. in L-head engine. First ordered by US Air Corps in 1938. Note air springs at front.

339

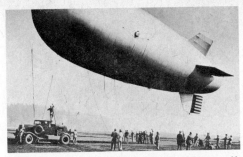

Truck, 5-ton, 4×4, Balloon Winch (Marmon-Herrington)
First procured in 1938, the largest order for these C90-BWS-4 balloon winch trucks was placed in 1940. Picture shows vehicle in action with an observation balloon. There was also a snowplough version (1937; with Sno-Go equipment).

Truck, 5-ton, 4×4, Cargo (USA/QMC) Hercules RXC 6-cyl., 115-bhp, 4F1R×2, wb 176 in, 274×96×122 in, 13,610 lb. Tyres 9.75-22. 529-cu. in. L-head engine. Brown-Lipe transmission. Wisconsin transfer. Also built as 4×2 with 501-cu. in Hercules RXB engine, both in 1931.

Truck, 6-ton, 4×4, Cargo (USA/QMC) Continental 16H 6-cyl., 120 bhp, 4F1R×2, wb 196 in, 303×101×125 in, 15,610 lb. Tyres 10.50-22. 611-cu. in. L-head engine. Wisconsin transmission. Double-reduction axles. Air brakes. Also 4×2 which was substantially similar. 1931.

Truck, 7½-ton, 4×4, Cargo (USA/QMC) Hercules HXG 6-cyl., 164 bhp, 4F1R×2, wb 196 in, 303×101×125 in, 16,590 lb. Tyres 12.00-20. 779-cu. in L-head engine. Basically similar to 6-ton version but heavier components, larger engine, etc. Built in 1932.

USA

TRUCKS, 5- to 10-TON
6×4 and 6×6

Makes and Models: Autocar (tanker, 6×4, 1939/40). Biederman 125/6WD (5-ton, 6×6, 1936). Clydesdale 300 (7½-ton, 6×6, 1936/37). Corbitt M1 (prime mover, 6×4, 1932), 12SD6 (7½-ton, 6×6, 1935). 54SD6 (7½-ton, 6×6, 1936), 33S6 (5-ton, 6×6, 1937), AA (6-ton, 6×6, 1939), 54SD6 (7½-ton, 6×6, 1939), 50SD6 (7½-ton, 6×6, 1939/40), etc. FWD (7½-ton, 6×6, 1937). GMC (8-ton, 6×4, from 1931). Hug 50-6 and 51-6 (7½-ton, 6×6, 1939/40). Indiana (see White). Mack T2 (prime mover, 6×4, 1929), NM (6-ton, 6×6, 1939). Marmon-Herrington T301-6 (5–6-ton, 6×6, 1933), TL31-6 (5–6-ton, 6×6, 1935), TH315-6 (5–6-ton, 6×6, 1936), etc. USA/QMC Class C (5-ton, 6×4, 1923), gas-electric (5-ton, 6×6, 1928), (6-ton, 6×4 and 6×6, 1931), TCSW (7½-ton, 6×4, from 1928), (7½-ton, 6×6, 1931), (9-ton, 6×4, 1931), (10-ton, 6×4, 1930), (10-ton, 6×6, 1930), etc. White T3 (prime mover, 6×4, 1931), Indiana 20X6 (7½-ton, 6×6, 1935–36), etc.

General Data: As can be seen from the above listing there was a good deal of activity in the field of heavy six-wheeled vehicles, particularly during the 1930s when several of the smaller truck makers were contracted for producing batches of 6×4 and 6×6 trucks and prime movers.

In addition the US QMC produced a number of experimental vehicles, after some World War I vintage 'Liberty' trucks had been converted and modified. The Air Corps procured several types of 6×6 vehicles from Biederman and Corbitt and the Artillery used heavy 6×4 prime movers made by Corbitt, GMC and White. Marmon-Herrington produced several types of 6×6 vehicles for the Ordnance Dept. and in addition supplied a considerable number to the Iranian (Persian) Army. Included in this section are several types of tank transporters or tank carriers. Until six-wheeled vehicles of this type appeared during the late 1920s the US Army had carried its tanks on a special platform version of the famous Mack 'Bulldog', a heavy duty truck with chain drive and solid rubber tyres which had been introduced in World War I and was still available in substantially the same form during the 1930s. The first American six-wheeled trucks with pneumatic tyres were built by Goodyear in 1918/19 to the design of Ellis W. Templin.

Vehicle shown (typical): Truck, 8-ton, 6×4, Artillery Prime Mover, T4 (GMC).

Technical Data:
Engine: GMC 6-cylinder, I-L-W-F, 707 cu. in, 156 bhp.
Transmission: 4F1R × 3.
Brakes: air.
Tyres: 9.75-22.
Wheelbase: 189½ in, BC 52 in.
Overall l × w × h: 288 × 96 × 97 in.
Weight: 19,325 lb.
Note: produced in 1931 for towing 3-in anti-aircraft gun. Later production (1935/36) had flat-fronted radiator grille. Also used for towing 8-in howitzer, low-loader trailers, etc.

Truck, 5-ton, 6×6, Field Service, E2 (Biederman) Hercules HXB 6-cyl., 148 bhp, 5F1R×2, wb 228 in, 340×92½×100 in approx. Chassis weight 13,950 lb. Tyres 9.75-20. Model 125/6WD chassis/cab of 1936–38. Price $5930. 707-cu. in L-head engine. Air brakes.

Truck, 7½-ton, 6×6, Cargo (Clydesdale) Hercules HXC 6-cyl., 164 bhp, 4F1R×3, wb 190 in, 320×96×125 in, 18,325 lb. Tyres 10.50-22. Model 300 chassis, supplied with and without winch in 1936/37 at $7460 and $6965 resp. 779-cu. in L-head engine. Air brakes.

Truck, 8-ton, 6×4, AA Prime Mover, M1 (Corbitt) Hercules 6-cyl., 156 bhp, 4F1R×3, wb 189¼ (BC 52) in, 306×96×94 in, 22,200 lb. Tyres 9.75-22. 779-cu. in L-head engine. Air brakes. Winch behind cab. Speed 40 mph. Produced by The Corbitt Co., Henderson, N.C., in 1932.

Truck, 6-ton, 6×6, Heavy Wrecker, M1 (Corbitt) During the late 1930s the US Army Ordnance Department procured a number of heavy wreckers, M1, with single swinging boom for recovery and general lifting purposes. Several manufacturers supplied chassis.

Truck, 5-ton, 6×6, Field Service, E2 (Corbitt) Similar in general configuration to the Biederman (qv) this vehicle was supplied to the US Army Air Corps in 1937. Model 33S6 chassis/cab cost $6445, had 252½-in wheelbase and 9.75-20 tyres.

Truck, 7½-ton, 6×6, Tractor, F1 (Corbitt) The Air Corps procured several of these tractor trucks for towing their standard Semi-Trailer, Fuel Servicing, 4000-Gallon, F1, often towing another unit with the aid of a dolly (thus becoming a full-trailer).

Truck, Tractor, 7½-ton, 6×6, Wrecking, C2 (Corbitt) Used by the Air Corps for salvaging crashed aircraft and for general towing. It had a fifth wheel for towing the Semi-Trailer, Wrecking, 40-ft, 12½-ton, C2 (for aircraft and general purpose hauling).

Truck, 7½-ton, 6×6, Tractor, F1 (Corbitt) Hercules HXC 6-cyl., 165 bhp, 5F1R×2, wb 180 in, 280×96×105 in. Model 50SD6 of 1939/40, used by Air Corps (later Army Air Forces) for towing 4000-gallon fuel-servicing semi-trailer. Note air springs at front.

Truck, 6-ton, 6×6, Cargo and Prime Mover (Corbitt)
Produced in 1939 for towing anti-aircraft artillery. Chassis had 9.75-22 tyres, 185-in wb and weighed 19,250 lb. Price was quoted as $7665. Later production had flat radiator grille. Note winch behind cab.

Truck, 7½-ton, 6×6, Cargo (Hug) Hercules HXC 6-cyl., 160 bhp, 4F1R×2, wb 182 in, 278×96×122 in, 21,500 lb. GVW 35,000 lb. Tyres 10.50-20. Air brakes. Winch behind cab. One of a few produced by the Hug Co. of Highland, Ill, in 1939–40 (Model 50-6).

Truck, 10-ton, 6×4, Tank Carrier (Mack) Developed from the 4×2 Model AC 4-cyl. 'Bulldog' Mack (which was also used for tank carrying) this 6-cyl. model of about 1930 had an additional rear axle, driven by secondary chains, and a winch. Shown with a Light Tank, T1E1.

Truck, 6-ton, 6×6, Machine Shop (Marmon-Herrington) Described as 'the first modern type mobile machine shop for the US Army Ordnance Dept.', this was also one of the first 6×6 military trucks produced by Marmon-Herrington. Chassis was Model T301-6 of 1933.

Truck, 6-ton, 6×6, Heavy Wrecker, M1 (Marmon-Herrington) Introduced in 1935 on a Model TL31-6 Hercules-engined all-wheel-drive Marmon-Herrington chassis this was one of the US Army Ordnance Department's first swinging-boom heavy wreckers of the type produced until 1945.

Truck, 6-ton, 6×6, Wrecking and Recovery (Marmon-Herrington) Used by the US Army Air Corps as a Wrecking and Recovery Truck this unit was based on a Model TH315-6 heavy duty 6×6 chassis. Note the small swinging-boom crane and the flashing light. Built in 1936.

Truck, 5-ton, 6×4, Cargo (USA/QMC 'Liberty') One of the first re-worked Standardized Class B 'Liberty' trucks of the post-WWI period was this six-wheeler. Still on solid rubber tyres the modification consisted of the relatively simple addition of an extra axle.

Truck, 5-ton, 6×4, Cargo (USA/QMC) Continental 4-cyl., 52 bhp, 4F1R, wb 172 in, width 84 in, height 122 in, 11,500 lb. 9.00-24 pneumatic tyres. Re-worked 'Liberty' truck, designated Class C. Designed and built in 1923 by Engineering Section, Motor Transport Division, US QMC.

Truck, 5-ton, 6×4, Cargo (USA/QMC) Basically similar to the Class C 'Liberty' shown on the previous page, this truck was built by the US QMC in 1924 for the US Marine Corps. Both vehicles had an articulating rear bogie with inverted semi-elliptic leaf springs.

Truck, 7½-ton, 6×4, Tank Carrier, TCSW (USA/QMC) Continental 15H 6-cyl., 105 bhp, 4F1R, wb 201 in, 301½ × 98½ × 105 in, 17,000 lb. 548-cu. in L-head engine. 'Liberty' gearbox. 48 produced by Quartermaster Corps in 1928. Some had closed cab and 'Liberty' pattern front end.

Truck, 7½-ton, 6×4, Tank Carrier, TCSW (USA/QMC) Continental 15H 6-cyl., 105 bhp, 4F1R, wb 201 in, 309 × 104 × 105 in, 18,000 lb. Pneumatic tyres, size 36 × 8. 12 built in 1930. TCSW = Tank Carrier Six Wheeled. Tanks were also carried on solid-tyred low-loader trailers, towed by heavy prime movers.

Truck, 5-ton, 6×6, Cargo, Gas-Electric (USA/QMC) Waukesha 6-cyl., 100 bhp, 8F8R, wb 152 in, 280 × 90 × 109 in, 15,000 lb. 40 × 8 tyres. Coleman front axle. Engine drove 40-kW generator, powering three 20-hp electric motors (Westinghouse), one for each axle. Produced experimentally in 1929.

345

Truck, 10–12-ton, 6×6, Gun Carrier (USA/QMC) Sterling LT6 6-cyl., 177 bhp, 4F1R×3, wb 186 in, 310×99×122 in, 18,700 lb. Hendrickson rear bogie with Wisconsin axles. Transporter for 155-mm GPG gun. Also 10-ton 6×6 and 6×4 Cargo variants. Built by QMC in 1930.

Truck, 7½–9-ton, 6×6, Cargo (USA/QMC) Hercules HXG 6-cyl., 164 bhp, 4F1R×2, wb 203 in, 303×101×125 in, 20,735 lb. Tyres 10.50-22. 779-cu. in L-head engine. Shown with 155-mm howitzer. Also made with 221-in wheelbase (length 357 in) and as 9-ton 6×4 (HXB engine). 1931.

Truck, 8-ton, 6×4, AA Prime Mover, T3 No. 4 (White) 6-cyl., 100 bhp, 4F1R×2, wb 197 (BC 46) in, 293½×98×92 in, 20,960 lb. GVW 39,920 lb. Tyres 40×8. Own 518-cu. in engine. Air brakes on all wheels. Speed 30 mph. Double-drum winch behind cab. 1931.

Truck, 7½-ton, 6×6, Cargo (White Indiana) Hercules HXC 6-cyl., 164 bhp, 4F1R×3, wb 190 in, 291×96×124 in, 19,250 lb. Tyres 9.75-22. Model 20X6, supplied in 1935 (with and without winch) and 1936 (shown; price $6760). Wisconsin/Timken-Detroit transfer case and driving axles.

USA
FIRE TRUCKS

The first standardized fire trucks of the US Army QMC in World War I were a ½-ton Dodge with special body made by Peter Pirsch & Sons (Truck, ½-ton, Fire Chief; equipped with Pyrene fire extinguisher, fire gong, chemical tank, fire lantern, hose basket with about 150 ft of hose, etc.), and three Trucks, 3–5-ton, Fire Pump, made by Ahrens-Fox, American LaFrance and Seagrave. There was also a Fire, Hose and Ladder Truck on the ¾-ton 4×2 GMC chassis. During the inter-war period the QMC built some 4×2 and 6×4 types at Holabird and others were acquired from specialist builders like General Fire, Howe, etc. Several were based on Ford/Marmon-Herrington 4×4 chassis.

Truck, 4×2, Fire (American LaFrance) One of two fire trucks used by the USMC (Marine Barracks, Parris Island, S.C., 1919). Equipment included ladders, scaling hooks, chemical tank, hoses, fire extinguishers, fire lanterns, searchlight, fire gong, etc.

Truck, ¾-ton, 4×2, Fire (GMC) Known officially as Truck, ¾-ton, Fire, Hose and Ladder, this vehicle was based on a 1917/18 GMC chassis. Engine: 4-cyl., 22.5 SAE-hp. Trans. 3-speed selective type. Chassis: 132-in wb, 35×5 pneumatic tyres, 56-in tread, 2900 lb.

Truck, 1½-ton, 4×4, Fire (Ford/Marmon-Herrington) Ford V-8-cyl., 80 bhp, 4F1R×2, wb 131½ in. Tyres 7.50-20. Marmon-Herrington Model B5 4×4 conversion of 1936 Ford Model 51 truck chassis. Range Fire Truck, supplied to Aberdeen Proving Ground. Base weight 6550 lb, price $3575.

Truck, 1½-ton, 4×4, Fire Crash (Ford/Marmon-Herrington)
Ford V-8-cyl., 85 bhp, 4F1R×2, wb 157 in. One of a Series of Crash Trucks supplied in 1937 to the US Navy for use at Naval Air Stations. Based on converted 1937 Ford Model 79 chassis/cab.

Truck, 1½-ton, 4×4, Air Crash Rescue (USA/QMC) Hercules 6-cyl., 71 bhp, 5F1R×2, wb 124 in, 183×82×79 in, 6160 lb. Tyres 7.00-20. 320-cu. in L-head (side-valve) engine. Clark transmission. Ross steering. Speed 50 mph. Built by QMC in 1932.

Truck, 5-ton, 4×2, Fire (USA/QMC) Known officially as Truck, Heavy, 750-gallon, 4×2 (2DT), Fire, this was one of several 'Liberty'-based vehicles manufactured at Holabird Quartermaster Depot, Baltimore, Maryland, during the 1930s.

Truck, 2½-ton, 6×4, Fire Crash (USA/QMC) Lycoming AEC 8-cyl., 130 bhp, 4F1R, wb 162 in, 257×90×72 in, 11,500 lb. Tyres 8.25-18. 420-cu. in L-head in-line engine. Speed 60 mph. Chemical fire apparatus, used at airfields, etc. Made by QMC at Holabird, 1932.

USA

TRACTORS, WHEELED (PRIME MOVERS)

Illustrated in this section are a few examples of wheeled tractors. There were several basic types, the most important being the agricultural/industrial tractor (the latter being a variant of the former but more suitable for road use), both with steering front wheels and propelled by large, wide rear wheels, and the special prime mover. Of the special wheeled prime mover there were few, their task usually being fulfilled by all-wheel drive and half-track trucks and tractors. The industrial type was employed in fair numbers for aircraft towing; these had solid rubber tyres, rather than the all-metal wheels of the agricultural type from which they were developed.

Tractor, Wheeled, 3×2 (Samson) The Samson 'Sieve-Grip' tractor was made from World War I until 1922. One is shown drawing a 75-mm gun during tests by the US Marines' 92nd Company, 10th Regt. at Quantico, Virginia, in 1918.

Tractor, Wheeled, 4×2 (John Deere) Agricultural tractor, made by John Deere about 1930. Illustrated in a book issued by the Field Artillery School at Fort Sill, this type was found to be 'not suitable for use in the FA except in emergencies'.

Tractor, Wheeled, 4×2 (McCormick-Deering) From 1924 the McCormick-Deering 10-20 farm tractor was available for industrial use, equipped with solid rubber-tyred wheels. It remained in production for many years and the US Navy used them for aircraft towing.

Truck, Heavy, 4×4, Prime Mover (Walter) Carrying registration number USA W53 this 6-cyl. Walter FH entered service in 1930. Tyre size was 44×10; the two behind the cab were 40×8 and could be added to the front wheels for extra traction on soft ground.

Tractor, 4×4, Prime Mover (Minneapolis-Moline) From 1938 Minneapolis-Moline produced several farm tractor-derived prime movers for military use. There were open- and closed-cab variants. A truck-type driven front axle was used. Note the full-width front rollers which prevented digging-in.

Tractor, 4×4, Prime Mover (Minneapolis-Moline) Some MM prime movers were tested by the National Guard at Camp Ripley in Minnesota, including this model with 2-wh. cargo/personnel trailer. They were christened 'Jeep'. Later some 6×6 variants appeared, as well as a purpose-built 4×4 aircraft tractor.

Tractor, 4×4, Prime Mover (Austin-Western) Pilot model of diesel-engined tractor, designed in 1939 and delivered in early 1940 by the Austin-Western Road Machinery Co. Tiller-steering operated hydraulically on both axles (four-wheel steer). Tyres 9.00-24, dual all round. Air brakes.

USA

CARS, TRUCKS and TRACTORS HALF-TRACK

Makes and Models: Autocar/Ordnance (APC, M2, 1939/40). Chevrolet/Cunningham (truck, 1934, 1937). Cunningham (car, T1, 1932), etc. Diamond T/Ordnance (APC, T14, 1939). Dodge/Chase (truck, 1922). Ford/Cunningham (trucks, 1933, 1935). Ford/Marmon-Herrington (trucks, T9, T9E1, M2, from 1936). Ford/Trackson (truck, T8, 1936). Fordson/Hadfield-Penfield (tractor). FWD/Holt Caterpillar (balloon winch truck, 1917). Garford/Holt Caterpillar (truck, 1917). GMC (trucks, T1, T4, T5, from 1933). Holt (tractors, 75 and 120 HP, from 1913/14). Jeffery/Holt Caterpillar (truck, 1917). Linn (trucks, T3, T6, from 1933). Lombard (tractor, 100 HP, from 1916). Mack/Christie (truck, 1920). Nash/Holt Caterpillar (truck, 1917). Packard/Holt Caterpillar (truck, 1917). White/Ordnance (APC, T7, 1938/39), etc.

General Data: The first American Half-track machines to be used for military service were the commercial Holt and Lombard. The former, in 75 HP four- and 120 HP six-cylinder variants, were widely used by the British in World War I; the Lombard mainly by the Russians. Holt (later to be known as Caterpillar) also produced track-bogies for conversion of 4 × 2 and 4 × 4 trucks. During the Twenties Christie Crawlers and others marketed half-track conversion kits for commercial and military use. Even the Ford Model T appeared as a 'snowmobile' half-track with skis replacing the front wheels. In 1931 the US Army purchased a French Citroën-Kégresse P17 *autochenille* and following tests a series of developments took place by Ordnance in conjunction with Cunningham, GMC and Marmon-Herrington which culminated in the well-known family of US half-tracks of World War II. Linn built a variety of heavier half-track trucks during the mid-1930s, which in general configuration were not unlike the earlier Lombards. Many experimental models were built, including conversions of conventional trucks, until the end of World War II (see also HALF-TRACKS, Olyslager Auto Library/Warne).

Vehicle shown (typical): Truck, 1¼-ton, Half-Track, Wire Laying, T4 (GMC)

Technical Data:
Engine: GMC 331 6-cylinder, I-I-W-F. 331 cu. in (3¾ × 5 in), 92 bhp at 2500 rpm.
Transmission: 4F1R × 2. Drive to front sprockets.
Brakes: hydraulic.
Tyres: 7.00-20.
Overall l × w × h: 204 × 70 × 96 in.
Weight: 8700 lb.
Note: 22 produced for Signal Corps in 1933/34. Payload 2400 lb. Cunningham rear bogie units. Max. speed 40 mph. Range of action 125 miles.

Tractor, Half-Track, Artillery, 15-ton (Holt Caterpillar 75)
4-cyl., 75 hp, 2F1R, 240 × 116 × 120 in, 24,100 lb approx. Used mainly by the British forces in World War I. Steering by front wheel and, for tight turns, by steering clutches. Rigid suspension. Also produced in Great Britain (Ruston).

Tractor, Half-Track, Artillery, 20-ton (Holt Caterpillar 120)
Similar to Holt 75 but with 120-hp 6-cyl. engine, hence greater weight (27,000 lb) and overall length (252 in). Both were made by The Holt Manufacturing Co. of Peoria, Ill. US Army received 267 and 154 respectively. British had about 1500 of both.

Tractor, Half-Track, Prime Mover (Lombard) Designed originally in 1901 by Alvin O. Lombard of Waterville, Maine, as steam-powered log hauler. In 1916 the Russian Government ordered 104 of these 100 HP IC-engined units for hauling artillery and supply trains.

Truck, Half-Track, Cargo (Garford/Holt Caterpillar) In addition to producing their 75 and 120 HP tractors, Holt produced half-track bogies for conversion of regular trucks, known examples being Garford (shown) and Packard 4 × 2 and FWD and Jeffery/Nash Quad 4 × 4. 1917/18.

Truck, Light, Half-Track, Cargo (Dodge/Chase) In 1922 a Dodge Brothers ½-ton truck of 1917 was fitted with additional rear wheels and Chase tracks. Similar but shorter tracks were used on the front wheels. Object was to improve off-road performance. A four-axle full-track conversion was also made (see page 313).

Tractor, Light, Half-Track (Fordson/Hadfield-Penfield) Tested at the Aberdeen Proving Ground but not accepted for military service was this standard Fordson farm tractor which was fitted with steel track adaptors made by Hadfield-Penfield Steel Company of Bucyrus, Ohio.

Truck, 5-ton, Half-Track, Cargo (Mack/Christie) Christie Crawlers, Inc., of Newark, N.J., devised this attachment to convert conventional trucks into half-tracks. Shown is the famous Mack Model AC, popularly known as the 'Bulldog', under test by the Ordnance Dept.

Truck, 5-ton, Half-Track, Cargo (USA/QMC) Continental 6-cyl., 105 bhp, 4F1R × 2, wb 160½ in, 280 (approx.) × 96 × 126 in, 14,000 lb. 40 × 8 front tyres. Air brakes. 548-cu. in L-head engine. Speed 35 mph. Built experimentally by the QMC in 1929 and known as 'Belted Six-Wheeler'.

Car, Half-Track, Reconnaissance, T1 (Cunningham) The first American half-track to use rear bogies patterned on the French Kégresse system was this pilot model built in 1932 by James Cunningham, Son & Co of Rochester, N.Y. It was modified/rebuilt in 1933 (T1E2) and again in 1934 (T1E3).

Car, Half-Track, Reconnaissance, T1E1 (Cunningham/RIA) Developed from the T1 shown on the left was this half-track reconnaissance car with three machine guns, radio, etc. 30 were built by Rock Island Arsenal in 1933. All T1 series half-track cars had Cadillac V8 engines.

Truck, 1-ton, Half-Track, T1 (GMC) 6-cyl., 92 bhp, 4F1R × 2, 190 × 70 × 78 in, 7600 lb. Payload 2000 lb. GMC Model 331 engine. Maximum speed 45 mph. Range of action 200 miles. Turning radius 20 ft. Designed and built for the Ordnance Department in 1933.

Truck, 1¾-ton, Half-Track, T5 (GMC) GMC Model 400 engine with 5F1R × 2 trans. Speed 40 mph. 222 × 74 × 90 in, 8300 lb. Payload 3600 lb. Designed as artillery prime mover in 1934. 24 produced. T5E1 had wider tracks, T5E2 had rubber block lubricated pin tracks (these were reworked T5s).

Truck, 5-ton, Half-Track, T6 (Linn) Hercules HXE 6-cyl., 174 bhp, 5F5R, wb 156 in, 246 × 92 × 98 in, 19,400 lb. Turning radius 33½ft. Tyres 9.75-20. 14-in tracks inside frame. One made in 1934. In 1936 an improved T6 appeared with 8-ton payload rating but weighing only 18,000 lb.

Truck, 8-ton, Half-Track, T3 (Linn) American LaFrance V-12-cyl., 222 (later 246) bhp, 4F1R, 268 × 93 × 78 in, 25,000 lb. Commercial Linn with special engine (WD 12). Tyres 9.75-20. Trailer load 31,000 lb. Speed 15 mph. Range 150 miles. Turning radius 35 ft. Several built, from 1933.

Truck, 1½-ton, Half-Track, T9E1 (Ford/Marmon-Herrington) Ford V-8-cyl., 85 bhp, 4F1R × 2, 229 × 82 × 85 in, 11,325 lb (gross). Modification of regular 1936 Ford Model 51 131½-in wb truck. First appeared (as T9 with four-wheel bogie) in July 1936. Also produced with front roller.

Truck, 1½-ton, Half-Track, M2 (Ford/Marmon-Herrington) Ford V-8-cyl., 85 bhp, 4F1R × 2. Modification of 1937 Ford Model 79 commercial chassis/cab, with military bodywork. This bogie design was used (in slightly modified form) on US armoured half-track vehicles throughout World War II.

USA
TRACTORS
FULL-TRACK

During World War I it was found that the Army required five sizes of tracked tractors, namely of 2½-, 5-, 10-, 15- and 20-ton capacity. For the latter two the 75 and 120 HP Holt Caterpillar were found suitable (see preceding section) but special designs were made for the others, by Rock Island Arsenal. Of these, only the 5- and 10-ton machines reached quantity production before the Armistice. During the 1920s and 1930s many more modern tractors were produced or purchased from the industry (in standard or modified form). From 1936 Marmon-Herrington built rubber-tracked light tractors for the US forces and for export. By 1940 most low-speed crawler tractors, light, medium and heavy types, were of commercial design (Allis-Chalmers, Caterpillar, International).

Tractor, Full-Track, 5-ton (Holt Caterpillar) In addition to half-track tractors (see preceding section) Holt produced several types of full-track models, one of which is seen here at Quantico, Virginia, in 1918. It belonged to the USMC, 10th Regiment.

Tractor, Full-Track, 5-ton, Armored (Ordnance/Holt) Holt 4-cyl., 56 bhp, 3F1R, 133 × 63 × 72 in, 10,725 lb. Clutch brake steering. Speed 7.4 mph. Armoured tractor with rigid suspension. 11,150 ordered; 3,480 delivered until 31 Jan. 1919. 459 shipped to France. Also as exp. SP gun.

Tractor, Full-Track, 10-ton, Armored (Ordnance/Holt) Holt 4-cyl., 75 bhp, 3F1R, 162 × 84 × 93 in, 21,500 lb. Clutch brake steering. Speed 5.2 mph. Articulated rigid suspension. 6,623 ordered; 2,014 delivered until early 1919. 628 shipped to France. Towed 155-mm gun.

Tractor, Full-Track, Light (Caterpillar) 4-cyl., 28 bhp, 3F1R, length 115 in, height 60½ in, weight 7500 lb. Drawbar pull 5645 lb. Speed 4.65 mph. Cruising radius 25 miles. Modified Caterpillar Twenty tractor of early 1930s at Fort Eustis, Virginia.

Tractor, Full-Track, 3½-Ton, T3 (Ordnance) Hercules 6-cyl., 100 bhp, 4F1R, length 120 in, height 52 in, weight 5700 lb. Speed 26 mph. Cruising radius (range of action) 150 miles. Controlled differential steering. Three-quarter front view shown. Mid-1930s.

Tractor, Full-Track, Heavy (Caterpillar) Typical heavy crawler tractor of the late 1930s (Caterpillar RD7). Weight (w/o winch) about 20,500 lb. 4-cyl. diesel engine, 61 drawbar hp. 60-in gauge with 18-in grouser track shoes. Front-mounted Hyster power winch.

Tractor, Full-Track, Crane (International) Modified IHC Model T35 'TracTracTor', fitted with 2-ton Roustabout swinging crane and rear-mounted power winch. Supplied in 1938. During World War II similar tractors were used by the Army Air Forces and Aviation Ordnance companies.

USA

COMBAT VEHICLES
WHEELED

Although many types of armoured and semi-armoured wheeled vehicles were produced in the US prior to 1940, the majority were experimental or improvised; very few reached quantity production. Some use was made of armoured cars at the Mexican border in 1916 but there was little activity until the 1930s when a long line of armoured cars and scout cars appeared (some, which had little more than armoured radiator shutters, are shown under 'Field Cars'). Many of these were designed by the Ordnance Dept., others were private ventures or produced for export. A random selection is illustrated here.

Gun Carrier, 3×2 (Duryea/Davidson) Basically a Duryea passenger runabout this car carried a Colt automatic gun with steel shield. Devised by Major (later Col.) R. P. Davidson in 1898 it was America's first combat vehicle. Two steam-driven cars followed in 1900.

Armored Car, 4×2 (Cadillac/Davidson) The United States' first true armoured car was also designed by Davidson, in 1915. Other Davidson designs included balloon destroyers, a medical car and a field kitchen (with electric cooker), all based on Cadillac car chassis.

Armored Car, 4×2 (King/AMCC) Built by the Armored Motor Car Company of Detroit in 1915 on 70-bhp 8-cyl. King car chassis. A few of several variants were subsequently made and used mainly by the USMC. Most had twin rear tyres. Note sand planks and spotlight.

Armored Car, 4×2 (Autocar) These vehicles (incl. support vehicles, see Canada), built in the United States by the Autocar Co., saw active service in France in 1918 with the first Canadian contingent. Based on 2-cyl. 2-ton truck chassis. One survived and is preserved in Canada.

Armored Car, 4×2 (White) Built in 1916, sponsored by some wealthy New Yorkers who formed the 1st Motor Battery of the N.Y. National Guard. Identical hulls were mounted on Mack and Locomobil (Riker) chassis, visible in background. Used on Mexican border.

Armored Car, 4×4 (Jeffery Quad) In 1914/15 the Thomas B. Jeffery Co. produced several armoured cars on their Quad chassis, which for this purpose was available with duplicate driving controls at the rear. Specimen shown had fixed turret and armour-protected wheels.

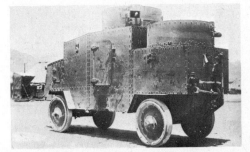

Armoured Car, 4×4 (Jeffery Quad) One of a number of Quad truck-based twin-turreted 'Armored Cars No. 1', used by the Army on the Mexican border in 1916. Length 200 in, height 110 in, weight 12,600 lb. 'Armored Cars No. 2' were smaller and based on 4×2 White truck chassis.

Armored Car, 4×4, T6 (USA/QMC) Franklin air-cooled 6-cyl. engine, 95 bhp. Weight 8700 lb. Speed 60 mph. Made in 1932 at Holabird Quartermaster Depot as one of a long line of experimental armoured and partly-armoured cars, starting with the T1 (Pontiac) in 1928.

Armored Car, 4×4, T11 (FWD) Cadillac V-8-cyl., 115 bhp, 5F1R, wb 114 in, length 177 in, height 86 in, wt 8350 lb. Pilot model, built to QMC specifications in 1932/33. In 1934 Marmon-Herrington was contracted for small-scale production. T11E1 had larger turret; T11E2 (1936) had Hercules engine.

Armored Car, 4×4 (Marmon-Herrington) Ford V-8-cyl., 85 bhp, 4F1R×2. Tyres 9.75-22 (bulletproof). Crew 5. Dual steering controls at rear. Weight 12,760–14.080 lb, depending on armour and armament. Commercial, 1934. Quantity of heavier version (TH310 ALF-1) with Hercules engine was sold to Iran.

Armored Car, 4×2 (Tucker) Packard V-12-cyl., 175-bhp, 5F2R, wb 109 in, 166×75×106 in, 10,750 lb. Named 'Tiger Tank' this very fast experimental vehicle was a private venture, designed by Preston Tucker in 1938. Noteworthy was the hydraulically-operated glass dome turret with 37-mm gun.

Scout Car, 4×4, T7/M1 (White Indiana) Hercules 6-cyl., 75 bhp, 4F1R×2, wb 131 in, 192×80×79 in, 7700 lb. Tyres 8.25-20. 282-cu. in L-head engine. Speed 50 mph. First of its type, based on Indiana 12X4 truck chassis in 1934. 76 more were built as Scout Car, M1.

Scout Car, 4×4, T9/M2 (Corbitt) Lycoming 8-cyl., 94 bhp, 4F1R×2, wb 132 in, 191×80×73 in, 7900 lb. Tyres 8.25-20. 280-cu. in L-head engine. Speed 50 mph. Next in line to T7 (T8 was not built). 20 built in 1935, followed by two as Scout Car, M2.

Scout Car, 4×4, T13 (Marmon-Herrington) Ford V-8-cyl., 85 bhp, 4F1R×2, wb 131 in, 190×75×70 in, 7710 lb. Tyres 8.25-20. Based on Ford/M.-H. chassis, 1937. 38 bought for National Guard. Note skate-mounted mg on all-round rail. Company produced scout cars from 1934, some for export.

Scout Car, 4×4, M2A1 (White) Hercules 6-cyl., 4F1R×2, wb 131 in. 200×79×78 in, 7810 lb. Developed from M2(E1) in 1937. After addition of radio redesignated M3 (1937–39) and following further improvements (wider body, etc.) became the well-known M3A1 of World War II.

USA

COMBAT VEHICLES TRACKED

Like wheeled combat vehicles of the pre-1940 era, US tracked vehicles (except tanks, which are not dealt with here) were rather large in number of types but small in number of units produced per type. Many were experimental models designed by the Ordnance Department and by a few private concerns like the Front Drive Motor Co. and US Wheel Track Layer Corp. (Christie) and Marmon-Herrington. Cunningham built a number of vehicles to Ordnance designs. John Walter Christie (1863–1944) was probably the best-known American private designer of combat (and other) vehicles and was particularly successful with tank suspension systems.

Gun Motor Carriage, Full-Track (Ordnance) Known as self-propelled caterpillar gun mount this was one of several types developed in 1917/18. It consisted of a 75-mm gun, mounted on a 2½-ton artillery tractor. The same gun was also mounted on the 5-ton model.

Gun Motor Carriage, Full-Track, Mk VII (Holt) Cadillac V-8-cyl., 70 bhp, 3F1R, 135 × 63 × 71 in, 10,600 lb. 75-mm Field Gun Carriage, M1916, based on modified undercarriage of 2½-ton artillery tractor. Track width 8 in. Max. road speed 9.5 mph. Two built in 1919.

Gun Motor Carriage, Full-Track (Christie) 6-cyl., 120 bhp, 4F4R, 236 × 111 × 80 in, 44,000 lb. Christie-designed mobile mount for 155-mm GPF gun, M1918. Centre wheels sprung, outer wheels rigid. Length of gun 236 in. Four built by Front Drive Motor Co. at Hoboken, NJ, in 1920.

Reconnaissance Tractor, Full-Track (Ordnance) 4-cyl., 15 bhp, 3F1R, 104 × 44 in, 1000 lb. Payload 500 lb. Road speed range 1.5–30 mph. Amphibious. Smallest in a range of experimental 'tractor caissons', designed by Colonel Chase about 1918/19. Others had payload ratings of 1½-, 3- and 3½-ton.

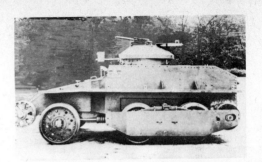

Combat Car, Convertible, T2 (Ordnance) Continental radial 7-cyl., 165 bhp, 4F1R, 177 × 75 × 89 in, 17,000 lb. One produced in 1931, originally known as Convertible Armored Car, T5. Used with or without tracks; max. speed 20 and 30 mph resp. One of many experimental models.

Combat Vehicle, Chassis, Convertible (Christie) Liberty V-12-cyl., 338 bhp, 4F1R, 204 × 84 × 72 in, 17,200 lb. One of J. Walter Christie's experimental designs, executed by the US Wheel Track Layer Corp, in 1928. Could run on wheels or tracks (70 and 42 mph resp.). Forerunner of Christie M1931 tank.

Combat Vehicle, Full-Track (Marmon-Herrington) Ford V-8-cyl. engined Model CTL3 for USMC, 1936/37. One of a range of M.-H. tractors and light tanks equipped with continuous-band steel-cable-reinforced rubber tracks, developed in co-operation with Goodrich. Crew 2. Engine at rear. 8200 lb.

OTHER COUNTRIES

This section covers a number of countries which used motor vehicles for their armed forces but did not have their own motor industry to furnish these vehicles. Some did have assembly plants or even limited manufacturing facilities but additional vehicles, particularly the special types, had to be imported. The emphasis is on transport vehicles and tractors and these were made and exported chiefly by the big manufacturing companies of Europe and North America. A complete listing is impossible to publish but many of the known types are mentioned and a random selection of typical examples is illustrated. Most nations did not commence motorization of their armies until the late 1930s and even then this was done on a relatively limited scale. Those countries which are not listed usually acquired standard commercial cars and trucks, particularly Fords and Chevrolets, to fill their transport needs. Ford and Chevrolet vehicles were attractive because they were relatively powerful, rugged and reliable at low cost. Their makers had adequate service and spares facilities all over the world and if required these trucks could be modified by specialist firms like Marmon-Herrington, DAF, etc., who offered various conversions e.g. all-wheel drive, tandem rear bogies, etc.

Argentina: US Dodge VF Series truck, converted to 4 × 4 by Marmon-Herrington in 1936. GVW 13,000 lb. 160-in wheelbase. 228 CID 92-bhp L-head Six engine. 32 × 6 tyres. Army also used 4 × 2 Chevrolets, Dodges, Fords, GMCs, Internationals, etc.

ARGENTINA: During the First World War Argentina acquired a quantity of FWD trucks from the United States; these were used for many years, eventually fitted with pneumatic tyres. During the 1930s many new US vehicles were acquired; these were chiefly commercial trucks and included Chevrolet, Dodge, Ford, GMC and International 4 × 2, Dodge/Marmon-Herrington 4 × 4, International 6 × 4, etc., as well as Harley-Davidson motorcycles and various types of passenger cars, incl. Dodge. From Europe came Belgian FN motorcycles, British Thornycroft 6 × 4 trucks/artillery prime movers (Tartar and Amazon), French Renault 6 × 4 trucks and Citroën-Kégresse half-track tractors. German Tempo 4 × 4 cars, etc. The overall variety of makes, models and countries of origin was probably unique. In 1938 Hispano-Argentina Fabrica de Automobiles (HAFADASA) produced a diesel-engined 6 × 6 truck for the Army.

Argentina: British Thornycroft Amazon Model WF/AC6/1 artillery tractor, one of a series delivered about 1938. In 1936 Thornycroft had supplied 12 lighter six-wheeled artillery tractors (Tartar) which were similar to the British WD normal-control 3-ton 6 × 4 type.

Bulgaria: Mercedes-Benz LO2750 2¾-ton commercial type trucks of 1935/36. These were available with Daimler-Benz OM65 diesel or M66 petrol engine, both 4942-cc units of 65 bhp. Wheelbase 4.25 m. Overall dimensions: 6.73 × 2.14 × 2.05 (cab). Weight 2950 kg.

China: Part of a fleet of 2000 US Dodge VH Series trucks, supplied in 1939. GVW 15,000 lb. 153-in wheelbase. 241.5 CID 99-bhp L-head Six engine with 5F1R transmission. Full floating hypoid rear axle. Tyres 34 × 7. Cabs and bodies were made locally.

BRAZIL: During the inter-war period the US producers Ford, General Motors and International Harvester established assembly plants in Brazil and these reputedly filled the main transport needs of the armed forces. It was not until after World War II (during which Brazil received many military vehicles from the US) that the Brazilian automotive industry was considerably expanded.

BULGARIA: Mainly German cars and trucks. In the mid- and late 1930s the Army acquired Mercedes-Benz cars (170V), as well as 4 × 2 and 6 × 4 trucks and in 1939 a fleet of 160 6 × 6 diesel trucks (*I.E. Lkw*) was ordered from MAN.

CHINA: Trucks used by the Chinese Army before (and during) World War II were mainly supplied by the United States and included Chevrolet, Dodge, Ford, GMC, International, Studebaker, etc. Marmon-Herrington supplied a quantity of track-laying tractors in 1936, after having designed a reconnaissance car for China two years earlier. German armoured cars were also used.

EGYPT: Military vehicles operated by the Egyptian Army since its formation (i.e. from the date of the Anglo-Egyptian treaty in 1936) were mainly of British origin, including Crossley, Ford, Guy, Morris-Commercial, etc. Chevrolets were supplied by General Motors' subsidiary in Alexandria. In 1938 the US Chrysler Corporation sold a large number of Dodge and Fargo light and medium trucks to Egypt; these were basically commercial chassis, equipped with oversize sand tyres. The order included more than 1000 units of 1000-lb and 3000-lb payload capacity. The Egyptian Army also operated at least one FWD Model B-based fire truck.

GREECE: During the First World War the Greek government used quantities of Americant rucks including Kissel (about 100 of these were ordered in 1914) and Peerless. Also acquired was a number of 3-wheeled Knox-Martin tractors with semi-trailers. Later Greece imported vehicles from several countries, including the legendary Pavesi articulated gun tractor from Italy. In the late 1930s there appeared German BMW motorcycles, Mercedes-Benz LG65/3 6 × 6 trucks (200 of these were supplied in 1938/39), British Bedford 1½-ton 4 × 2 trucks, etc.

INDIA: The Indian forces, both during World War I and the interwar period used almost exclusively British vehicles (Crossley, Karrier, Leyland, Morris-Commercial, Thornycroft, etc.) and several of these are covered in the chapter on Great Britain. In view of the special local operating conditions many of the vehicles used in India (supplied through the India Office in London) were modified in various respects and some were specially designed and built. In some instances (especially armoured cars) bodywork was built locally on imported chassis. This became customary particularly after 1939 when large numbers of Canadian chassis (notably Chevrolet and Ford) were assembled and bodied in India. Most of these vehicles were subsequently used in the fighting in the Middle East.

IRAN (PERSIA): For the motorization of their country in the 1930s the Iranian authorities bought a relatively large number of vehicles from the Marmon-Herrington company in the United States. These included reconnaissance, scout and armoured cars, 4×4 and 6×6 trucks and prime movers, etc. The first order came in 1932/33, followed by many more until 1940. French Laffly six-wheelers were also imported during the same period.

IRELAND (EIRE): Early Irish Army vehicles were either imported and bodied locally or assembled from imported components. The latter were mainly Fords, which were assembled by Henry Ford & Son Ltd at Cork. At the beginning of World War II (during which Eire remained neutral) the two major types of Irish Ford military vehicles were simple open MG carriers on light commercial chassis and armoured cars on shortened 1½-ton truck chassis. Amongst imported British vehicles of the 1920s and 1930s were BSA motorcycles, Leyland 4×2 trucks, Morris-Commercial Model CDSW 6×4 artillery tractors, etc. During the 1920s there were armoured cars based on Lancia, Peerless, Rolls-Royce and other chassis. Ex-WWI British Crossley trucks were also in service for many years. In the 1930s Ireland acquired a number of Landsverk 6×4 armoured cars from Sweden (*qv*). Many of these Irish vehicles were in service until long after World War II.

MEXICO: Mainly US vehicles, including Dodge, Ford, GMC, etc. These were commercial trucks, usually with special bodywork. Fords with Marmon-Herrington all-wheel drive were supplied in the late 1930s and Germany sold the twin-engined Tempo G1200 cross-country car.

Colombia: 3½-ton 4×2 Brockway truck, used by the armed forces of the Republic of Colombia in the early 1920s. Like other Latin American countries, Colombia imported most of her commercial and military trucks from the United States.

Egypt: The Chrysler Corporation exported many military vehicles under the marque name Fargo. This is one of 21 Model FJ4 133-in wb 1½-ton chassis supplied to Egypt in 1938. It had a 228 CID 92-bhp 6-cyl. engine, 4F1R transmission and 10.50-16 tyres.

Greece: Among US vehicles purchased by Greece in 1914 was a number of these 3-wheeled chain-drive Knox-Martin tractor units. The star above the steering gearbox indicated to the driver the position of the front wheel (which had almost 90° lock).

India: Vehicles used in India during the 1930s were practically all of British origin. Illustrated is a Humber Snipe 80 Tourer which was supplied to Army Headquarters in 1933. It had a 76-bhp 3.5-litre L-head Six engine with 4F1R gearbox.

Iran: From 1933 the Persian Army acquired many Marmon-Herrington vehicles. This Model A30S Reconnaissance Car, one of the first, carried personnel and a motorcycle. Based on the same 149-in wb chassis was a tractor for Bofors AA guns.

Iran: Marmon-Herrington Model A30-6 6×6 Cargo and Prime Mover trucks. Some 30 of these were supplied in 1935. Other Marmon-Herrington 6×6 trucks for Iran included heavy DSD400-6 and DSD800-6 artillery tractors, etc., delivered in 1940.

Ireland: Machine Gun Carriers, based on 1941 style 112-in wb ¾-ton Ford Commercial chassis (Model designations: 1NC for 40-bhp Four, 1GC for 90-bhp Six, 11C for 90-bhp V8, 19C for 100-bhp V8). Chassis were assembled at Cork.

Ireland: Armoured Car Mark IV, built in Ireland (Eire) on shortened Ford V8 1½-ton truck chassis. It had a turret-mounted Vickers MG. 21 were made in 1941 (following 14 Mk IIIs) and some were still in service with Irish UN forces in the Congo 25 years later.

Mexico: Dodge truck of 13,000 lb GVW rating, supplied by the US Chrysler Corp. in 1937. Bodywork (with folding troop seats) was similar to that used by the US Army. 92-bhp 228 CID L-head Six engine. 133-in wheelbase. 32 × 6 tyres.

Mexico: 1939 Ford V8 122-in wheelbase 1-ton truck chassis (Model designations: 92Y with 60-bhp, 91Y with 85-bhp, 99Y with 95-bhp V8 engine), converted to four-wheel drive by Marmon-Herrington. Four-door all-steel command car bodywork.

New Zealand: Gun Portee, carrying Fordson tractor (industrial type) and field gun. The chassis was a six-wheeler conversion of the 1936 Ford Model 51 1½-ton truck. Similar trucks were used in South Africa, with different bodywork for carrying gun and limber.

Norway: Combination motortricycle and gun carriage produced in 1900 by Cudell & Co. Motoren- und Motorfahrzeug-Fabrik of Aachen, Germany (1898–1908). Cudell was one of Germany's pioneer car manufacturers and this three-wheeler was patterned on the French De Dion.

NETHERLANDS EAST INDIES: Mainly American Chevrolet and Ford, many of which were converted to six-wheelers by DAF in the Netherlands (with the DAF Trado system, *qv*). Marmon-Herrington of the USA supplied all-wheel drive Ford trucks as well as small and medium full-track gun tractors, etc. Before 1930 there were American Fageol, Garford and Wichita trucks and various other civilian types, including motorcycles and cars.

NEW ZEALAND: Like other British Commonwealth countries New Zealand for its armed forces used vehicles mainly from Great Britain and Canada. During the 1930s a number of 6 × 4 trucks was supplied by Ford, Guy and Leyland. The latter were of the Terrier type and employed as GS trucks, searchlight carriers, artillery tractors, etc.

NORWAY: The Norwegian Army before World War II used relatively few motor vehicles and these were imported from various countries including the United States (e.g. Chevrolet), Great Britain, Sweden, etc. Most transport vehicles were regular commercial trucks.

PORTUGAL: Military vehicles used by the Portuguese Army were imported from various countries and used both in Portugal and its colonies in Africa. Included were French Peugeot cars (WWI), British Albion, Dennis and Hallford trucks (WWI), US FWD Model B and Kelly-Springfield trucks (WWI), German Daimler 4 × 4 tractor/truck (60 PS *Zugwagen*, 1908/09; later converted to half-track), etc. In the late 1930s Praga of Czechoslovakia supplied a number of their full-track artillery tractors (Model T VI P, with truck type body and cab).

RUMANIA: Mainly imported trucks from other European countries, including Thornycroft 3-ton 6 × 4 (1929) from Great Britain, Austro-Daimler ADGR 2–3-ton 6 × 4 (1936) from Austria, Praga RV 2-ton 6 × 4 (1936), Tatra T27 3-ton 4 × 2 (1938), Tatra T93 2-ton 6 × 6 (1940) and Praga T IV R full-track artillery tractors (1937) from Czechoslovakia, BMW motorcycles, Tempo G1200 cross-country cars, MAN 6 × 6 (*Einheitsdiesel I.E. Lkw*) and various other types from Germany in the late 1930s etc. Rather unusual was a series of American Ford V8 COE trucks of about 1940 with Marmon-Herrington all-wheel drive; these were used as prime movers for AA guns.

369

Portugal: About 1908 the German firm of Daimler in Marienfelde produced this heavy truck with four-wheel drive for the Portuguese government. It was employed in Angola (Port. W. Africa) and fitted with rear track bogies for improved traction in sand.

Portugal: Czech Praga Model T IV P (or T4P) full-track artillery tractor. CKD/Praga supplied similar tractors to other countries, incl. the Netherlands East Indies, Peru, Rumania, Sweden and Turkey. Most had a truck type body and cab, as shown.

Rumania: American Ford V8 COE trucks of about 1940, with Marmon-Herrington all-wheel drive, stub axle-mounted support wheels amidships and special bodywork with single rear door. Used for towing AA guns. Rumania used chiefly European vehicles.

Rumania: Czech Tatra Model T93 2-ton 6×6 truck, supplied in 1940. Air-cooled 70-bhp 3980-cc (80×99 mm) V8 engine with 4F1R×2 transmission. Tyres 6.00-20. Weight 3150 kg. Speed 70 km/h. Also with other body styles.

Spain: This huge 'mobile blockhouse' was one of 24 supplied in 1914 by Schneider, the French armaments firm, to the Spanish Army for use in Morocco. It was probably based on a 24/40 CV Brillié-Schneider Paris type bus chassis.

Spain: During the Civil War of 1936–39 the Spanish used a wide diversity of motor vehicles. Shown is an improvised armoured vehicle, used by the Republicans. It was locally made and based on an ex-USA 'Liberty' truck chassis of 1917–18.

SOUTH AFRICA: Prior to the outbreak of World War II most if not all military vehicles used in South Africa were imported from Great Britain and North America. By 1940 large numbers of Chevrolet and Ford light and medium trucks were in service.

SPAIN: Although Spain did have a small motor industry (notably Hispano-Suiza) the country relied almost entirely on imported vehicles for its armed forces. In 1912 Daimler of Germany supplied a large 4 × 4 artillery tractor (see Germany) and a few years later Benz/Gaggenau designed a number of military vehicles in conjunction with the Spanish Government. Thornycroft of Britain, FWD of the USA and some other firms also supplied trucks at about the same time. In 1920 Benz delivered two four-wheel drive tractors (Model VRZ; 4-cyl. Model S120 engine, 8154 cc (120 × 180 mm), 55 bhp at 1000 rpm). During the civil war of 1936–39 a very large variety of vehicles was used by both sides. Many of these were commercial types of American, British, French, German, Italian and Soviet origin: Bedford, Chevrolet, Dodge, Ford, GAZ, Krupp, MAN, OM (Autocarretta), Pavesi (P4 tractors), Phänomen, ZIS, etc. The relatively few vehicles which survived after this period were kept in military and civilian service until long after World War II (in which Spain remained neutral). Some were still operational in the late 1960s.

TURKEY: In World War I Turkey used mainly German vehicles, known makes including Benz cars and DAAG trucks. Czech Praga R 2-ton trucks were also employed. During the 1930s vehicles of a variety of origins were acquired for military service, including six-wheeled trucks from France (Laffly), Germany (Büssing-NAG III GL6) and the USSR (ZIS-6), regular 4 × 2 commercial trucks from the USA (Ford V8, 1935) and full track artillery tractors from Czechoslovakia (Praga T6, 1936 and T7, 1938). Some of the latter were still in use after World War II, having been re-engined with Deutz air-cooled diesels.

YUGOSLAVIA: Mainly imported vehicles from other European countries, including Czechoslovakia (e.g. Praga RN Trucks, 1937) and France.

Spain: US Dodge T200 Series 4 × 4 artillery prime mover truck with winch. 12,000-lb GVW rating. 150 supplied in 1938. 92-bhp 228 CID L-head Six engine with 4F1R × 2 transmission. Body platform 9′ × 6′6″. Tyres 32 × 6. Wheelbase 159 in.

Spain: 1937 Chevrolet artillery tractors with Dutch DAF Trado rear bogies, supplied to Spain. The Trado balancing beams between frame and rear wheels made these 6 × 4 vehicles excessively wide (frame and axle could be narrowed but at considerable extra cost).

Turkey: During the 1930s Germany exported various types of *Geländegängige Lkw* (cross-country trucks) which were also used by her own *Wehrmacht*. This is a Büssing-NAG Model III GL6 supplied to Turkey about 1933/34. It had a 90-bhp 6-cyl. petrol engine.

Turkey: The Turkish Republic, under Kemal Atatürk, was one of the very few countries to employ Soviet military vehicles. Illustrated are some ZIS-6 6 × 4 trucks (derived from Russia's ZIS-5 4 × 2), led by a German DKW 500-cc sidecar combination. c. 1937.

INDEX

AC 177
Adler 11, 115, 119, 125–127, 133, 134, 139, 165, 169
AEC 175, 183, 192, 200, 206–208, 214, 217, 219, 223, 279
Aero 34, 36
Ahrens-Fox 347
Albion 175, 185, 192, 193, 199–201, 207, 208, 212, 222, 224, 369
Alco 295
Alfa Romeo 239, 241, 256, 259
Allis-Chalmers 298, 356
Alvis-Straussler 175, 217, 237
American Austin 304, 306
American LaFrance 347, 355
AMO 277, 279, 280
Ansaldo 25, 239, 260, 261
ARA 113
ARGENTINA 364
Argus 129, 132
Ariel 175
Aries 78, 101
Armstrong Siddeley 217, 218, 224, 233, 234, 265
Armstrong Whitworth 228
Arrol-Johnston 187
Atkinson 239
Atterbury 295, 314
Audi 115
Austin 25, 122, 175, 180, 181, 184, 185, 187, 199, 201, 229, 237, 268, 276, 279, 283
Austin-Western 350
AUSTRALIA 6, 7

AUSTRIA 8–26
Austro-Daimler 8–14, 16–26, 94, 155, 275, 369
Austro-Fiat 8, 9, 11–13, 23, 44, 235, 236
Austro-Tatra 8, 9, 35
Autfit 34
Autocar 32, 192, 193, 294, 298, 309, 328, 340, 351, 359
Automatic 295
Auto Union 115, 125, 128, 129, 131, 161, 165
Aveling & Porter 213, 215
Avery 320

Balachowsky et Caire 89, 90
Bantam 298, 305
BD 35
Bedelia 76
Bedford 147, 175, 186, 187, 191, 192, 195, 199, 200, 206, 267, 365, 371
BELGIUM 27–31
Belsize 192, 200
Benelli 240
Benz 115, 116, 119, 121, 136, 139, 145, 149, 151, 155, 157, 164, 167, 264, 371
Bergmann 119, 139
Berliet 31, 65, 68, 71, 73, 74, 78, 79, 84, 89, 94, 99, 103, 104, 268
Berna 8, 200, 201, 269, 290, 291, 293
Berna-Perl 8, 11, 12
Bernard 65, 92, 99, 104
Bethlehem 334

Bianchi 239–241, 243, 246, 250, 253, 260
Biederman 298, 328, 335, 340–342
BMW 115, 117, 118, 122–124, 129, 168, 268, 365, 369
Bock & Holländer 8, 9
Bofors 284, 287
Borgward 115, 169
Botond 236
Bovy 27
Boydell 174, 215
Bray 174, 215
BRAZIL 365
Breda 239, 250, 255–257, 260
Bremer 167, 172
Briscoe 32
British Berna 200, 201
British Quad 200
Brockway 298, 334, 366
Brooke 174
Broom & Wade 215
Brossel 27, 29, 30
Brush 174
BSA 175, 176, 268, 366
Buick 32, 184, 187, 245, 268, 290, 298, 301, 307, 309
BULGARIA 365
Burford 192, 199, 223, 224, 231, 279
Burrell 174, 215
Büssing (-NAG) 8, 12, 115, 116, 136–139, 142, 143, 147, 149, 154, 155, 157, 161, 163, 166, 170, 172, 173, 235, 289, 371, 372

Cadillac 32, 294, 296, 298, 301, 303, 304, 309, 310, 354, 358
Caldwell-Vale 6
Caledon 200, 279
CANADA 32, 33
Carden-Loyd 227, 229, 232, 274
Case 33
Caterpillar 6, 17, 223, 256, 298, 300, 351, 352, 356, 357
Ceirano 239, 250, 253, 260
CGT 102
Chalmers 32
Charron 102
Chatillon-Panhard 89, 91, 94
Chenard-Walcker 23, 65, 84, 85
Chevrolet 6, 7, 27, 29, 63, 147, 230, 239, 265, 266, 268, 270–272, 284, 285, 298, 301, 303–305, 307, 308, 311–313, 317, 319, 323, 351, 364–366, 369, 371, 372
CHINA 365
Chiyoda 263, 266
Christie 351, 353, 362, 363
Chrysler 32, 84, 262, 317, 365, 368
Citroën 23, 31, 64, 65, 68, 69, 75, 77, 84, 85, 105–108, 110, 111, 223, 268, 274, 351, 364
CKD 34, 60
Clayton 213, 215, 223, 226
Clement-Talbot (see Talbot)
Cleveland 300

Cleveland (Cletrac) 112, 298
Clydesdale 200, 298, 323, 340, 341
Clyno 176, 177
Coleman 298, 328, 329, 338
COLOMBIA 366
Commer 27, 175, 182, 187, 191, 192, 195, 199–201, 222
Corbitt 298, 323, 324, 328, 329, 335, 340–343, 361
Cottin-Desgouttes 68, 79
Crochat 79
Crosley 298, 304, 306
Crossley 33, 175, 178, 184, 187, 188, 192, 193, 197, 200, 207–209, 222–225, 230, 231, 233, 267, 365, 366
Csepel 235, 236
Cudell 115, 369
Cugnot 64
Cunningham 298, 307, 351, 354
CWS 273
Cyklonette 119
CZ 34–36
CZECHOSLOVAKIA 34–61

DAF 268, 270–272, 364, 372
Daimler (D) 8, 11, 17, 115, 116, 119, 136, 139, 145, 149, 150, 155, 156, 158, 161–164, 167, 172, 204, 217, 218, 366, 370, 371
Daimler (GB) 175, 178, 183, 184, 186, 187, 192, 199, 200, 202, 215, 216, 268
Daimler-Benz (see Mercedes-Benz)
Danubius 11

Dasse 27
DAT 264
Davidson 294, 301, 358
De Dion-Bouton, 65, 75, 80, 100, 115, 119, 274, 369
Delahaye 65, 80, 84, 101, 105, 108, 268
Delaunay-Belleville 65, 80, 277
Delco 300
Demag 116, 168, 169, 171, 284
Denby 314, 315
DENMARK 62
Dennis 6, 175, 192, 199, 200, 202, 205, 222, 369
Deutz 155, 158
Dewald 81
Diamond T 298, 311, 312, 317, 328, 329, 334, 351
Diatto 27, 241, 246, 247, 250, 252
Dinos 139
Dixi 122, 123, 139, 165
DKW 116, 117, 122, 123, 268, 372
DMT 227
Dodge 6, 27, 84, 89, 175, 239, 266, 298, 301, 304, 306, 307, 309, 311, 313, 317, 323, 324, 347, 351, 353, 364–366, 368, 371, 372
Douglas 6, 176, 268, 269
Drednot 32
Driggs-Seabury 295
Duplex 298, 320
Dürkopp 11, 119, 120, 139, 140, 149, 151, 155, 158, 167
Dür-Wagen 172

Duryea 358
Dux 139, 140

EGYPT 365, 366
Ehrhardt 139, 141, 155, 158, 161, 162, 164, 217, 218
EIRE (see IRELAND)
Eisenach 115
Electric 295
Excelsior 268
Eysink 268

Fafnir 119
Fageol 269, 369
Famo 116, 171
Fargo 6, 298, 301, 303, 304, 311, 317, 323, 324, 365, 366
Faun 116, 135, 139, 149, 154, 155, 160
FBW 292
Federal 266, 295, 298, 309, 318, 323, 324, 328, 332
Fiat 8, 11, 25, 60, 83, 84, 192, 200, 235, 238, 239, 241–251, 253–261, 268, 273–276
FINLAND 63
FN 27–31, 268, 364
Foden 174, 175, 213, 223
Ford 6, 7, 27, 29, 31, 32, 62, 63, 70, 144, 178–181, 184, 186–188, 190, 192, 200, 209, 222, 230, 233, 239, 245, 262, 264, 268–272, 274, 276–278, 281, 284, 285, 288, 296, 298, 301–309, 311–313, 317–319, 323, 325, 328–330, 347, 348, 351, 355, 361, 364–366, 368–371

Ford (D) 116, 129, 144
Ford (F) (see Matford)
Fordson 7, 175, 192, 196, 199, 200, 209, 217, 351, 353, 369
Foster 215, 216
Fowler 115, 155, 174, 215
Framo 116, 122, 123
FRANCE 64–114
Franklin 323, 325, 360
Frera 240
Fross-Büssing 8, 11, 12
FWD 6, 89, 200, 217, 219, 223, 225, 279, 294–296, 298, 320–323, 325, 338, 340, 351, 352, 360, 364, 365, 369, 371

Garford 192, 269, 279, 309, 314, 315, 328, 330, 334, 351, 352, 369
Garner 175, 192, 196, 197, 199, 200, 206, 217, 236
Garrett 213, 215
GAZ 277, 278, 281–283, 371
General Fire 347
General Motors 6, 32, 63, 84, 262, 317, 328, 365
GERMANY 115–173
Gilera 240
Gillet 27, 28
Glasgow 217
GMC 27, 29, 32, 84, 89, 93, 100, 184, 245, 269, 271, 274, 292, 296, 298, 307–309, 311, 312, 314, 315, 318, 320, 323, 328, 330, 335, 336, 338, 340, 347, 351, 354, 364–366
Gnome-Rhône 66, 67
Gotfredson 33

374

Gräf & Stift 8–12, 17, 19, 21, 26, 44
Graham 265
Gramm (-Bernstein) 32, 192, 295, 334
GREAT BRITAIN 174–234
GREECE 365, 367
Guy 175, 187, 191, 192, 194, 199, 200, 207, 209, 217, 218, 220, 223, 225, 365, 369
Guzzi 240, 260

HAFADASA 364
Hall 176
Halley 192, 194, 199, 200, 202
Hallford 6, 199, 200, 202, 369
Hanomag 116, 122, 124, 125, 155, 160, 171
Hansa (-Lloyd) 116, 119, 139, 141, 144, 149, 163, 168
Hardy 200, 207
Harley-Davidson 262, 268, 298–300, 364
Heavy Aviation 296, 315, 332
Hendee (see Indian)
Hendrickson 33, 298
Henney 307
Henschel 166, 135, 138, 148, 149, 152
Hillman 175, 182
Hispano (-Suiza) 364, 371
Hiziri 264
Holt 6, 17, 112, 171, 214, 223, 226, 282, 351, 352, 356, 362
Holverter 181, 192
Horch 116, 119, 125, 126, 128, 129, 131–133, 139,

141, 155, 159, 284
Hornsby 174, 215, 216, 227, 228
Hotchkiss 65, 69–72, 74, 94–96, 100, 107
Howe 347
Howey-Wiley 306
Hudson 6, 263, 268, 298, 302, 310, 321
Hug 298, 340, 343
Humber 175, 180, 367
HUNGARY 235–237
Hupp 32
Hurlburt 279

INDIA 366, 367
Indian 268, 298–300
Indiana 298, 323, 327, 328, 340, 346, 361
International 298, 304, 307, 308, 311, 313, 328, 336, 356, 357, 364, 365
IRAN 366, 367
IRELAND 366, 368
Isotta Fraschini 239, 241, 250, 252, 256, 260, 261
Isuzu 263, 266
Itala 241, 246, 248, 250
ITALY 238–261
Itar 35

JAG 280–282
JAP 177
JAPAN 262–267
Jawa 34–36
'Jeep' 304
Jeffery 32, 33, 78, 79, 89, 93, 94, 199, 295, 320–322, 351, 352, 359
John Deere 298, 349
Jowett 180, 187

Kaelble 116, 155, 160
Karrier 175, 192, 199, 200, 203, 207, 210, 212, 217, 218, 366
Kégresse 23, 31, 64, 67, 105, 111, 223–225, 256, 274, 277, 351, 354, 364
Kelly-Springfield 27, 32, 78, 192, 200, 269, 295, 332, 369
King 307, 308, 358
Kissel 321, 365
Klöckner-Humboldt-Deutz 116, 284
Knox 83, 214, 365, 367
KO-HI 267
Kommunar 282
Komnick 119
Krauss-Maffei 116, 168, 270, 171
Kriéger 101
Krupp 116, 129, 136–138, 144, 148, 152, 155, 158, 162, 166, 171, 173, 235, 236, 371
Kurogane 262, 265

Lacre 6, 199
Laffly 14, 64, 65, 71, 72, 74, 84, 86, 94–96, 102, 104, 366, 371
Lanchester 174, 187, 188, 229
Lancia 60, 192, 239, 241–243, 246, 248, 250, 254, 260, 261, 279, 366
Landsverk 58, 272, 284, 287–289, 366
Landwehr 8, 17, 20
Lanz 116, 155, 159, 173

Latil 27, 30, 64, 65, 71, 72, 79, 84, 86, 87, 89, 90, 92–94, 97, 101, 105, 217, 275, 322
Laurin & Klement 9, 11, 34, 35, 42, 44, 45, 264
Lauster 163
Lehaitre 67
Leyland 33, 175, 186, 192, 199, 200, 203, 209, 211–213, 222, 231, 366, 369
LGOC 183, 192
'Liberty' (see USA/QMC)
Licorne 65, 71
Light Aviation 296, 314, 315
Linn 298, 351, 355
Lippard-Stewart 295
Locomobile 200, 279, 302, 333, 359
Lombard 282, 351, 352
Long Distance 294, 314
Lord Baltimore 295
Lorraine 65, 71, 74, 94, 98, 113
LUC 139, 141

Mack 200, 295, 296, 298, 314, 328, 332, 340, 343, 351, 353, 359
Maffei 167, 168
MAG 9, 11, 235
Magirus 116, 136, 138, 139, 145, 148, 149, 152, 155, 159, 166
Mais 295
MAN 55, 116, 135, 138, 142, 145, 153, 154, 365, 369, 371
Manfred Weiss 230, 235, 237
Mann 213, 215

Mannesmann-Mulag 133, 139, 149, 163
Marienfelde 8, 11, 115, 149, 150
Marmon 323
Marmon-Herrington 29, 31, 298, 304, 306, 311–313, 323–326, 328, 330, 331, 338–340, 343, 344, 347, 348, 351, 355, 356, 360, 361, 363–370
Marshall 215
Marta 11, 235
Martell 223, 225, 231
Martini 269, 290
Matchless 175
Matford 65, 70, 76, 84, 86
Maudslay 175, 192, 200, 203
Maybach 116, 129
McCormick-Deering 349
McLaren 215
Mercedes 119, 121, 133, 155, 277
Mercedes-Benz 116, 122, 125–130, 132, 133, 135, 136, 138, 145–149, 153, 154, 156, 168, 170, 171, 289, 365
Mercier 67
Merryweather 222
Metallurgique 27
MEXICO 366, 368
Miesse 27, 29
Militor 296, 320, 322
Milnes-Daimler 174, 204
Minneapolis-Moline 298, 350
Minerva 27, 31, 268
Mitchell 321
Mitsubishi 263
Monet-Goyon 66
Moreland 298, 333

Morris 6, 22, 175, 181, 186, 223, 225, 231
Morris-Commercial 175, 181–183, 185, 187, 190–192, 196–200, 206, 217–220, 223, 226, 230, 365, 366
Motobécane 66
Motosacoche 268, 290
Müller 155

Nacke 119, 139, 142, 149
NAG 116, 119, 121, 134, 139, 142, 146, 149, 151, 155, 156
NAMI 277
Napier 174, 178, 184, 187–189, 192, 193, 199, 200, 279
Nash 79, 89, 275, 296, 298, 320, 322, 351, 352
National 321
Nazarro 241, 246
Neckarsulm 117
Nesselsdorfer 9, 11, 19, 34, 35, 44, 56
NETHERLANDS 268–272
NETHERLANDS EAST INDIES 369
NEW ZEALAND 369
Nimbus 62
Nissan 265
Norton 175
NORWAY 369
NSU 115–118, 139

OAA 264
Ogar 35, 36, 61
Oldsmobile 304, 330
OM 239, 246, 249, 250, 252, 254, 371

Omnia 268
Opel 116, 119, 120, 122, 125, 128–130, 133–135, 137, 139, 146, 147, 161, 165
Ordnance (USA, RIA) 299, 300, 351, 356, 357, 362, 363
Orion-Wagen 172
Oshkosh 298, 338
Overland 234

PABZ 8, 11
Packard 32, 78, 192, 200, 264, 269, 277, 279, 295, 298, 307, 314, 315, 333, 351, 352
Pagefield 200, 204
Paige 314, 315, 321
Panhard 27, 65, 81, 84, 87, 89, 91, 100, 103, 108, 110, 111
Pavesi 30, 89, 217, 218, 233–235, 239, 241, 257, 259–261, 275, 284, 286, 290, 365, 371
Pavesi-Tolotti 256, 258
Peerless 6, 200, 204, 223, 227, 229, 321, 332, 365, 366
PERSIA (see IRAN)
Peugeot 65, 66, 68, 76, 81, 84, 87, 95, 102, 105, 112, 113, 276, 369
Phänomen 116, 119, 125, 129, 131, 133, 134, 137, 371
Phänomobil 119
Phelon & Moore (Panther) 176
Pidwell 112

Pierce-Arrow 78, 83, 200, 204, 229, 269, 314, 316, 334
Plymouth 284, 298, 301, 303
Podeus 139, 142, 155, 159
Pohl 155
POLAND 273–276
Polski Fiat 273–276
Pontiac 298, 304, 305, 360
Porsche 8, 17, 20–22, 56
PORTUGAL 369, 370
Praga 9, 11, 34, 35, 37, 39, 40, 42–46, 49, 51, 56–61, 235, 284, 369, 370, 371
Premier 321
Presto 269
Protos 119, 134
Puch 8, 9–11
Purrey 78
PZInz. 273–275

Quad (see Jeffery, Nash)

Rába 9, 11, 17, 35, 37, 44, 235–237
RAF 9, 34
Rapid 241
Reinickendorf 17
Renault 64, 65, 68–71, 73, 75–77, 81–84, 88, 89, 91, 94, 98, 101, 102, 112–114, 268, 269, 364
René Gillet 66, 67
Reo 32, 199, 298, 311, 323, 326
Republic 314, 315, 334
Rheinmetall 161, 173
Richard & Hering 139
Riker 200, 333, 359
Rikuo 262
Riley 175, 181

Roadless 223, 225, 226, 231, 234
Robey 215
Rochet-Schneider 65, 84
ROF 228
Röhr 35, 122, 124
Rokko 264
Rolls-Royce 6, 25, 175, 179, 187, 189, 229, 230, 277, 366
Romfell 237
Rover 175, 179, 184, 203
Royal Enfield 175
RUMANIA 369, 370
Russell 32, 33
RUSSIA 277–283
Ruston 213, 215, 223, 226, 352

SAG 151
Saint Chamond 112
Sampson 294, 295, 314
Samson 349
Sankyo 262
Sarolea 27, 28
Saurer 8, 10–13, 23, 24, 26, 65, 82, 84, 105, 142, 169, 200, 254, 269, 276, 290–293
Sava 31
Scammell 98, 175, 206, 214, 217, 221, 266
Scania-Vabis 284–286
Scat 241, 250
Scemia 75
Schneider 75, 89, 91, 108, 112, 113, 264, 371
Scott 176
SD 199, 233
Seabrook/Standard 200
Seagrave 347

Selden 334
Selve 129, 132
Sentinel 56, 213, 223
Serpollet 115, 119
Sheffield-Simplex 229, 279
Siddeley Deasy 178, 184
Siemens-Schukert 155
Simms 177
Sinclair 322
Singer 186, 219
Sisu 63
Skoda 17, 19, 21, 34–36, 39, 40, 43, 44, 47, 49, 51–53, 56–58, 60, 61, 171, 284, 287
Soller 256
Somua 65, 108, 109, 111
SOUTH AFRICA 371
Spa 200, 239, 241, 245, 246, 248–250, 252, 254, 257, 259, 260
SPAIN 371, 372
Spyker 268, 269
Standard (GB) 175
Standard (US) 332
Star 184, 192
Sterling 298, 334
Stevens 200, 205
Stewart-Crosbie 216
Steyr 8–12, 14–16
Stoewer 35, 116, 119, 122, 124, 125, 139, 143, 151
Straker (-Squire) 174, 184, 192, 200, 205
Straussler 92, 104, 175, 181, 192, 196, 199, 200, 217, 219, 229, 230, 235–237
Studebaker 6, 11, 32, 84, 178, 186, 268, 295, 298, 302, 309, 314, 328, 365
STZ 282

Sumida 263, 267
Sunbeam 175, 178, 179, 184
SWEDEN 284–289
SWITZERLAND 290–293
Syracuse 299

Talbot 175, 178, 183, 187, 189, 192
Tarrant 6
Tasker 213, 215
Tatra 8, 19, 34, 35, 38–41, 43, 44, 48–50, 53–61, 71, 74, 94, 98, 124, 266, 369, 370
Tempo 116, 122, 124, 284, 364, 366, 369
Terrot 66
TGE 263, 266
Thompson 233
Thornycroft 7, 62, 174, 175, 192, 194, 198–200, 205, 207, 211–213, 215, 217, 221, 222, 232, 264, 266, 364, 366, 371
Tidaholm 284, 288
Tilling-Stevens 44, 175, 200, 207
Toyota 262, 263, 265
Triangel 62
Triumph (D) 115, 117
Triumph (GB) 175–177
Trojan 175, 186, 187, 217, 221
Tucker 360
TURKEY 371, 372

Unic 65, 75, 107, 110
United 332
Ursus 274, 276
US, USA/QMC 296, 297, 316, 323, 327, 328, 331,

332, 334, 335, 337–340, 344–348, 353, 360, 371
USA 294–363

Vauxhall 175, 179, 184, 186
Verlie 295, 332
Velocette 175
Vickers (-Armstrongs) 31, 176, 181, 214, 227, 229, 231, 232, 234, 267

Victoria 117, 118
Volkswagen 165
Volvo 284–287, 289
Vomag 116, 135, 139, 143
Voran 129, 132
Vulcan 175, 185, 187, 192, 194, 198, 200, 212, 223

WAF 8, 11, 12
Wallis & Steevens 213, 215
Walter (CS) 35
Walter (US) 93, 298, 322, 350
Wanderer 116, 119, 125–127
Wantage 213
Ward LaFrance 298, 319
Waverly 309, 310
White 32, 78, 83, 94, 99, 102, 200, 279, 294, 295, 298, 302, 307, 309, 310, 314, 316, 319, 323, 327, 328, 332, 334, 340, 346, 351, 359, 361
Wichita 269, 369
Willème 65, 94
Willys (-Overland) 27, 192, 298
Winton 294, 309, 310
Wolseley 174, 175, 178, 181, 184, 187, 189, 192, 200,

204, 215, 216, 229, 232, 234, 268
Wolseley-Siddeley 184

YaG (see JAG)
YUGOSLAVIA 371

Z 34, 35, 39, 40, 61
Zetor 61
ZIS 277, 278–282, 371, 372
Zündapp 116, 117
Züst 241, 250, 252